Polyoxometalate Chemistry

Some Recent Trends

World Scientific Series in Nanoscience and Nanotechnology

Series Editor-in-Chief: Frans Spaepen (*Harvard University, USA*)

Vol. 1 Molecular Electronics: An Introduction to Theory and Experiment
Juan Carlos Cuevas (*Universidad Autónoma de Madrid, Spain*) *and Elke Scheer* (*Universität Konstanz, Germany*)

Vol. 2 Nanostructures and Nanomaterials: Synthesis, Properties, and Applications, 2nd Edition
Guozhong Cao (*University of Washington, USA*) *and Ying Wang* (*Louisiana State University, USA*)

Vol. 3 Molecular Cluster Magnets
edited by Richard Winpenny (*The University of Manchester, UK*)

Vol. 4 Plasmonics and Plasmonic Metamaterials: Analysis and Applications
edited by Gennady Shvets (*The University of Texas, Austin, USA*) *and Igor Tsukerman* (*The University of Akron, USA*)

Vol. 5 Inorganic Nanomaterials from Nanotubes to Fullerene-Like Nanoparticles: Fundamentals and Applications
Reshef Tenne (Weizmann Institute of Science, Israel)

Vol. 6 Plasmon Resonances in Nanoparticles
Isaak D. Mayergoyz (University of Maryland, USA)

Vol. 7 Scanning Probe Microscopy for Energy Research: Materials, Devices, and Applications
edited by Dawn A. Bonnell (The University of Pennsylvania, USA) and Sergei V. Kalinin (Oak Ridge National Laboratory, USA)

Vol. 8 Polyoxometalate Chemistry: Some Recent Trends
edited by Francis Secheresse (Université de Versailles-St Quentin, France)

Volume 8

World Scientific Series in **Nanoscience and Nanotechnology**

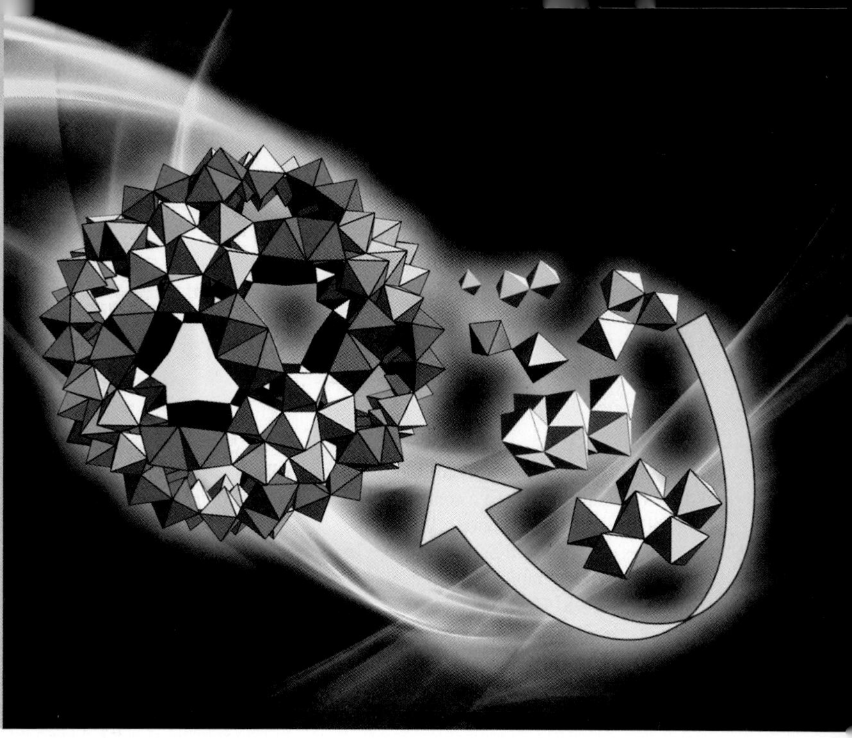

Polyoxometalate Chemistry

Some Recent Trends

Editor

Francis Sécheresse
Université de Versailles-St Quentin, France

World Scientific

NEW JERSEY • LONDON • SINGAPORE • BEIJING • SHANGHAI • HONG KONG • TAIPEI • CHENNAI

Published by

World Scientific Publishing Co. Pte. Ltd.
5 Toh Tuck Link, Singapore 596224
USA office: 27 Warren Street, Suite 401-402, Hackensack, NJ 07601
UK office: 57 Shelton Street, Covent Garden, London WC2H 9HE

Library of Congress Cataloging-in-Publication Data
Polyoxometalate chemistry : some recent trends / editor, Francis Secheresse, Université de Versailles-St Quentin, France.
pages cm -- (World Scientific series in nanoscience and nanotechnology ; volume 8)
Includes bibliographical references.
ISBN 978-9814458979
1. Polyoxometalates. I. Secheresse, Francis.
QD474.P645 2013
546'.3--dc23
2013008277

British Library Cataloguing-in-Publication Data
A catalogue record for this book is available from the British Library.

In-house Editor: Rhaimie Wahap

Typeset by Stallion Press
Email: enquiries@stallionpress.com

Printed in Singapore by Mainland Press Pte Ltd.

PREFACE

Polyoxometalate (POM) chemistry remains a very attractive and active field, an illustration being that in 2011, more than 1000 articles have been published covering fundamental and applied research, synthesis and reactivity, physicochemistry and biochemistry and many other domains. This means that writing a book to give an up-to-date information on POM chemistry is simply impossible. For this reason, some recent outstanding results obtained by young researchers, covering molecular- and nanochemistry, have been selected and assembled in this book.

The aim of this book is to highlight in the domains of the synthesis of new POMs with a specific attention to polyoxothioanions, and also to provide some novelties and perspectives in selected domains such as nanochemistry, macro-ions self-assembly, *modus operandi* in syntheses, magnetism, luminescence and nanochemistry.

The stability of POM-protected metal nanoparticles is closely dependent on several parameters such as the electrostatic forces typical of anion-stabilized colloids, direct bonding between POMs and the metal surfaces, and the formation of numerous ionic and hydrogen bonds between POMs bonded closely to one another on the metal surface. This is presented in Chap 1.

The behavior of macro-ions in solution is shown to be quite different from the simple ionic solution or colloidal systems. Their self-assembling into a single-layered, spherical, hollow vesicle structure, namely the "blackberry" structure is developed in Chap 2.

The case of "*one-pot*" reactions often reported in POMs synthesis can hide very complex solution speciation processes arising from the presence of a highly dynamical situation, that depends on a multitude of factors such as pH, ionic strength, reaction time, temperature, counterions, concentration of starting materials, presence of electron donors and their redox potential. This is treated in Chap 3.

Examples of spin clusters with tunable interactions are given and single molecule magnet based on POMs are specifically tackled in Chap 4.

Besides paramagnetic transition metal centers and lanthanoid ions encapsulated in archetypal lacunary polyoxoanions, magnetically functionalized Kleperates are described, their discovery tracing back to nearly 15 years. These domains at the border of physics and chemistry are presented in Chaps 5 and 6.

The host-guest concept revealed very productive for the elaboration of POMs, the softness of sulfur compared to oxygen opening on fluxional systems with high electron localization. Fluxionality opens to mobile units and giant structures presenting internal cages described in Chap 7.

Francis Sécheresse
Université de Versailles-St Quentin
December 2012

CONTENTS

LIST OF COLOR PLATES

COLOR PLATE 1

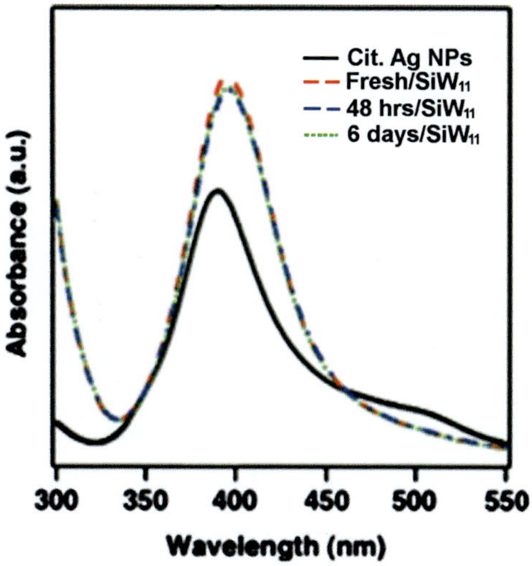

Fig. 9 (Chapter 1). UV-vis spectroscopic analysis of the displacement of citrate by α-$SiW_{11}O_{39}^{8-}$ on Ag NPs. After addition of α-$SiW_{11}O_{39}^{8-}$ to a solution of citrate-stabilized Ag NPs (solid black curve), absorbance due to the surface plasmon resonance (SPR) increased to new values (dashed red curve) and then remained constant (blue and green curves). (Reprinted from Ref. 87 with permission from Springer Science & Business Media.) (See page 15.)

COLOR PLATE 2

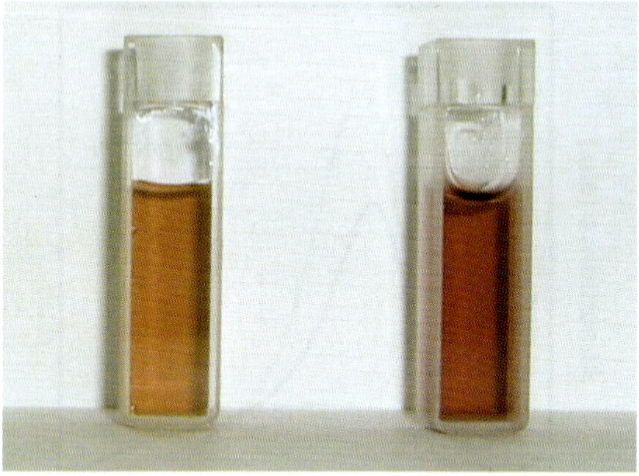

(a)

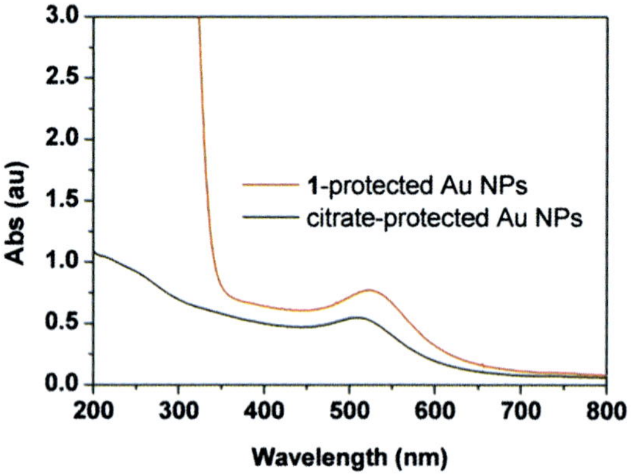

(b)

Fig. 10 (Chapter 1). (a) Photographs and (b) spectra of citrate-protected Au NPs and α-$K_9AlW_{11}O_{39}$-protected Au NPs. The cuvette at the left in panel (a) contains citrate-protected Au NPs. Addition of α-$K_9AlW_{11}O_{39}$ (**1**) to a final concentration of 2.5 mM, causes a visible color change (cuvette at right). Corresponding UV-visible spectra are shown in panel (b). Reprinted from Ref. 61 with permission from the Elsevier. (See page 16.)

COLOR PLATE 3

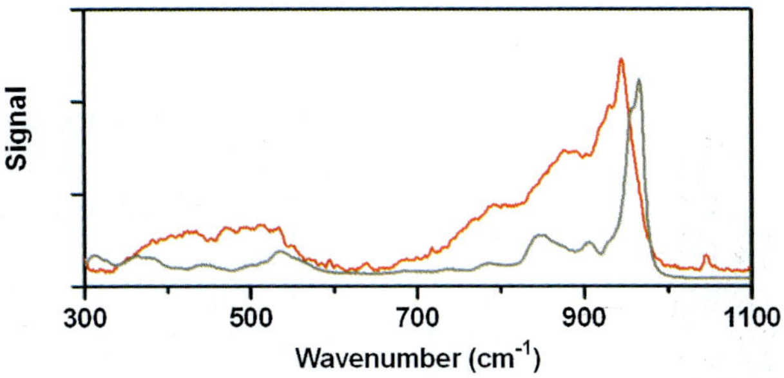

Fig. 13 (Chapter 1). Raman spectrum of solid α-$K_9AlW_{11}O_{39}$ (gray line) and the SER spectrum of an aqueous solution of α-$K_9AlW_{11}O_{39}$-protected Ag(0) NPs (red line). (Reprinted from Ref. 50 with permission from the American Chemical Society.) (See page 21.)

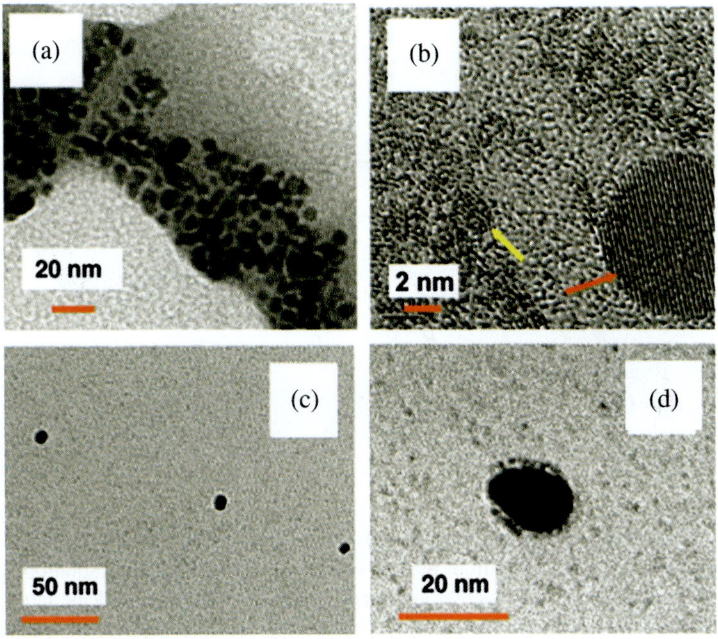

Fig. 14 (Chapter 1). (a) TEM image of Ag(0) (dark) embedded in amorphous $K_9AlW_{11}O_{39}$ (gray). (b) HRTEM image of the dry sample showing α-$K_9AlW_{11}O_{39}$ (yellow arrow) and a Ag(0) particle (red arrow). (c) Cryo-TEM image of individual α-$K_9AlW_{11}O_{39}$— protected NPs with no evidence of aggregation. (d) α-$AlW_{11}O_{39}^{9-}$ — protected Ag(0) particles "trapped" in cryogenically frozen water–glass. Freely solvated POMs appear as small dark objects in the background. (Reprinted from Ref. 50 with permission from the American Chemical Society.) (See page 22.)

COLOR PLATE 4

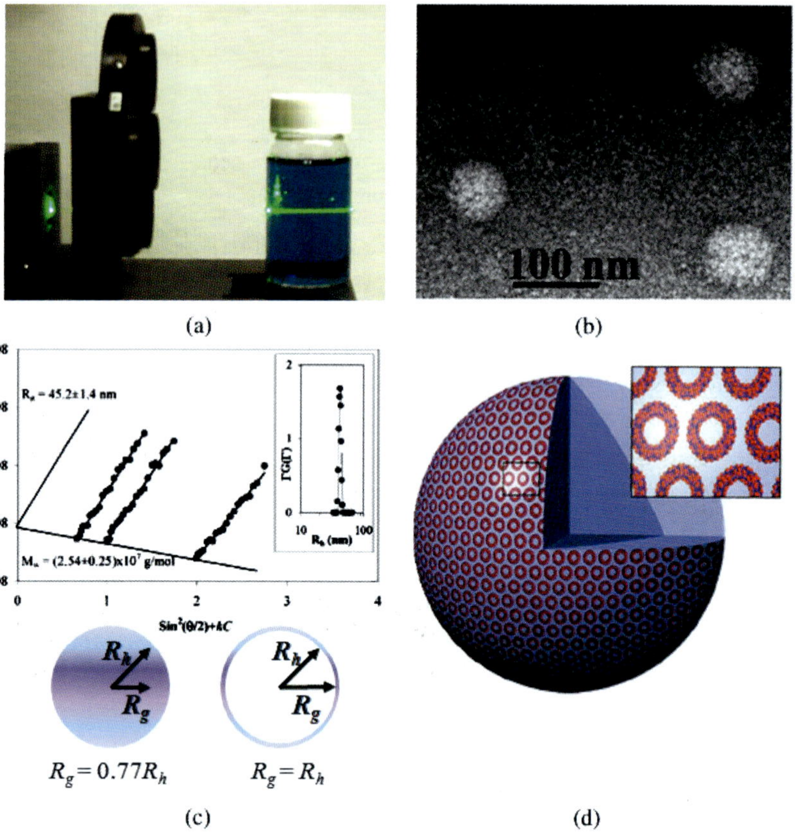

Fig. 4 (Chapter 2). (a) Tyndall effect of a $\{Mo_{154}\}$ aqueous solution when a green laser (532 nm) passes through the solution. (b) TEM image of the blackberry structure formed by $\{Mo_{154}\}$ macroanions. (c) Zimm plot of a series of $\{Mo_{154}\}$ aqueous solutions reveals the radius of gyration R_g and weight averaged molecular weight (M_w) of the large blackberry structure. (d) A cartoon illustrates the blackberry structure. (Reprinted with permission from Ref 13. Copyright 2003 Nature Publishing Group.) (See page 61.)

COLOR PLATE 5

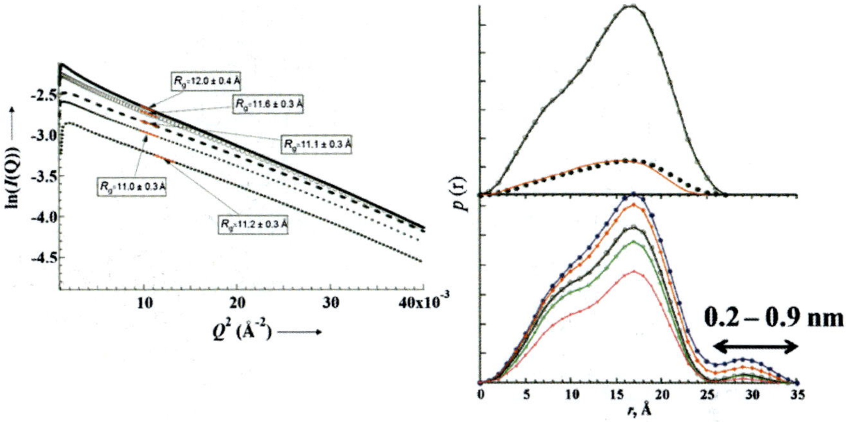

Fig. 8 (Chapter 2). *Left*: Guinier plots for 0.26 mM {Mo72V30} macroions in water and water/acetone mixed solvents containing various amount of acetone (in vol %) (solid line: 75% acetone/water, open circles: 65% acetone/water, dashed line: 45% acetone/water, dotted line: 10% acetone/water, open squares: in pure water). *Right top*: Distance distribution functions based on calculated and experimental scattering curves for $\{Mo_{72}V_{30}\}$ as obtained by using an indirect Fourier Transform of the primary SAXS data. (closed circles: 0.1 mg/mL, or 0.013 mM $\{Mo_{72}V_{30}\}$, open circles: 0.052 mM $\{Mo_{72}V_{30}\}$, red line: $\{Mo_{72}V_{30}\}$ calculated). *Right bottom*: Experimental distance distributions for 0.26 mM $\{Mo_{72}V_{30}\}$ in water and acetone/water mixed solvents with various acetone content (in vol %). (green: 75% acetone/water, blue: 65% acetone/water, red: 45% acetone/water, light blue: 10%acetone/water, magenta: in pure water).[32] (Reprinted with permission from Ref 32. Copyright © 2009 Wiley InterScience.) (See page 69.)

COLOR PLATE 6

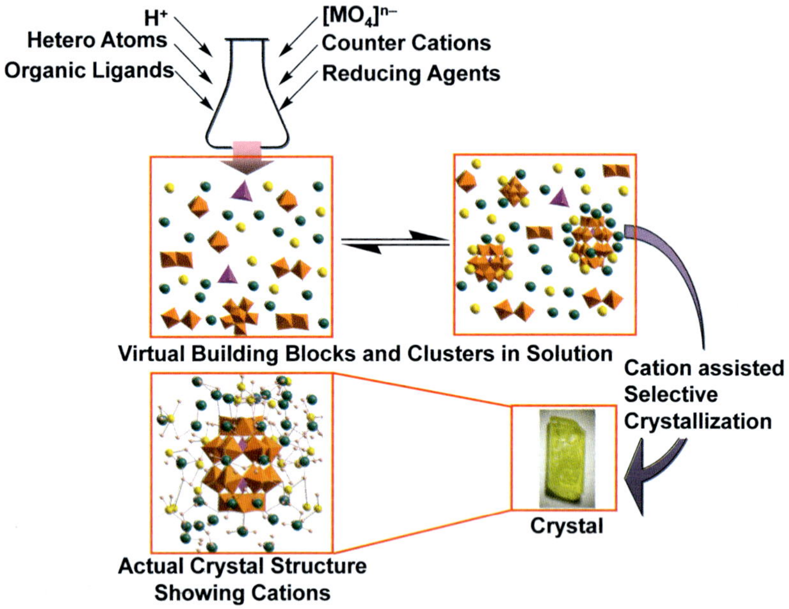

Fig. 2 (Chapter 3). Schematic representation of the traditional "one-pot" synthesis of POM clusters leading to the formation of various structural archetypes in solution highlighting the role of cations in selective crystallization of a particular product. Yellow and green spheres represent different cations. The top left shows a range of "virtual" geometrically possible units and the top right shows the cations complexing the clusters and pulling certain structures out of solution. (See page 103.)

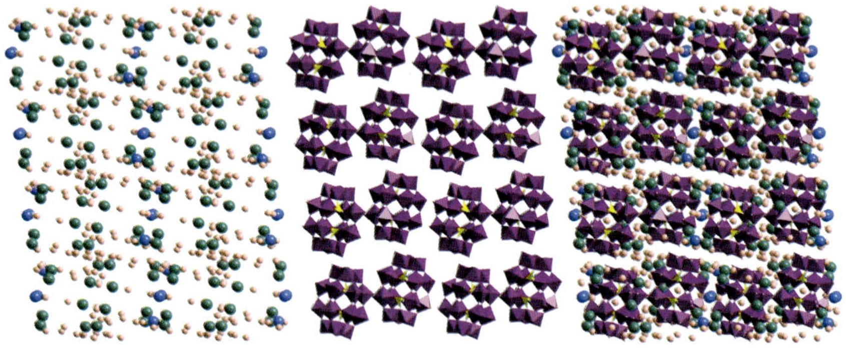

Fig. 3 (Chapter 3). Packing diagrams depicting the separate and combined arrangements of cations and POM cluster anions in a crystal lattice. *Left*: only cations (blue and green spheres) and solvent water molecules (rose spheres); *Middle*: only POM clusters (violet and yellow polyhedra) and *Right*: POM cluster anions, counter cations and solvent water molecules together in a crystal lattice. (See page 104.)

COLOR PLATE 7

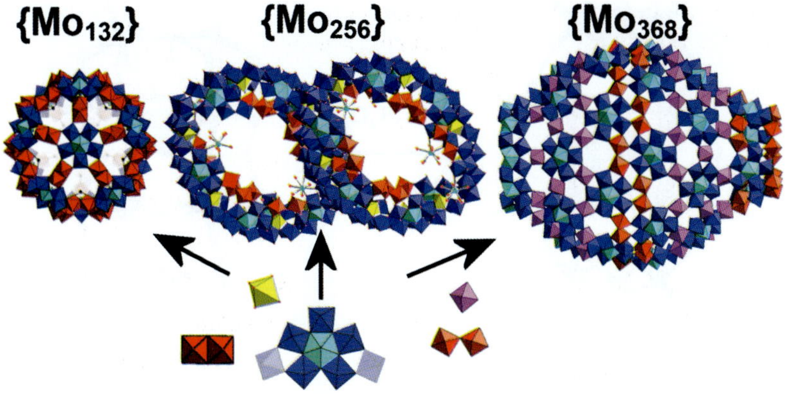

Fig. 4 (Chapter 3). $\{Mo_1\}$, $\{Mo_2\}$, $\{(Mo)Mo_5\}$ etc. building blocks self-assemble according to the reaction conditions leading to various structurally diverse Mo-blue clusters.[5,6] (See page 107.)

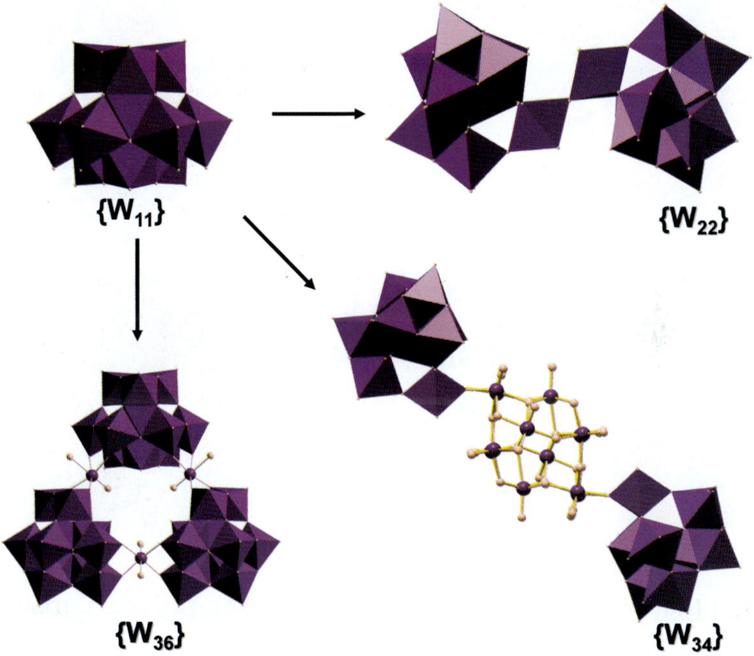

Fig. 7 (Chapter 3). Formation of three isopolytungstate formed by the different linking modes of same basic building block $\{W_{11}\}$. $\{WO_6\}$ units (violet polyhedra), W, (violet spheres), O (rose spheres). (See page 113.)

COLOR PLATE 8

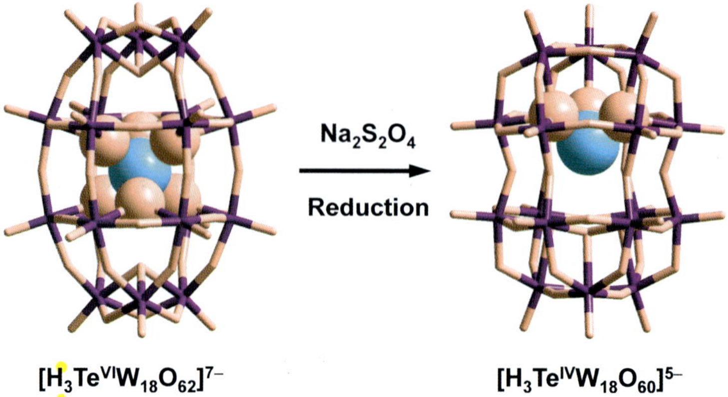

Fig. 8 (Chapter 3). Structural transformation as a result of the reduction of Te^{VI} to Te^{IV} inside a tungsten based molecular cage. (See page 116.)

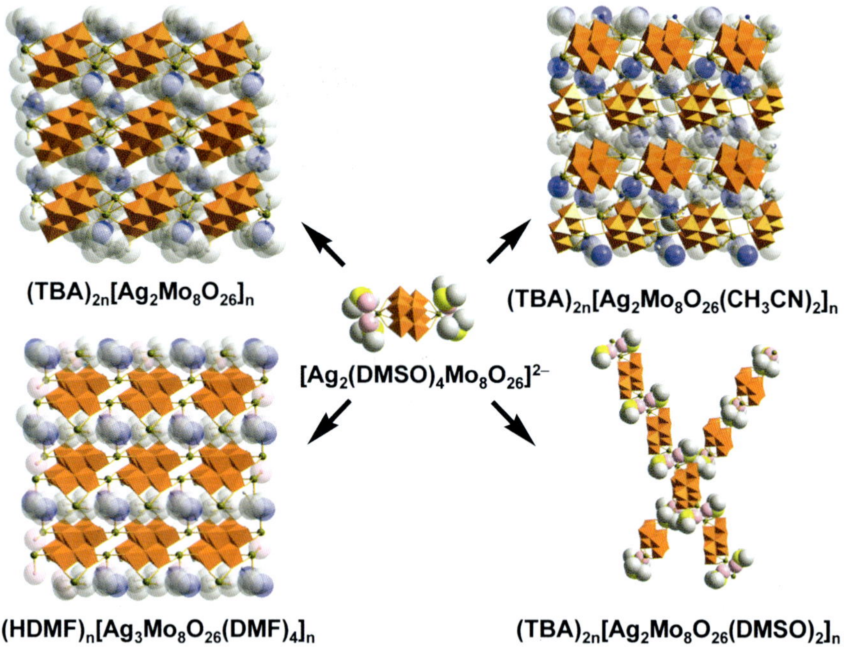

Fig. 10 (Chapter 3). Summary of the structures isolated from the silver-octamolybdate reaction system. The solvents and counterions are represented in space fill models to highlight their influence in determining the overall topologies achieved which include monomeric structures, 1D chains, grid and 2D network as shown. Color scheme: $\{Mo_8\}$ polyhedra orange, O (rose), C (grey), N (blue), S (yellow). (See page 119.)

COLOR PLATE 9

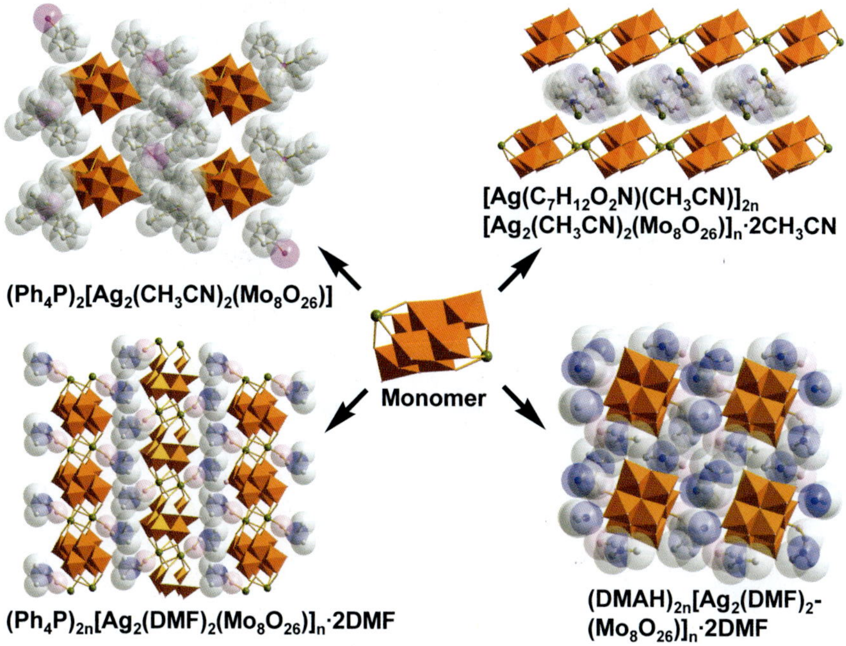

Fig. 11 (Chapter 3). More examples of silver-octamolybdate based architectures isolated as a result of solvent and counter ions influencing self-assembly. Color scheme: $\{Mo_8\}$ polyhedra (orange), O (rose), C (grey), N (blue), P (pink). (See page 120.)

COLOR PLATE 10

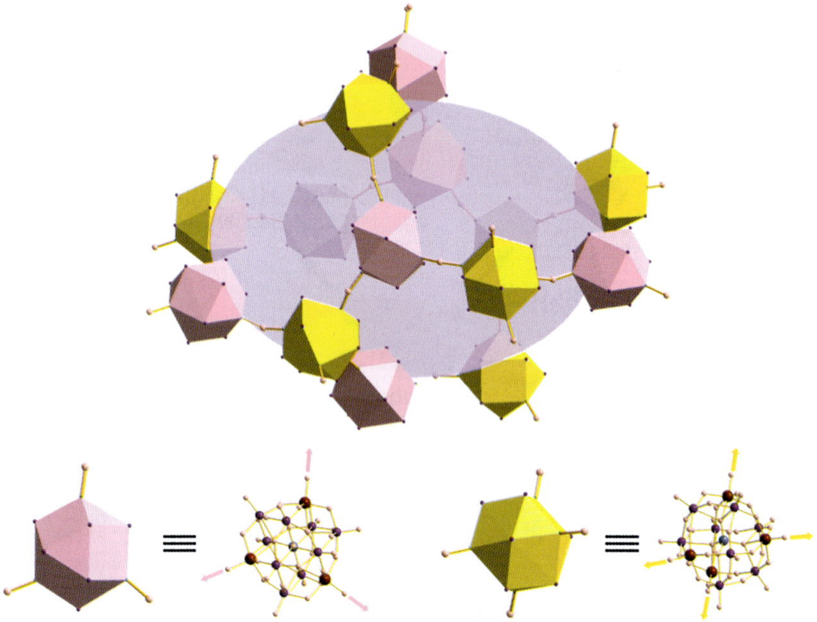

Fig. 13 (Chapter 3). Illustration of the nanosized pockets in $[(C_4H_{10}NO)_{40-}(W_{72}Mn_{12}O_{268}Si_7)n]$ highlighted by the purple ellipsoid (dimensions: ~ 2.7\2.4\1.3 nm). The trigonal (rose) and tetrahedral (yellow) building block connectors are shown at the bottom; rose and yellow arrows highlight the connecting modes. (See page 125.)

COLOR PLATE 11

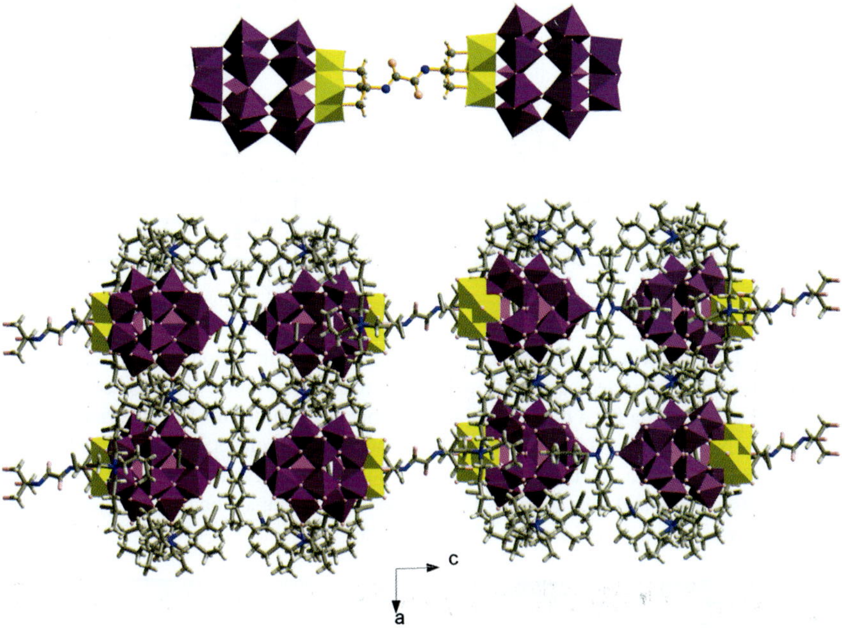

Fig. 16 (Chapter 3). Combined ball and stick and polyhedral representation of the "inorganic–organic–inorganic" hybrid cluster anion derived by grafting linear bis-(tris) ligands onto the V_3-capped Wells-Dawson-type cluster $TBA_5H_4[P_2V_3W_{15}O_{62}]$ (above) and the packing diagram (below) of the same compound showing the selective gathering of TBA cations around dumb-bell heads leading to voids in the crystal lattice. C (grey), H (white), N (blue), O (rose), W (violet polyhedra), V (yellow polyhedra), P (pink polyhedra). (See page 131.)

COLOR PLATE 12

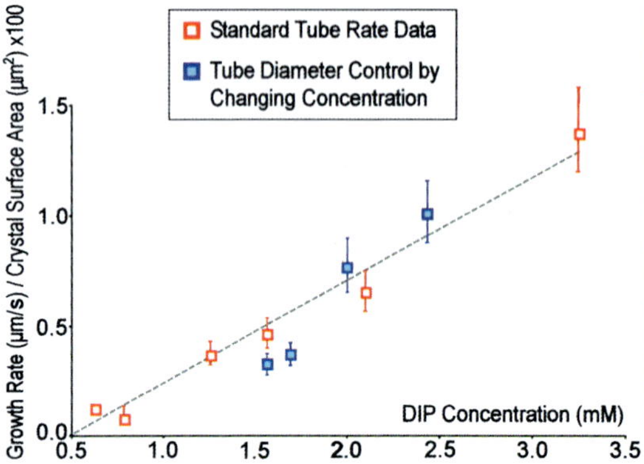

Cation conc. (mM)	Speed/Surface μm/s/μm^2 x100
0.4	0.002
0.6	0.012
0.8	0.007
1.3	0.037
1.6	0.046
2.1	0.065
3.3	0.138

Fig. 20 (Chapter 3). Plot of growth rate/crystal surface area against the overall concentration of DIP, showing how the diameter is related to the cation concentration. Red squares show the linear decrease in growth rate as the concentration is decreased, and blue squares show rate data for a tube in which the concentration of DIP-Me is varied while it is growing. Note: The crystal surface area is considered as it relates directly to the limiting rate at which water can enter the membrane surrounding the crystal, and therefore the rate of POM dissolution and tube growth. This allows direct comparison of tubes grown from crystals of varying sizes and shapes. (See page 135.)

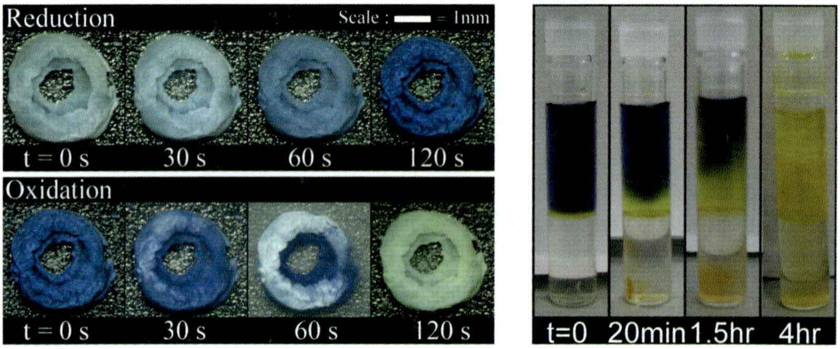

(a) (b)

Fig. 27 (Chapter 3). (a) A schematic representation of one cell being grown inside another. A salt containing a large organocation (green) is injected into a salt solution containing a large POM anion (blue) forming a membrane. A second POM that contains another POM reagent (red) is injected into the first cell, producing a second membrane. Adding an external reagent (yellow) can then cause a reaction in the inner cell after a diffusion time. (b) Time-lapse images are shown below with the growth of a DIP-Me cell in $\{PMo_{12}\}$, followed by the construction of a $\{PW_{12}\}$ cell inside the encapsulated DIP-Me droplet. The $\{PW_{12}\}$ cell contains potassium permanganate and when hydrogen peroxide solution is added to the outer $\{PMo_{12}\}$ solution, discoloration of the $\{PW_{12}\}$ cell is observed after a few minutes. (See page 147.)

COLOR PLATE 13

(a)

(b)

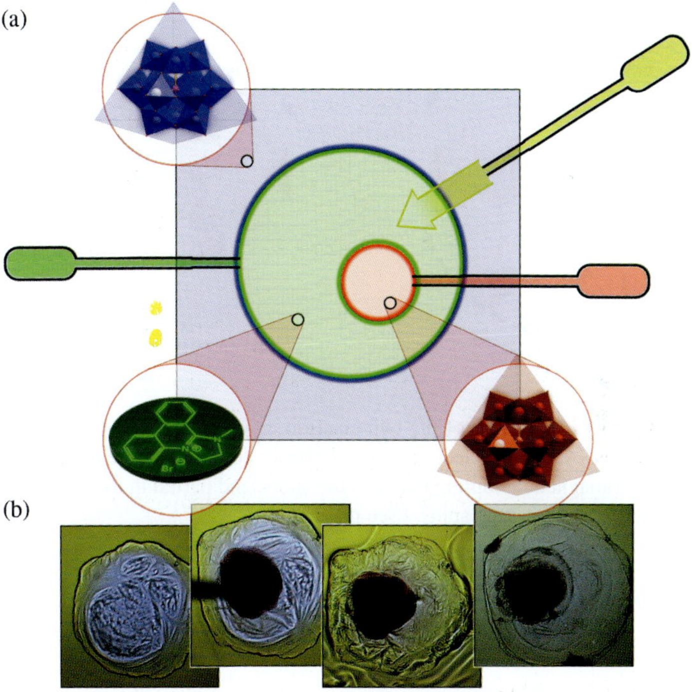

Fig. 28 (Chapter 3). (a) Addition of dilute sodium dithionite (reducing agent) to a washed membrane in water, results in the reduction of the POM over 120 seconds. After washing with deionized water, addition of hydrogen peroxide solution (30% v/v) to the same system causes the POM to become reoxidized over approximately 120 seconds. This process is fully reversible and can be repeated up to three times without significant membrane degradation. (b) Diffusion of H_3O^+ cations through a $\{PW_{12}\}$-DIP-Me membrane showing the change in color of the bromocresol purple in the upper DIP fraction from purple to yellow indicating the lowering of the pH in this region. (See page 148.)

COLOR PLATE 14

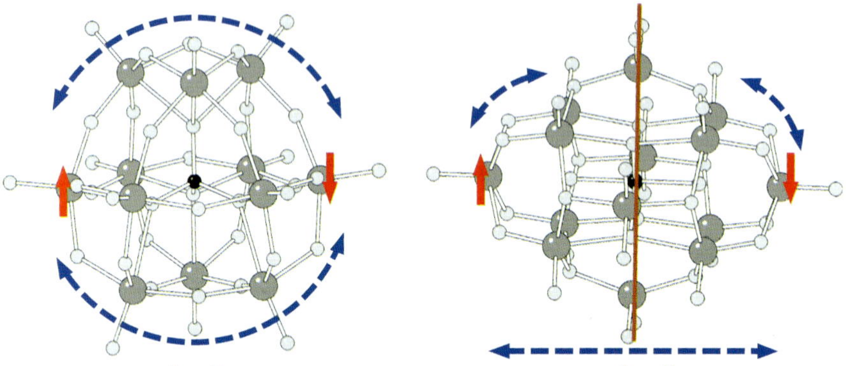

Fig. 8 (Chapter 4). Structure of the two electron reduced $[PW_{12}O_{40}]^{5-}$ and $[GeV_{14}O_{40}]^{8-}$ anions. Blue arrows show the electron delocalization: complete in the case of W_{12} giving a strong antiferromagnetic coupling between the spins, and partial in the case of V_{14} giving a weak interaction between the spins. (See page 166.)

COLOR PLATE 15

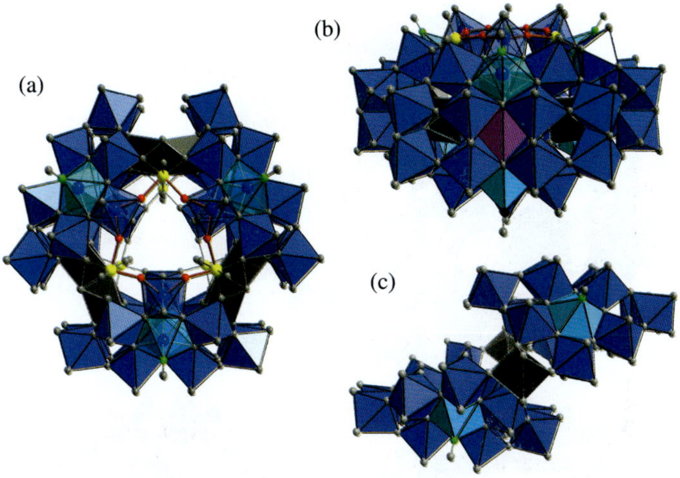

Fig. 1 (Chapter 5). Structure of the $[H_3Mo_{57}V_6(NO)_6O_{183}(H_2O)_{18}]^{21-}$ (= $\{Mo_{57}V_6\}$)-type cluster anion, determined from single-crystal X-ray diffraction analysis of its ammonium salt, $(NH_4)_{21}\{Mo_{57}V_6\}\cdot 65H_2O$. The Mo atoms and MoO_n polyhedra of the $\{Mo_{17}(NO)_2\}$ groups are shown in blue (the MoO_7 pentagonal bipyramids in light blue); the $\{Mo^V_2\}$ groups as dark gray polyhedra. O: light gray, N: green, V: yellow spheres. (a) View along the molecular C_3 axis in central projection highlighting the frontal of the two $(\text{–V–O–Mo–O–})_3$ pores by brown bonds and red O atoms. The coordination polyhedra of the Mo positions belonging to this pore as well as to the adjoined seven-coordinated Mo sites are shown transparent. This pore motif represents the general superexchange pathway of the M_3 groups in both the $\{Mo_{57}M_6\}$ structures as well as the $\{Mo_{72}M_{30}\}$-type Keplerates. (b) Side view approx. perpendicular to the view in (a), showing the symmetrical structure of the arched $\{Mo_{17}(NO)_2\}$ building block which consists two $\{Mo_8(NO)\}$ groups joined by an additional MoO_6 octahedron (purple). (c) Structure of the $\{Mo_{36}(NO)_4\}$ anion consisting of two $\{Mo_{17}(NO)_2\}$ groups fused to two MoO_6 octahedra (dark grey). (See page 175.)

COLOR PLATE 16

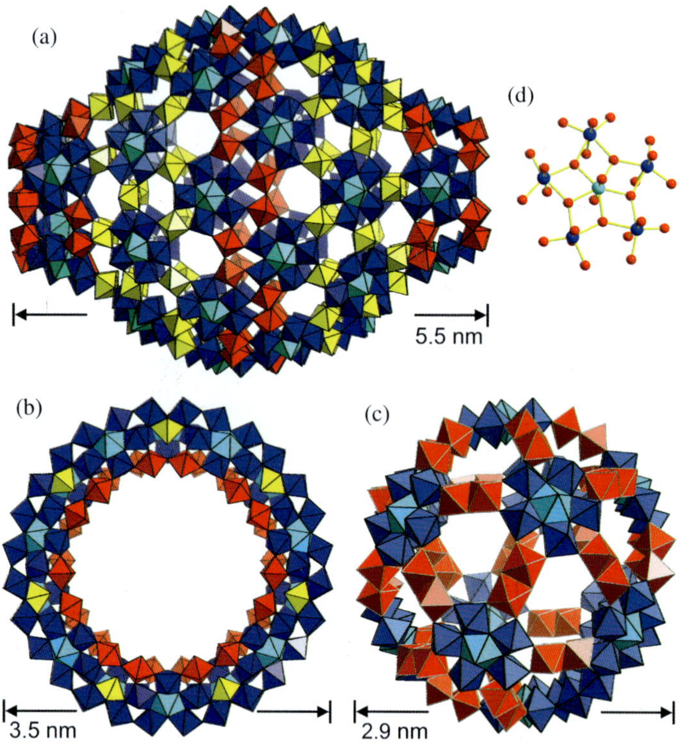

Fig. 2 (Chapter 5). Comparison of the mixed-valent polyoxomolybdate(V/VI) cluster anions $\{Mo_{368}\}$ (a), $\{Mo_{154}\}$ (b), and $\{Mo_{132}\}$ (c) which are all based on the pentagonal $\{(Mo)Mo_5\}$ building block (light/dark blue) that is displayed in ball-and-stick representation in (d). Red: $\{Mo_2\}$-type groups, yellow: mononuclear Mo linker groups. Additional ligand groups (sulfate in the case of $\{Mo_{368}\}$, carboxylates in the case of $\{Mo_{132}\}$) are omitted for clarity. (See page 177.)

COLOR PLATE 17

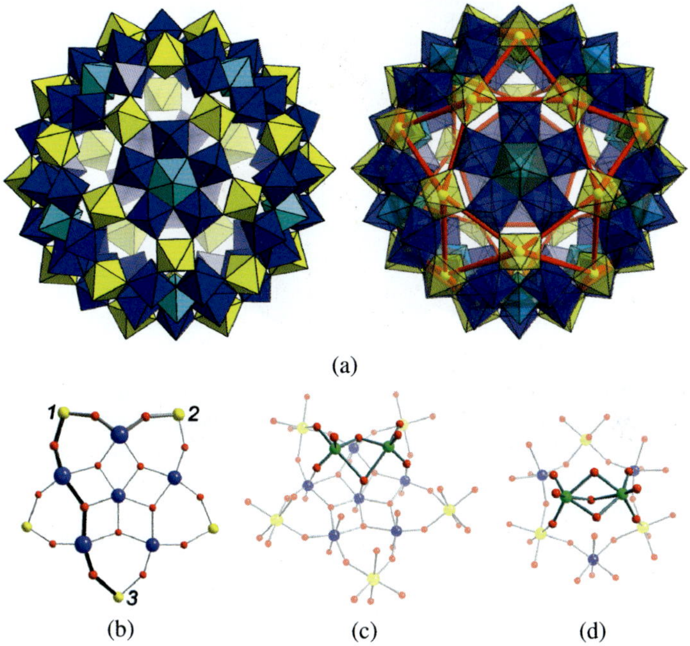

Fig. 3 (Chapter 5). (a) Polyhedral representation of the (charge-neutral) $\{Mo_{72}Fe_{30}\}$ cluster (left) and with its inscribed Fe_{30} icosidodecahedron (red lines, right). Acetate and $\{Mo_2\}$-type ligands coordinating to Fe and Mo positions from the inside of the cluster sphere are omitted. (b) Ball-and-stick representation of a substructure of the $\{Mo_{72}Fe_{30}\}$ cluster consisting of an $\{(Mo)Mo_5\}$ group with five surrounding Fe centers (Mo: blue, Fe: yellow, O: red). The nearest-neighbor exchange pathway between Fe positions 1 and 2 via a –O–Mo–O– bridge is indicated by thick gray bonds, the next-nearest neighbor exchange pathway to Fe position 3 by thick dark gray bonds. (c, d) Structural details and coordination modes of the $[Mo_2O_7(H_2O)]^{2-}$ and the $[H_2Mo_2O_8(H_2O)]^{2-}$ ligands, (L′ and L″), respectively. (See page 179.)

COLOR PLATE 18

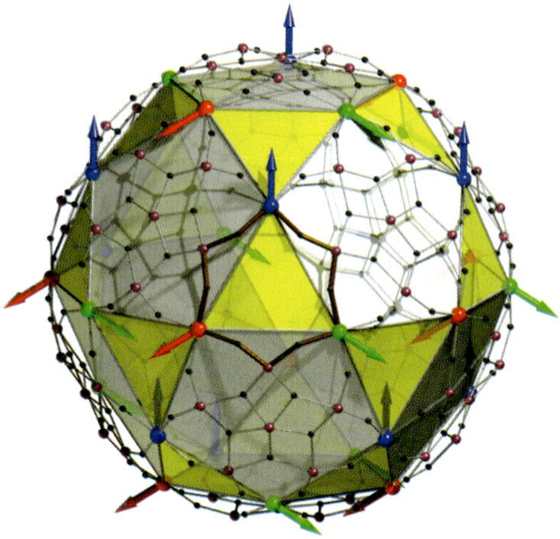

Fig. 5 (Chapter 5). Superposition of the highly symmetrically frustrated classical ground state configuration onto the $\{Mo_{72}Fe_{30}\}$ cluster structure (terminal oxygen positions and ligands omitted). The ten classical spin vectors of each of the three sublattices (red, green, blue) assume a relative orientation of 120°. Note that the three-colorability of the spin icosidodecahedron results in a configuration where any given spin site is surrounded by four nearest neighbors, all of which belong to different sublattices. One of the 20 triangular $(\text{–Fe–O–Mo–O–})_3$ pores is highlighted by thick brown bonds (cf. Fig. 1). (See page 181.)

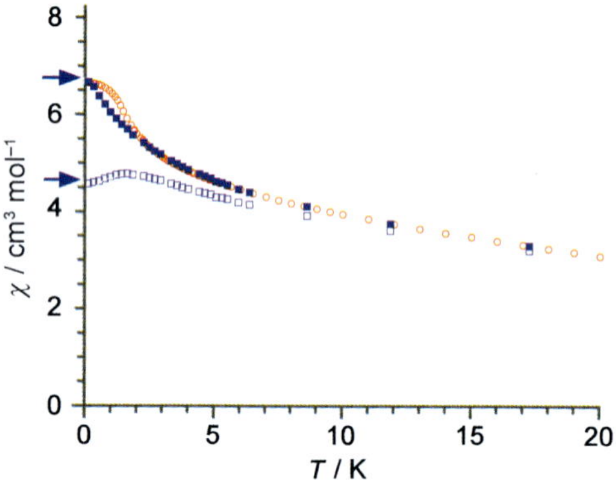

Fig. 6 (Chapter 5). Temperature dependence of the molar susceptibility χ of $\{Mo_{72}Fe_{30}\}\cdot$ca.$150H_2O$ at 0.5 Tesla (redrawn from ref. 28), showing that χ approaches a finite value at 0 K. Red circles: experimental data, blue open squares: classical Monte Carlo simulational data, blue solid squares: sum of simulational data plus contribution of paramagnetic impurity. Lower blue arrow: low-temperature limit χ_0 of the classical Heisenberg model. Upper blue arrow: value of χ_0 plus the impurity correction extrapolated to $T = 0$ K. (See page 183.)

COLOR PLATE 19

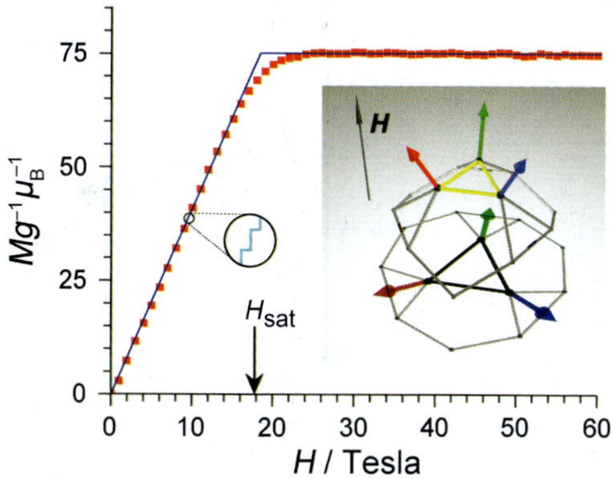

Fig. 7 (Chapter 5). Field dependence of the molar magnetization M of $\{Mo_{72}Fe_{30}\}$·ca.150H_2O at 0.46 K. Red squares: experimental data, dark blue graph: linear increase expected for the classical spin vector model until the system saturates at $H_{sat} = 17.7$ Tesla. The equidistant quantum spin level crossing steps are indicated by the magnified circular inset. Note that the actual experimental magnetization data density was much higher than shown here and would have allowed identification of the expected level crossing steps. Inset: Simplified representation of the 'umbrella' spin phase (of classical spin vectors) is shown on a section of the Fe_{30} spin array. This spin phase is characterized by a continuous alignment of the spin vectors of the three (red, green, blue) sublattices with an external magnetic field vector, while the relative 120° angular orientation is retained in the projection perpendicular to the H vector. (See page 184.)

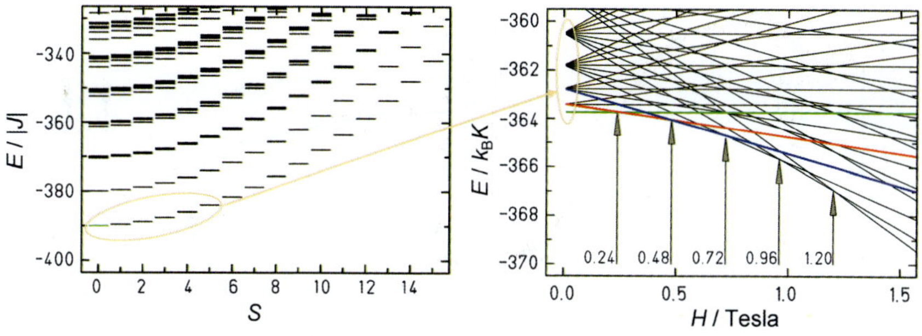

Fig. 8 (Chapter 5). *Left*: Lowest and some excited parabolic "rotational bands", i.e. sets of spin multiplet eigenvalues, as a function of the total spin quantum number S resulting from the approximate quantum model for $\{Mo_{72}Fe_{30}\}$. Note that the ground state is a singlet (green). This model results in a singlet ground state (green level). *Right*: Zeeman splitting caused by an external field separates the M_S substates. The lowest substates intersect at equal field increments of 0.24 Tesla, causing spin level crossing (compare magnified steps in Fig. 7). (See page 185.)

COLOR PLATE 20

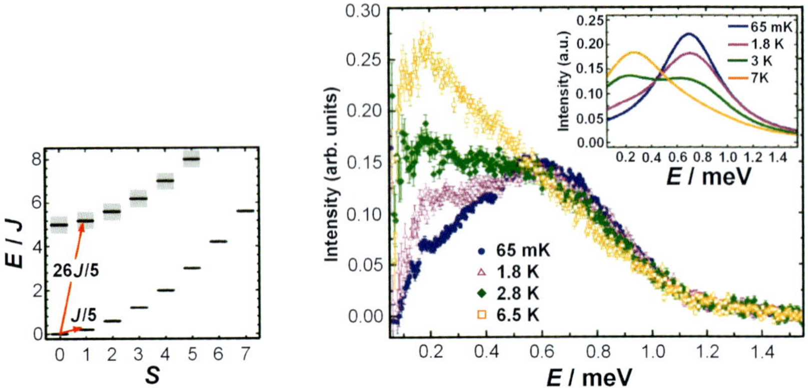

(a) (b)

Fig. 9 (Chapter 5). (a) Energy levels calculated for the lowest and the first excited rotational band for $\{Mo_{72}Fe_{30}\}$ (compare Fig. 8). Red arrows indicate allowed transitions from the $S = 0$ ground state to states characterized by $S = 1$. The shading of the energy levels of the first excited rotational band represents potential deviations from the ideal single-J model due to the fact that the high degeneracy of these levels can be partially lifted by a variety of perturbations. (b) Temperature evolution of the magnetic contribution of the inelastic neutron scattering intensity of deuterated $\{Mo_{72}Fe_{30}\}$, compared to the theoretical scattering patterns for different temperatures (inset), calculated using the quantum rotational band model and assuming identical probabilities for all allowed transitions. (See page 187.)

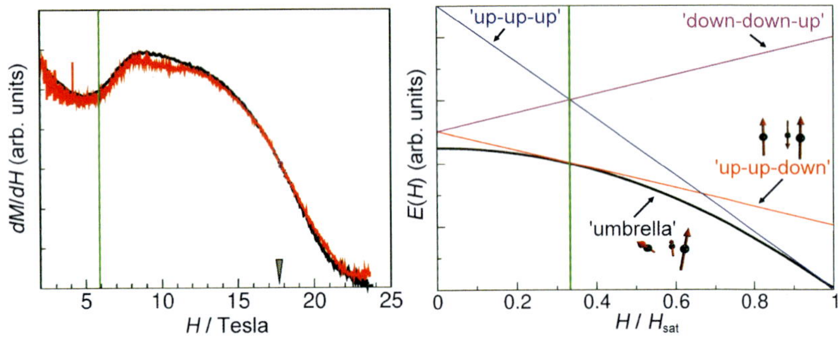

Fig. 10 (Chapter 5). *Left*: Field dependence of the experimental differential susceptibility dM/dH of $\{Mo_{72}Fe_{30}\}$ at 0.42 K. Black and red graphs indicate increasing and decreasing external field in the pulsed-field measurements, respectively. A local minimum is visible slightly below one-third of the saturation field H_{sat} ($H_{sat}/3$, indicated by a vertical green line; a gray arrow indicates H_{sat}). *Right*: The competing spin phases in $\{Mo_{72}Fe_{30}\}$ (all of them characterized by three sublattices). While the 'umbrella' spin phase is the most stable phase below saturation, the 'up-up-down' spin phase approaches the energy of the 'umbrella' phase in the vicinity of $H_{sat}/3$. Thus the 'up-up-down' spin phase is thermally populated at finite temperatures, which reduces the overall differential susceptibility. (See page 188.)

COLOR PLATE 21

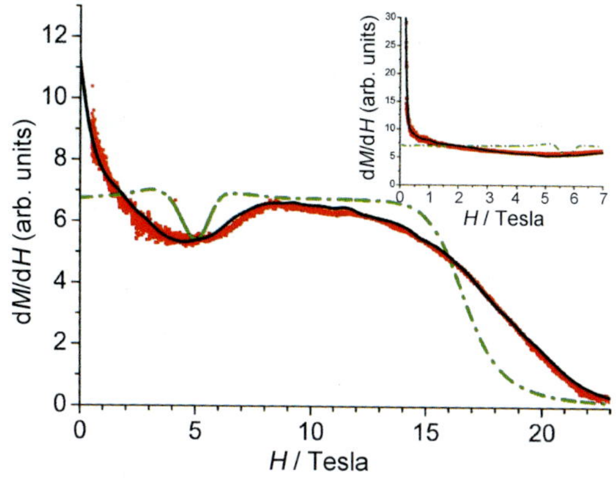

Fig. 11 (Chapter 5). Experimental (red dots) and simulated (black graph: multi-*J* model, green dashed graph: single-*J* model) field dependence of d*M*/d*H* for $\{Mo_{72}Fe_{30}\}$ at 420 mK and 60 mK (inset). Multi-*J* simulations were performed for an ensemble of 100 Fe_{30} spin clusters, where the average *J* (i.e. averaged over all 60 nearest-neighbor Fe···Fe contacts per molecule) was allowed to vary by ±15% over the entire ensemble, while the spread of individual J values within each $\{Mo_{72}Fe_{30}\}$ molecule was allowed to vary by ±40%. (See page 189.)

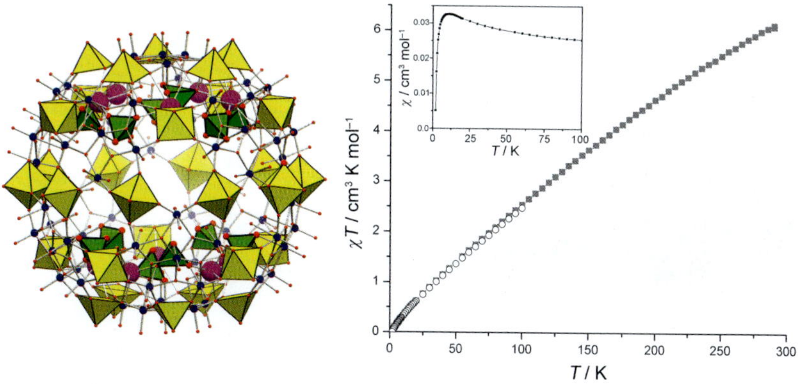

Fig. 12 (Chapter 5). *Left*: Structure of the $\{Mo_{72}V_{30}\}$ polyanion, emphasizing structural features that cause deviations from an ideal icosahedral geometry of the metal skeleton: The two coplanar $(KSO_4)_5$ rings (K: large purple spheres, sulfate: green tetrahedra) that coordinate to the inner surface of the cluster, specifically to 10 of the 12 $\{(Mo)Mo_5\}$ building blocks. Also note that ten outer V positions reside in a square-pyramidal environment, while the remaining 20 V centers adopt an octahedral coordination geometry. Mo: blue, O: red spheres. *Right*: Experimental susceptibility data for the $\{Mo_{72}V_{30}\}$-based compound (grey squares: 0.1 Tesla, open circles: 1.0 Tesla, corrected for diamagnetic and temperature-independent paramagnetism contributions). The immediate onset of χT even at 2 K and the maximum in χ vs. T at approx. 11 K (inset) indicate the presence of low-lying multiplet states that can be thermally populated even at low temperatures. (See page 194.)

COLOR PLATE 22

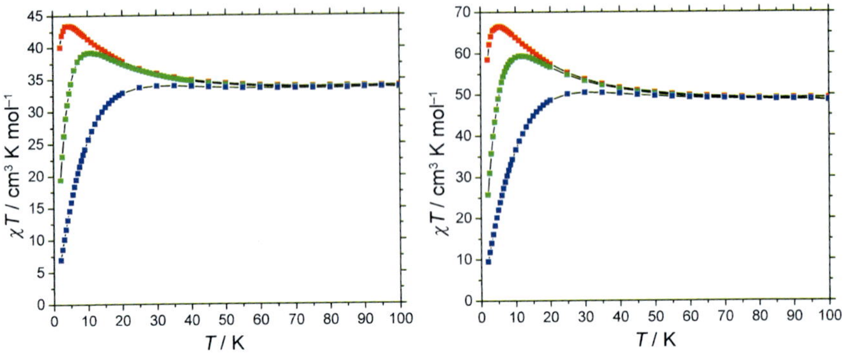

Fig. 13 (Chapter 5). Field-induced suppression of the ferromagnetic coupling as visible from the field dependence of χT curves for $K_{23}Na_4[\{Mo^{V/VI}_6O_{21}(H_2O)_3(SO_4)\}_{12}(V^{IV}O(H_2O))_{15}(Fe^{III}(H_2O)_2)_7(Mo^{V}O(H_2O))_8]\cdot 150H_2O$ (left) and $K_{21}Na_2[\{Mo^{V/VI}_6O_{21}(H_2O)_3(SO_4)\}_{12}(V^{IV}O(H_2O))_{11}(Fe^{III}(H_2O)_2)_{11}(Mo^{V}O(H_2O))_8]\cdot 150H_2O$ (right). 0.1 Tesla: red, 1.0 Tesla: green, 5.0 Tesla: blue data points. For 0.1 Tesla, the maxima are reached at 4.5 K for the $\{Mo_{72}V_{15}Fe_7Mo_8\}$ and at 5.5 K for the $\{Mo_{72}V_{11}Fe_{11}Mo_8\}$ compound, where the χT values correspond to $S=9$ and $S=11$ total spin states, respectively. (See page 195.)

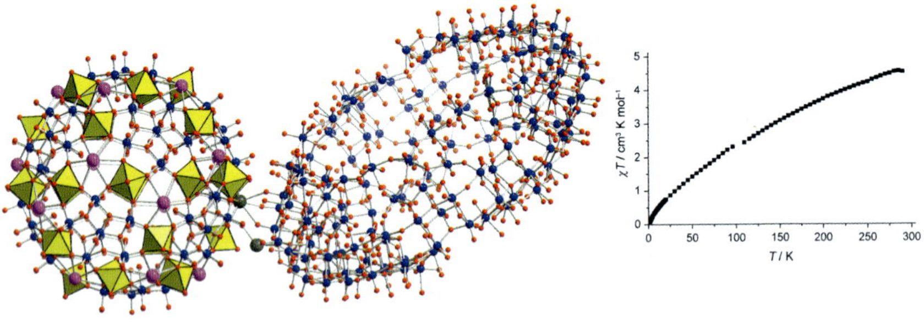

Fig. 14 (Chapter 5). *Left*: Simplified structure of the "ring-sphere" assembly $\{Mo_{72}V_{30}\}K_2\{Mo_{142}\}$ as it occurs in the solid state. Potassium cations coordinated to the Keplerate moiety as purple spheres, the two K^+ positions interlinking the cluster sphere and the cluster ring are shown as grey spheres. V: yellow octahedra, Mo: blue, O: red spheres. The sulfate groups in the Keplerate sphere are fully disordered and are not shown. *Right*: Temperature dependence of χT of $K_{41}Na_{11}(VO)_4\{Mo_{72}V_{30}\}K_2\{Mo_{142}\}\cdot$ca.$500H_2O$ at 0.1 Tesla (corrected for the contributions of the four vanadyl cations disordered in the crystal lattice). (See page 196.)

COLOR PLATE 23

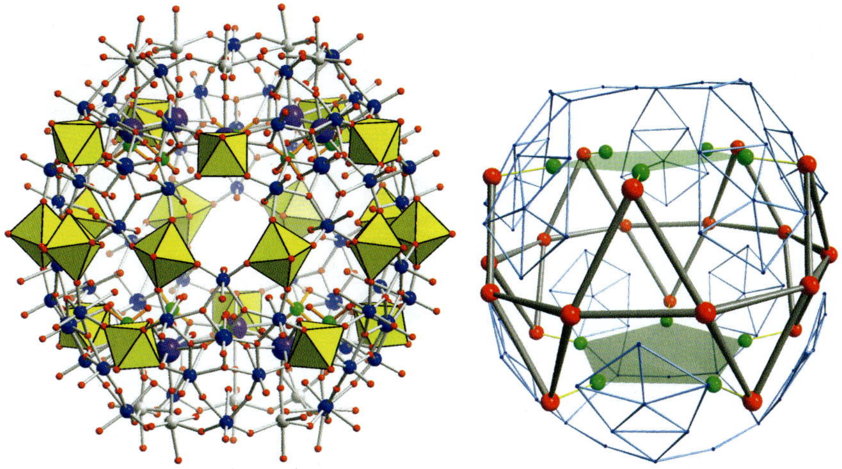

(a) (b)

Fig. 15 (Chapter 5). (a) Structure of the $\{Mo_{75}V_{20}\}$ cluster anion highlighting the two $\{NaSO_4\}_5$ rings similar to the pentagonal $\{KSO_4\}_5$ rings in $\{Mo_{72}V_{30}\}$ (V: yellow octahedra, Na: purple, S: green, Mo: blue, O: red spheres; Mo positions with occupancy factors of 0.5 are shown as gray spheres). (b) Simplified presentation of the metal skeleton of the $\{Mo_{75}V_{20}\}$ cluster from which the 10 alternating V_3-type isosceles triangles (red spheres, thick gray bonds) can be recognized. (See page 197.)

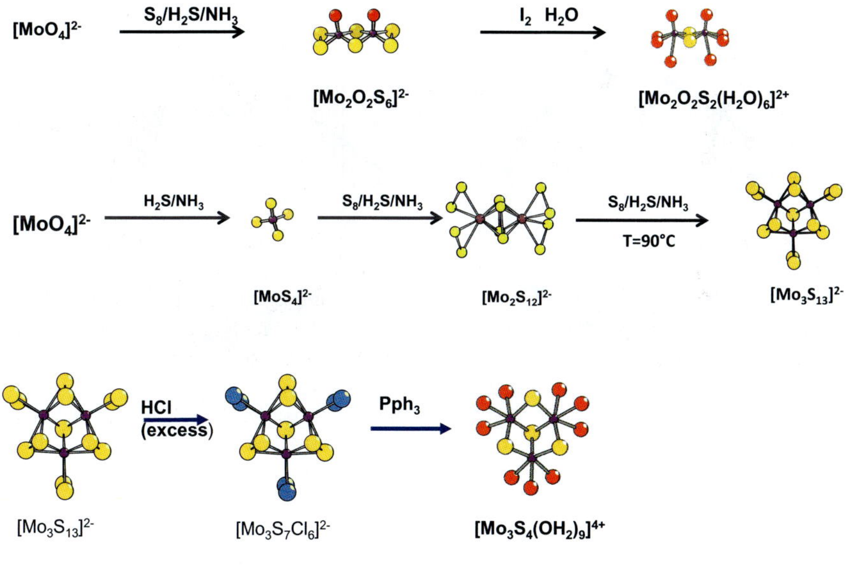

Fig. 6 (Chapter 7). Formation pathway of thiometalates derived from the $[MoO_4]^{2-}$ precursor (S in yellow, O in red). (See page 253.)

COLOR PLATE 24

(a) $M_3S_4(H_2O)_6$ group

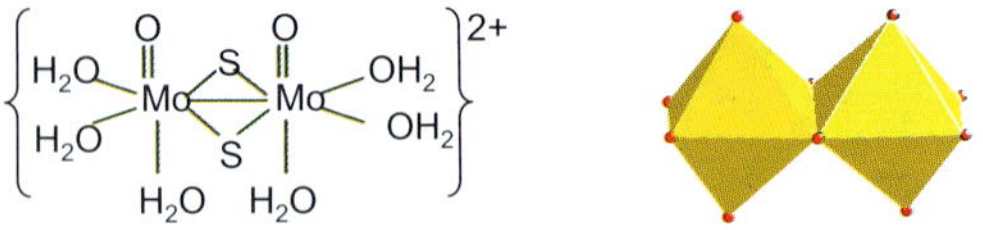

(b) $M_3S_4(H_2O)_9$ group

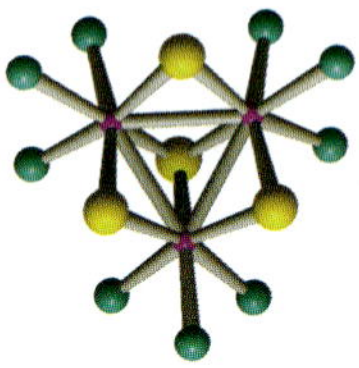

(c) M_3O_{13} group

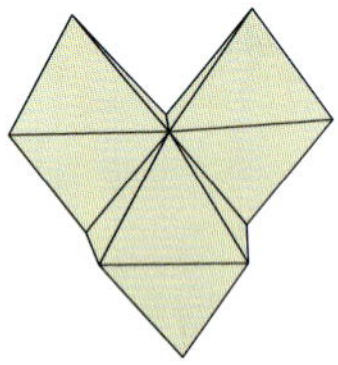

Fig. 7 (Chapter 7). Representation of the $[Mo_2O_2S_2(H_2O)_6]^{2+}$ and $[Mo_3S_4(H_2O)_9]^{4+}$ thio-precursor used in self condensation process (a) and in coordination to vacant POMs (a and b.) In grey M^{VI} and in yellow M^{V} metal cations. (See page 254.)

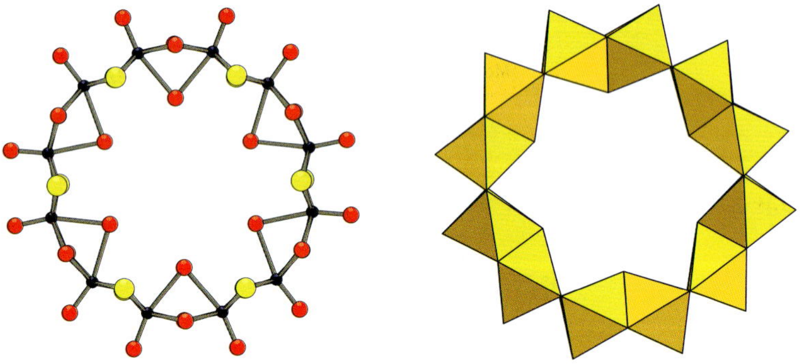

Fig. 8 (Chapter 7). Molecular structure of $[Mo_{12}O_{12}S_{12}(OH)_{12}(H_2O)_6]$ in ball and stick (Mo in blue, S in yellow, O in red) and polyhedral representations (Mo^{V} as yellow polyhedra). (See page 256.)

COLOR PLATE 25

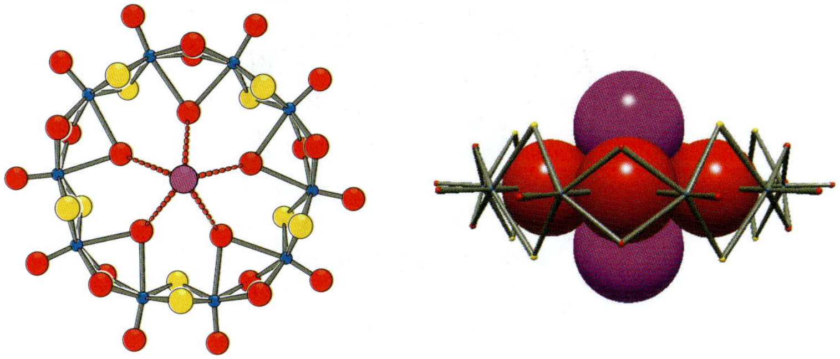

Fig. 9 (Chapter 7). Representations of $[X_2Mo_{10}S_{10}O_{10}(OH)_{10}(H_2O)_5]^{2-}$ in ball and stick and in space filling models (Mo in blue, S in yellow, O in red and X in purple). (See page 258.)

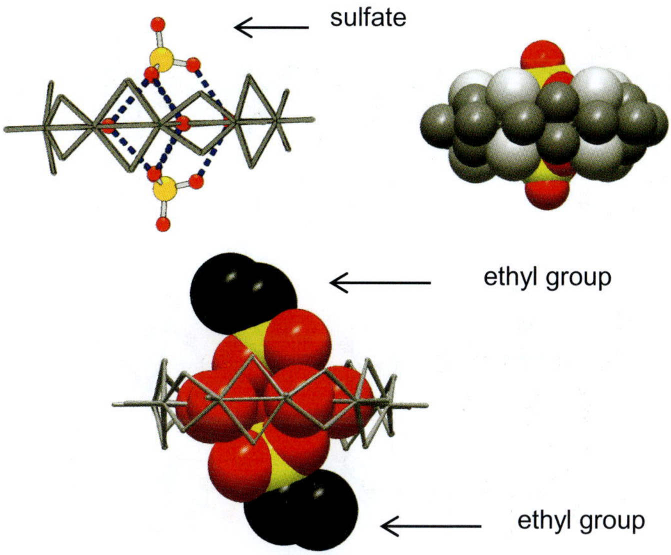

Fig. 11 (Chapter 7). Mo_{10}-ring stabilized by sulphate and ethylsulfonate groups (O in red). (See page 259.)

COLOR PLATE 26

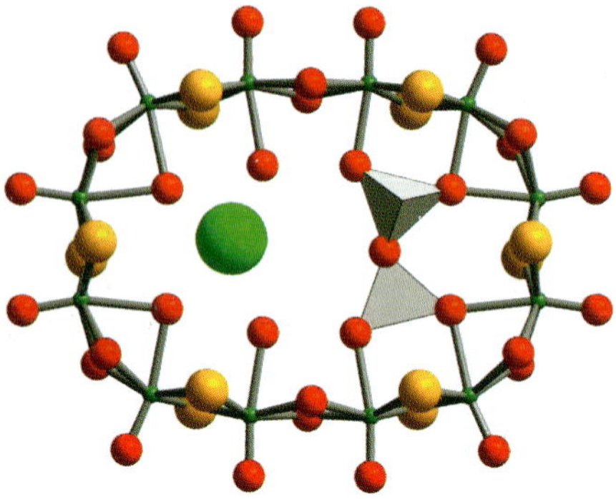

Fig. 12 (Chapter 7). View of the molecular structure of $[Cl(P_2O_7)Mo_{12}O_{12}S_{12}(OH)_{12}(H_2O)_4]^{5-}$ (Mo in blue, S in yellow, O in red and Cl in green). (See page 260.)

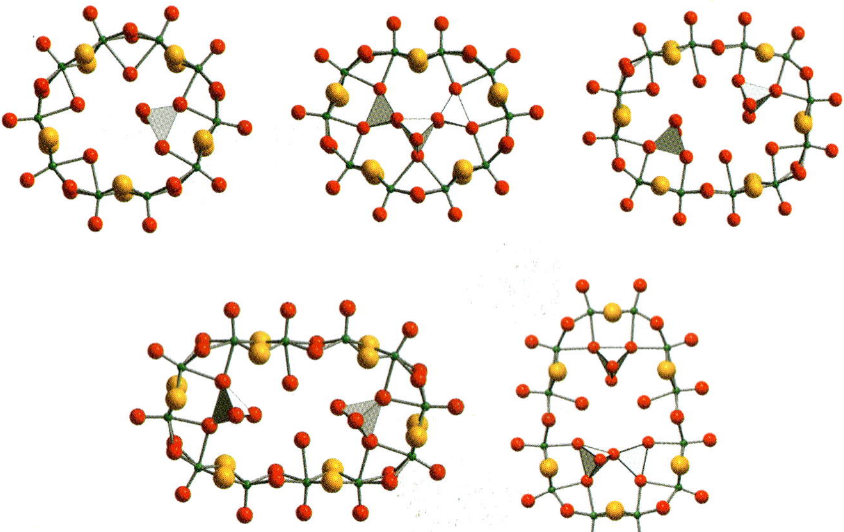

Fig. 14 (Chapter 7). Ball and stick representations of: top, $[(HPO_4)Mo_{10}O_{10}S_{10}(OH)_{10}(H_2O)_3]^{2-}$, $[(P_3O_{10})_2Mo_{10}O_{10}S_{10}(OH)_{10}]^{5-}$, $[(HPO_4)_2Mo_{12}O_{12}S_{12}(OH)_{12}(OH)_{2}]^{6-}$, and below, $[(HPO_4)_2Mo_{12}O_{12}S_{12}(OH)_{12}(H_2O)_2]^{6-}$ and $[(HPO_4)(P_2O_7)Mo_{12}S_{12}O_{12}(OH)_{12}(H_2O)_2]^{6-}$, (Mo in blue, S in yellow, O in red and phosphate and diphosphate guests in polyhedral models). (See page 262.)

COLOR PLATE 27

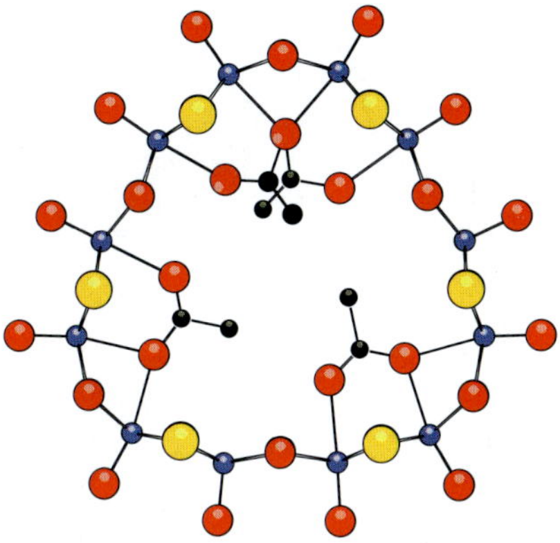

Fig. 15 (Chapter 7). Representation of the triangular shaped $[Mo_{12}O_{12}S_{12}(OH)_{12}(CH_3COO)_3]^{3-}$ (Mo in blue, S in yellow, O in red and C in black). (See page 263.)

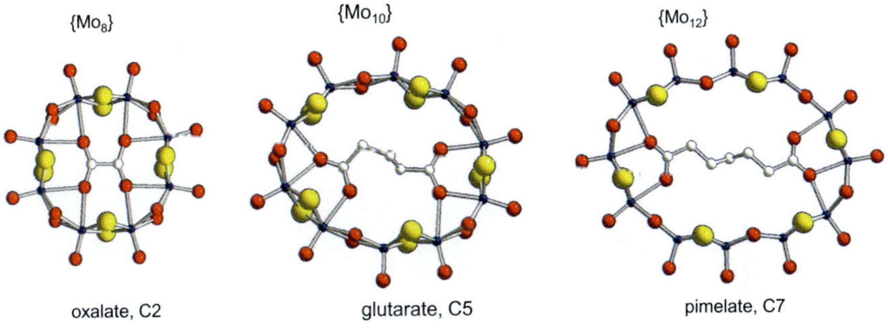

Fig. 17 (Chapter 7). Molecular structures of $\{Mo_8Ox\}^{2-}$, $\{Mo_{10}Glu\}^{2-}$, and $\{Mo_{12}Pim\}^{2-}$ showing the templating effect on the size and shape of the host-guest compounds (Mo in blue, S in yellow, O in red and C in black). (See page 266.)

COLOR PLATE 28

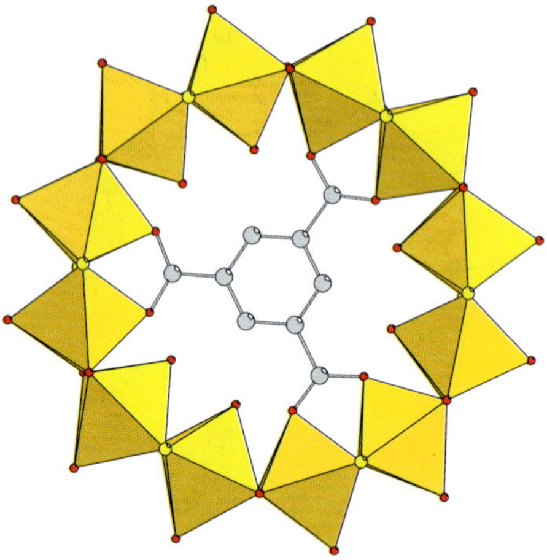

Fig. 18 (Chapter 7). Representation of the perfectly circular $[Mo_{12}O_{12}S_{12}Trim]^{3-}$ (Mo in blue, S in yellow, O in red and C in black). (See page 267.)

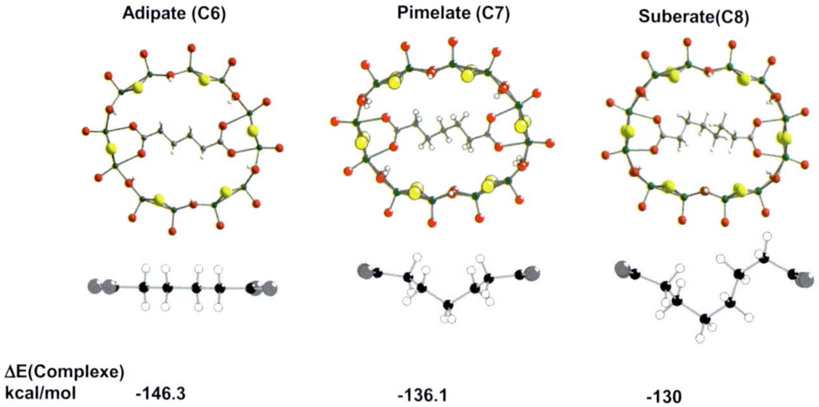

Fig. 19 (Chapter 7). Top, view of the distorsion generated on increasing the length of the inner alkyl chain (Mo in blue, S in yellow, O in red and C in black), below in black, view of the optimized DFT geometries of the alkyl chains with the corresponding energies. (See page 268.)

COLOR PLATE 29

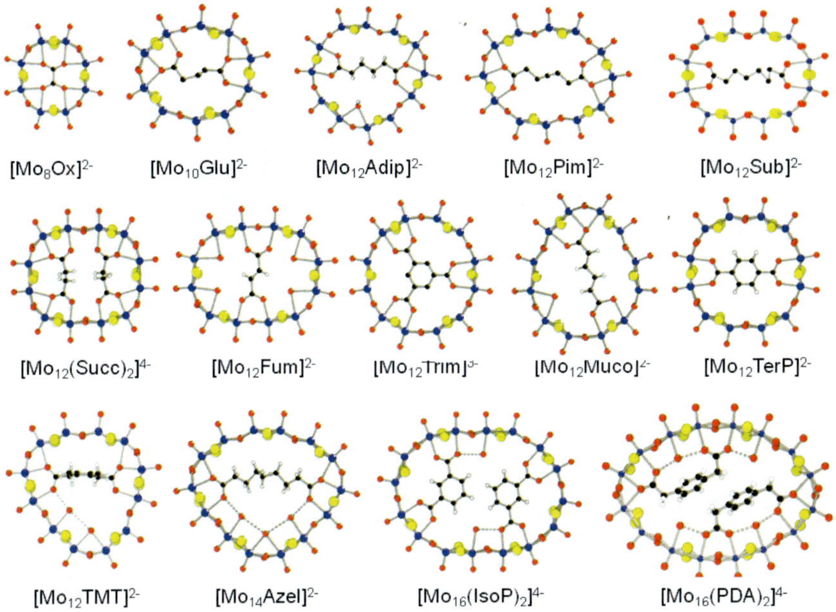

Fig. 20 (Chapter 7). Host-guests compounds obtained with flexible and/or rigid dicarboxylate anions (Mo in blue, S in yellow, O in red and C in black).(Copyright © Wiley 2011.) (See page 268.)

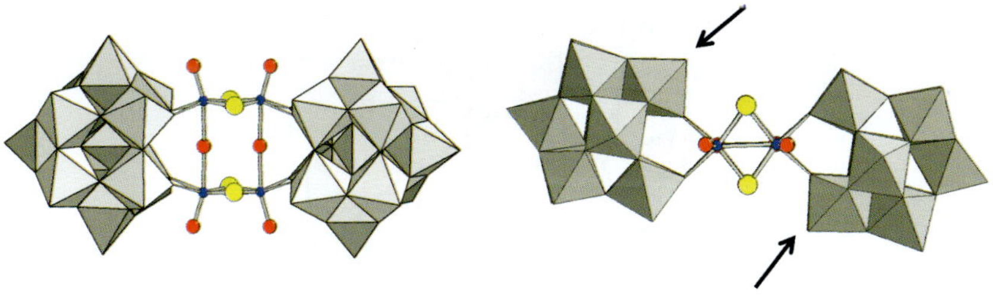

(a) (b)

Fig. 27 (Chapter 7). (a) View of the transoid isomer $[(PW_{11}O_{39})_2(Mo_4S_4O_4(OH_2)_2)]^{10-}$; (b) the two equivalent octaedra maked with arrows are 180° staggered relatively to the two $\{PW_{11}\}$. The central cluster $\{H_4Mo_4S_4O_6\}$ is represented in ball and stick model, (Mo in blue, S in yellow, O in red), a coordination polyhedra representation is used for the two $\{PW_{11}\}$. (See page 279.)

COLOR PLATE 30

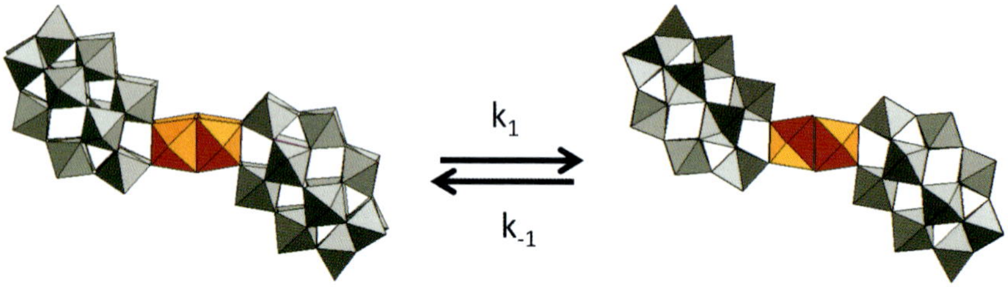

transoid *cisoid*

Fig. 28 (Chapter 7). Representation of the structural change resulting from the *transoid-cisoid* isomerization of $[(P_2W_{17}O_{61})_2(Mo_4S_4O_4(H_2O)_2)]^{16-}$ in solution (Mo^{VI} in grey, Mo^{V} in yellow). (See page 280.)

Fig. 31 (Chapter 7). Polyhedra representation of $[(H_2P_2W_{15}O_{56})_4\{Mo_2O_2S_2(H_2O)_2\}_4\{Mo_4S_4O_4(OH)_2(H_2O)\}_2]^{28-}$ (Mo^{VI} in grey polyhedra, Mo^{V} in yellow polyhedra). (See page 283.)

COLOR PLATE 31

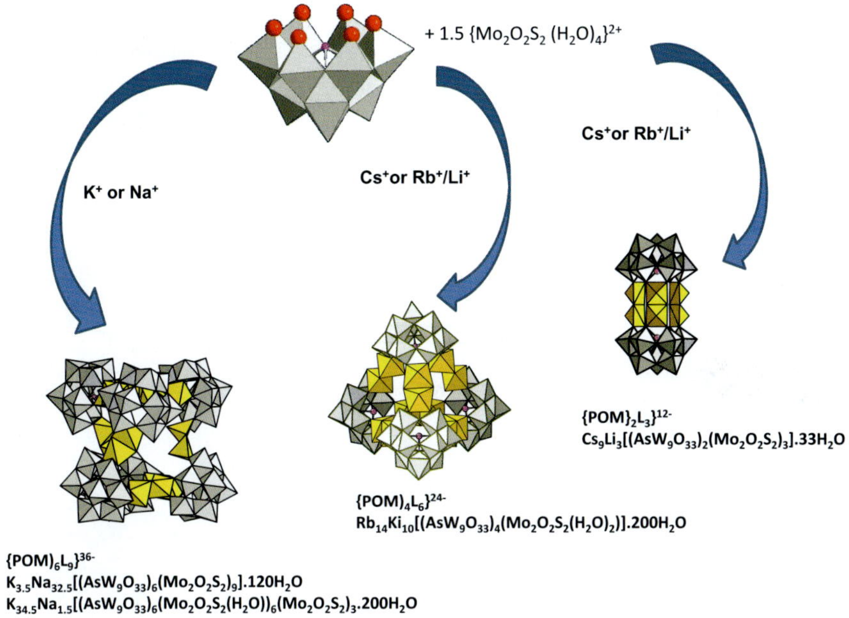

Fig. 32 (Chapter 7). Diagrammatic representation of the conditions of formation of $\{POM_2L_3\}^{12}$, $\{POM_4L_6\}^{24-}$ and $\{POM_6L_9\}^{36-}$, with their molecular geometries and chemical compositions, (M^{VI} in grey, M^{V} in yellow). (See page 285.)

COLOR PLATE 32

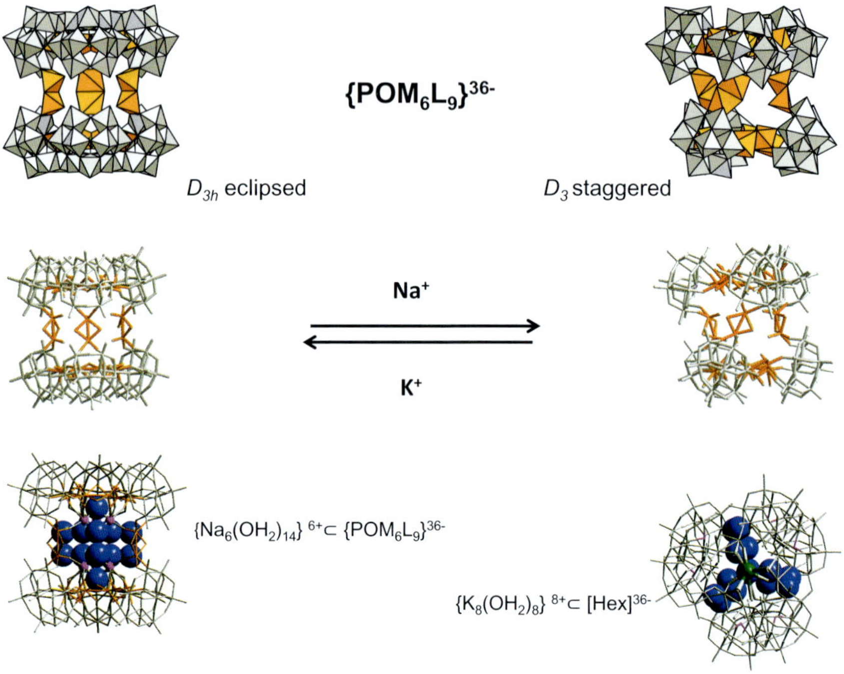

Fig. 33 (Chapter 7). Representations of the two conformations eclipsed and staggered of $\{POM_6L_9\}^{36-}$ in stick and polyhedra models. View of the aquo clusters $[Na_6(H_2O)_{14}]^{6+}$ and $[K_8(OH_2)_8]^{86+}$ trapped in the respective structures, (M^{VI} in grey, M^{V} in yellow, Na^+ and K^+ cations as blue spheres). (See page 286.)

COLOR PLATE 33

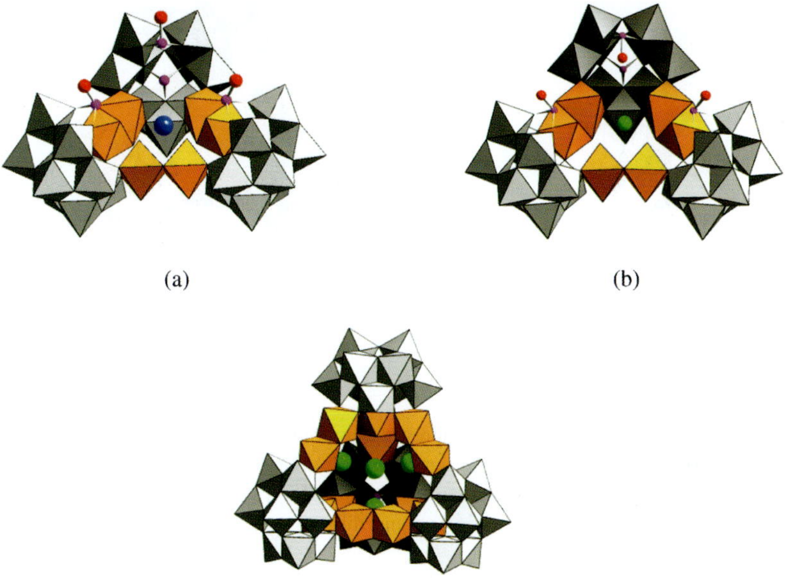

(a) (b) (c)

Fig. 36 (Chapter 7). Views in polyhedra models of (a) (up, up, up) and (b) (up, up, down) isomers of $[(HAs_2W_9O_{34})_3(Mo_2O_2S_2(OH_2)_2)_3]^{15-}$; (c) $[(\beta\text{-}AsW_9O_{33})_4(Mo_2O_2S_2(OH_2)_2)_6]^{24-}$ obtained after oxidation by iodine and reaction with $\{Mo_2O_2S_2\}$, (M^{VI} in grey, M^{V} in yellow, cations as colored spheres). (See page 289.)

(a) (b) (c)

$[Mo_8S_8O_8(OH)_8Mo(W)(O)OH_2)]^{2-}$

$[Mo_{40}S_8O_{120}]^{24-}$

$[\{W_6O_{21}(H_2O)_6\}_{12}\{Mo_2O_2S_2(CH_3COO)\}_{20}\{Mo_2O_2S_2(H_2O)_2\}_{10}]^{32-}$

Fig. 37 (Chapter 7). Representations in polyhedra of (a) $[M_9S_8O_{10}(OH)_{10}(H_2O)]^{2-}$, (b) $[Mo_{40}S_8O_{120}]^{24-}$ and (c) common backbone of the oxo $[Mo_{132}O_{372}(SO_4)_{30}(OH_2)_{72}]^{72-}$ and thio $[Mo_{132}S_{60}O_{312}(SO_4)_{15}(OH_2)_{102}]^{42-}$ Kleperates (M^{VI} in grey, M^{V} in yellow). (See page 290.)

COLOR PLATE 34

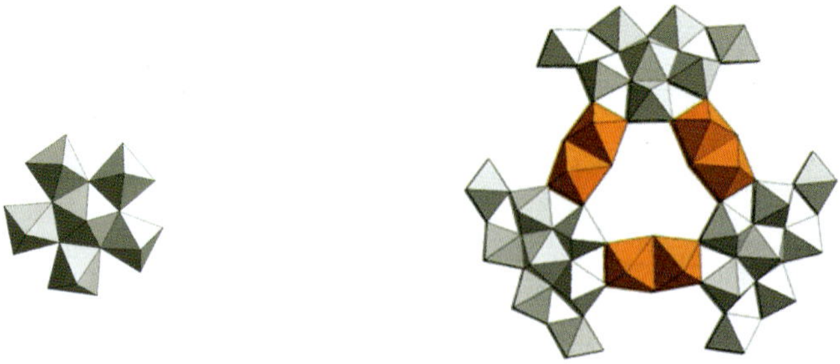

(a) (b)

$\{Mo_6O_{21}\}^{6-}$ $\{Mo_6O_{21}\}_3\{M_2O_2X_2\}_3$

Fig. 38 (Chapter 7). (a) Polyhedra representation of the $\{M_6O_{21}\}^{6-}$ pentagon$^-$, M = Mo, W, formed of a central MO_6 pentagon surrounded by five MO_6 octahedra; (b) view of the structural motif formed of 3 pentagons (in grey) and 3 $\{Mo_2O_2X_2\}$ X = O or S linkers ensuring the structural connectivity in oxo and thio Mo_{132} Kleperates (M^{VI} in grey, M^{V} in yellow). (See page 291.)

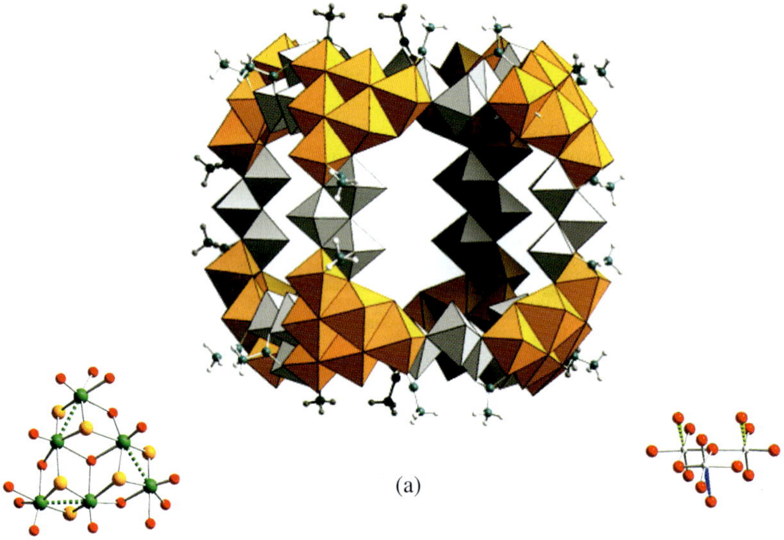

(a)

(b) (c)

Fig. 39 (Chapter 7). (a) Polyhedra representation of $[H_{32}Mo_{84}S_{48}O_{200}(AcO)_{24}]^{46-}$; $\{Mo_6S_6O_6(H)\}$ units located at the corners of the cube (in yellow); $\{Mo_3O_{11}\}$ linkers located at the edges of the cube (in grey); (b) ball and stick model of $\{Mo_6S_6O_6(H)\}$ units; (c) ball and stick model of $\{Mo_3O_{11}\}$ linkers (Mo in green, S in yellow, and O in red). (See page 293.)

COLOR PLATE 35

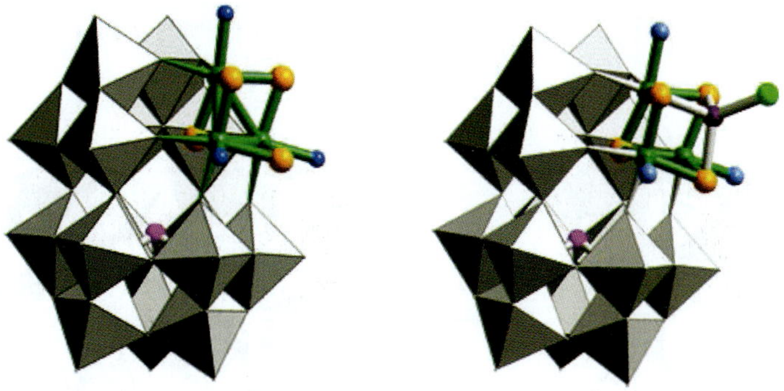

(a) (b)

Fig. 40 (Chapter 7). Representations of (a) $[AsW_{15}Mo_3S_4(H_2O)_3O_{53}]^{9-}$ showing the substitution of a $W_3O_{13}\}$ group by Mo_3S_4 and (b) $[AsW_{15}Mo_3S_4CuCl(H_2O)_3O_{53}]^{9-}$, As^{III} in pink, W^{VI}, in grey, Mo^{IV} in small green spheres, S in yellow, Cu in dark purple, oxygen of terminal water molecules in blue. (See page 295.)

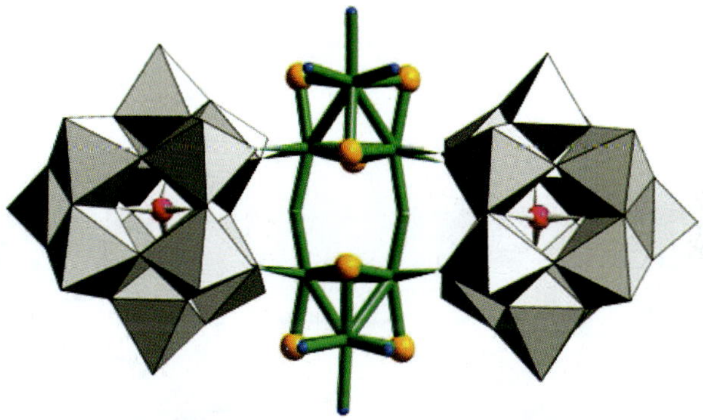

Fig. 41 (Chapter 7). Stick representation of $[(SiW_{11}O_{39})Mo_3S_4(H_2O)_3(\mu\text{-}OH))_2]^{10-}$ showing the two Mo_3S_4 groups blocked between two POMs subunits. W^{VI} in grey polyhedra, two Mo_3S_4 groups in ball and stick model. (See page 296.)

COLOR PLATE 36

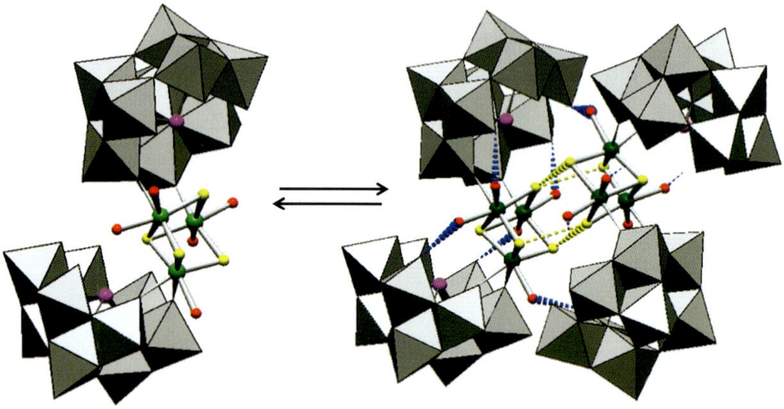

(a) (b)

Fig. 42 (Chapter 7). Representations of (a) the $[(H_4AsW_9O_{33})_2(Mo_3S_4(H_2O)_5)]^{6-}$ "monomer" and (b) the $[(H_4AsW_9O_{33})_2(Mo_3S_4(H_2O)_5)]_2^{12-}$ "dimer". W are grey octahedra, Mo are green spheres, S in yellow and O in red. Hydrogen bonds as blue dotted lines and S-S contacts in yellow ones. (See page 297.)

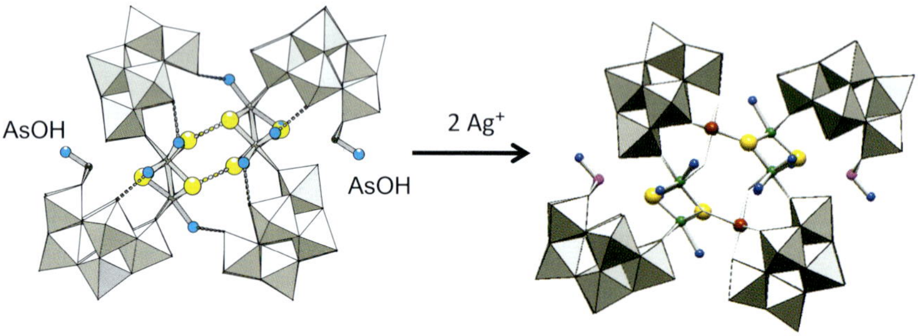

(a) (b)

Fig. 43 (Chapter 7). (a) Representation of $\{[(H_2As_2W_9O_{34})(H_4AsW_9O_{33})(Mo_3S_4(H_2O)_5)]_2\}^{14-}$ showing the two AsOH groups grafted on AsW9 units; (b) view of $\{[Ag_2(H_2As_2W_9O_{34})(H_2AsW_9O_{33})(Mo_3S_4(H_2O)_5)]_2\}^{16-}$, Ag^+ ions are dark purple spheres linked to sulfur atoms of Mo_3S_4 and oxygen atoms of {AsW9} POMs units. (See page 298.)

Chapter 1

POLYOXOMETALATE-PROTECTED METAL NANOPARTICLES: SYNTHESIS, STRUCTURE AND CATALYSIS

YIFENG WANG
School of Chemistry and Chemical Engineering, Shandong University, Jinan 250199, PR China

IRA A. WEINSTOCK
Department of Chemistry, Ben Gurion University, POB 653, Beer Sheva, 84105 Israel

1. Introduction

Metal-oxygen cluster–anions (polyoxometalates, or POMs)[1,2] constitute a large and rapidly growing class of discrete molecular structures with applications ranging from catalysis[3,4] to functional materials.[5] POMs are inexpensive, minimally or non-toxic, negatively-charged clusters comprised of early-transition metals, usually in their d^0 or d^1 electronic configurations (e.g., V(V), Mo(VI) and Mo(V), or W(VI)), bridged by oxygen atoms (formally O^{2-}, or occasionally HO^-, ions). Representative POMs range from small isopolyanions, such as $W_6O_{19}^{2-}$ (<1 nm in diameter), slightly larger Keggin or Wells-Dawson heteropolyanions (e.g., $PMo_{12}O_{40}^{3-}$ and $P_2W_{18}O_{62}^{6-}$), and partially reduced "wheel-like" oxomolybdate nanoclusters[6,7] that contain up to 176 Mo atoms (4.1 nm in diameter[8]). These and related polyoxoanions, and derivatives prepared by incorporation of main-group, transition-metal,[9,10] or f-block[11–14] cations, are used as molecular models of magnetic oxides,[15] and in applications from catalysis[3,16–20] and electron transport in fuel cells,[21,22] to the design of functional-nanocomposite materials.[23] These and thousands of other POMs can be prepared in quantity via either traditional (serial and kinetically controlled) or thermodynamic (one-pot) syntheses in water. As a class, POMs possess extensive and reversible redox chemistries,[24–26] which is central to their use in many applications.[25,27–30] Their reduction potentials, acidities, and other key properties relevant to catalysis and materials science, can be extensively yet readily altered by the elemental composition of the POM cluster itself. In addition, because many redox-active POMs are well defined and stable in solution, they are deployed as physicochemical "probes" of electron-transfer processes.[31–34]

As molecular anions, it is not surprising that polyoxometalates can stabilize colloidal metal nanoparticles in solution. And, given the ease with which POMs can be converted from water-soluble to organic-solvent soluble forms (as a function of their counter cations), they have been used with considerable success to stabilize metal nanoparticles in both media. While individual bonds between the oxide ligands of polyoxometalates and metal(0)-nanoparticle surfaces appear to be considerably weaker than those, for example, between alkanethiol ligands and gold(0) surfaces, data from numerous publications demonstrate that POMs can be far more effective than simple organic or inorganic ions such as citrate or phosphate.

Studies of POMs on planar metal and graphite surfaces, by Anson,[35] Nadjo and Keita,[36–38] Klemperer,[39] Gewirth[40–42] and Barteau,[43,44] and others,[45] demonstrated the formation of POM monolayers with well-defined packing geometries. During that time, Finke[46–49] demonstrated that POMs stabilized Ir(0) and Rh(0) nanaoparticles in organic solvents through a combination of the electrostatic mechanisms common to anion-stabilized colloids and the steric mechanisms typical of, for example, alkanethiol-protected gold(0) nanoparticles. The combined electrostatic and steric mechanism was proposed in response to data indicating direct bonding between POMs and the metal(0) nanoparticle surface. These direct interactions are entirely consistent with the formation of highly stable POM monolayers on planar surfaces.

More recently, Weinstock used cryogenic transmission electron microscopy (cryo-TEM) to image POM-stabilized Ag(0) and Au(0) nanoparticles in water.[50,51] Well-defined POM monolayers were observed covering the nanoparticles. The close distances between POMs on the metal(0) surfaces suggested that POM counter-cations must be extensively incorporated as critical structural components of the POM monolayer. This role is analogous to that of counter-cations in POM monolayers on planar surfaces, and between oxomolybdate macro-ions in single-walled POM vesicles.[52] Hence, the stability of POM-protected metal(0) nanoparticles likely derives from several phenomena, namely, electrostatic forces typical of anion-stabilized colloids, direct bonding between POMs and the metal(0) surfaces, and the formation of numerous ionic and hydrogen bonds between POMs bound closely to one another on the metal(0) surface. This unique class of metal-nanoparticle structures is the subject of this chapter.

The chapter is divided into three sections, which concern the synthesis, structure and reactivity of POM-stabilized metal(0) nanoparticles. Unlike most nanoparticle-protecting ligands, POMs are redox active, and their reduced forms can reduce metal cations to colloidal metal(0), which are then stabilized by the oxidized POM anions. These methods are described, along with reductions of metal cations to metal(0) nanoparticles by traditional reducing agents, but in the presence of POM ligands. The second section, on the structures of POM-stabilized metal(0) nanoparticles, includes important observations from many groups, along with a detailed summary of new findings obtained with the aid of cryo-TEM imaging. Unlike alkanethiol-Au(0) nanoparticles, the POM-protected metal(0) particles are soluble analogs for the metal(0)/metal-oxide interfaces in heterogeneous metal-oxide supported metal(0) catalysts. Because of this, and given the redox behavior of the POMs themselves, POM-protected metal(0) nanoparticles have been investigated for use as catalysts for selective oxidations and other reactions. Results from those studies, reviewed in the last section of this chapter, indicate that POM-protected metal(0) nanoparticles can serve as active catalysts with unique selectivities, and in some cases, unprecedented rates and stabilities. Recent information concerning the detailed structures of POM-protected metal(0) nanoparticles provide new options for establishing structure–reactivity relationships for this versatile class of nano-scale materials.

2. Synthesis of Polyoxometalate Stabilized Metal(0) Nanoparticles

2.1. *Control over shape and size*

Methods for controlling and selecting particle size are of paramount interest for optimizing and standardizing nanoparticle properties, and remain an issue of tremendous importance in nanochemistry. This section will focus on size and shape control of POM-stabilized metal(0) nanoparticles and related nanostructures. Interested authors are also encouraged to read a recent review by Louis Nadjo,[53] where reduced POMs are used as both the reducing and stabilizing reagent.

2.2. Ir *and* Ru *nanoparticles by* H_2 *reduction*

Finke and co-workers carried out systematic work on the synthesis of POM-protected noble-metal nanoparticles and their applications in catalytic hydrogenation. Nearly monodispersed Ir(0) (30 ± 4 Å, ~900 Ir; 20 ± 3 Å, ~300 Ir) and Rh(0) (40 ± 4 Å, ~2700 Rh) nanoparticles were synthesized by H_2-reduction of polyoxoanion-supported M (M = Ir^{I} and Rh^{III}) complexes in acetone (Eq. 1 and Fig. 1).[46,48,49,54–58] The POM ligand serves to control nucleation and growth, and small (several nm) Ir(0) and Rh(0) nanoparticles with polydispersities of less than 15% were reproducibly prepared. Both nanoparticles are highly active olefin-hydrogenation catalysts. The active catalyst was isolable. These nanostructures were exhaustively characterized by a variety of methods, including transmission electron microscopy (TEM), ultracentrifugation, solution molecular-weight measurements, fast-atom bombardment mass spectrometry and (importantly) elemental analysis. These methods led to empirical formulas such as, for example, $[Ir_{\sim900}(P_4W_{30}Nb_6O_{123})_{\sim60}](Bu_4N)_{\sim660}Na_{\sim330}$ for 30 ± 4 Å Ir(0) NPs, and $[Ir_{\sim300}(P_4W_{30}Nb_6O_{123})_{\sim33}](Bu_4N)_{\sim300}Na_{\sim233}$ for 20 ± 3 Å Ir(0) NPs.[46,58]

$$(Bu_4N)_5Na_3[(1,5\text{-cyclooctadiene})Ir^{I}\cdot P_2W_{15}Nb_3O_{62}] + H_2$$

$\longrightarrow$ Ir(0) +  + H^+ + $P_2W_{15}Nb_3O_{62}^{9-}$ (1)

The zeta potential (ζ) of solutions of the 30 ± 4 Å Ir(0) nanoparticles was ca. –40 mV,[46] consistent with a negative net charge on the colloidal particles, and Finke suggested that the metal particles were stabilized by an "electrosteric" — combined DLVO-type electrostatic plus steric-stabilization — mechanism.[59]

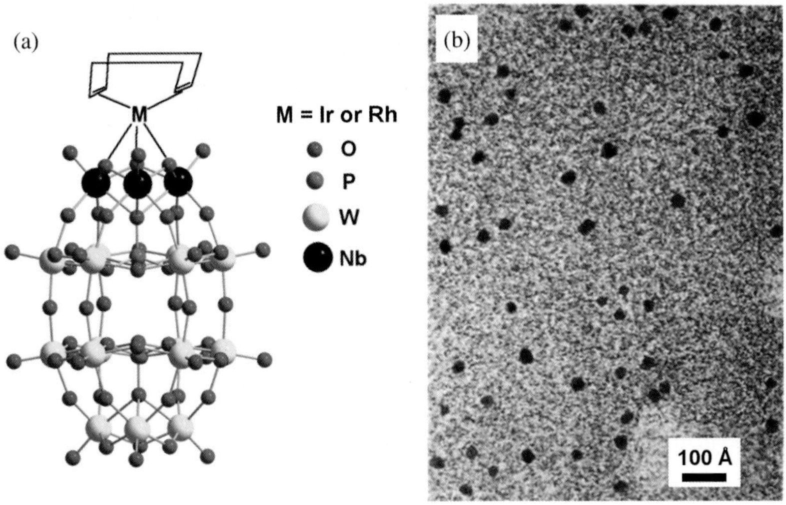

Fig. 1. (a) Ball and stick representation of $[(1,5\text{-cyclooctadiene})Ir \cdot P_2W_{15}Nb_3O_{62}]^{8-}$. (b) TEM image of the $Ir_{\sim 900}$ NPs. (Panel (b) is reprinted from Ref. 46 with permission from the American Chemical Society.)

The net charge of the colloidal Ir(0) particles can be estimated using the equation:[60]

$$q = 4\pi\varepsilon r\zeta \tag{2}$$

where ε is the permittivity of the solvent ($\varepsilon = 3.32 \times 10^{-10}$ F m^{-1} for acetonitrile at 298 K), r is the radius of the nanoparticle including the ligand shell ($r \approx 3.5$ nm, assuming 2.0 nm thickness of the $P_4W_{30}Nb_6O_{123}{}^{16-}$ shell), and ζ is the zeta potential value. For the 30 Å Ir(0) nanoparticles, Eq. 2 gives, $q = 5.8 \times 10^{-19}$ C, i.e., ~3.6e. Therefore, in acetonitrile the Ir(0) nanoparticles are negatively charged, and each 30 Å nanoparticle carries a net charge of 3.6-. At the time of Finke's work,[46,48,49,54–58] direct cryo-TEM imaging of POM anions on metal(0) cores had not been achieved (see Sec. 3.2).[50–52,61] Therefore, the specific organization of the POMs and the metal(0) nanoparticles was not clear.

The size of the Ir(0) nanoparticles was influenced by the choice of POM complex used to bind Ir(I). For example, using $(Bu_4N)_5Na_3[(1,5\text{-cyclooctadiene})Ir^I \cdot P_2W_{15}Nb_3O_{62}]$ as the precursor, the Ir(0) NPs had an average size of 30 ± 4 Å, while using $(Bu_4N)_5Na_3[(1,5\text{-cyclooctadiene})$

$Ir^{I} \cdot Si_2W_{15}Nb_3O_{62}$], an average size of 23 ± 3 Å was obtained. The H_2 pressure affected the rate of gas-to-liquid diffusion, and primarily impacted monodispersity, rather than average size.[48] While providing remarkably narrow polydispersities, this *in situ* protocol did not appear to provide an efficient means for controlling nanoparticle size and shape.

2.3. *Reactions of metal salts with reduced polyoxometalate anions*

2.3.1. *Synthesis of nanoparticles*

Since Finke's discovery of catalytically active POM-stabilized Ir(0) NPs, a wide variety of metal(0) nanostructures in water and other solvents stabilized by polyoxometalate anions have been prepared. One widely applicable technique is to react metal salts with reduced polyoxometalates. Herein, the polyanions serve as both reductant and stabilizer. A first example of this method took advantage of the photocatalytic activity of some polyoxometalates. In 2002, Papaconstantinou *et al.* prepared α-$PW_{12}O_{40}^{3-}$-stabilized Ag(0) NPs by reaction of Ag^+ with photochemically-reduced forms of α-$H_3PW_{12}O_{40}$. The POM was reduced by UV irradiation in the presence of isopropanol in water (Eqs. 3 and 4)[62]

$$\alpha\text{-}PW_{12}O_{40}^{3-} + \tfrac{1}{2}CH_3C(H)OHCH_3 \xrightarrow{h\nu} \alpha\text{-}PW_{12}O_{40}^{4-} + \tfrac{1}{2}CH_3C(O)CH_3 + H^+ \quad (3)$$

$$\alpha\text{-}PW_{12}O_{40}^{4-} + Ag^+ \longrightarrow \alpha\text{-}PW_{12}O_{40}^{3-} + Ag^0_{colloidal} \quad (4)$$

This method has been used to prepare POM-stabilized M(0) NPs, where M = Au,[62] Ag,[62] Pt,[62] Pd,[62–64] and others. It has been used in both one-pot and two-step protocols, in post-synthesis modifications (e.g., the incorporation of Ag(0) and Au(0) NPs into films pre-embedded with the POM photocatalyst/stabilizer[65]), and as a "UV-switchable reducing agent" for engineering Au-core/Ag-shell nanoparticles and other bimetallic nanostructures.[66,67] More generally, reduced polyoxometalates are readily obtained by a variety of means, and can then be reacted with a wide range of metal salts.[68–73] For example, narrowly disperse solutions of M(0) NPs, where M(0) = Ru, Rh, Pt, and Ir, stabilized by $PV_2Mo_{10}O_{40}^{5-}$, were prepared by stoichiometric chemical reduction of the molybdovanadophosphate (e.g., by Zn(0)), followed by reactions of the reduced POMs with transition metal salts.[71] When supported on alpha-alumina, some of these (Ag and Ru) are effective catalysts for the "direct" O_2-epoxidation of alkenes in the liquid phase.[71] Nadjo recently prepared POM-stabilized noble metal NPs by reacting pure salts of reduced POMs[74–76] such as $H_7[\beta\text{-}P(Mo^V)_4(Mo^{VI})_8O_{40}]$,[76] $(NH_4)_{10}[(Mo^V)_4(Mo^{VI})_2O_{14}(O_3PCH_2PO_3)_2(H_2OPCH_2PO_3)_2]$,[76] $[\varepsilon\text{-}P(Mo^V)_8(Mo^{VI})_4O_{36}(OH)_4[La(H_2O)_{2.5}Cl_{1.25}]_4]$, with noble-metal salts. Isolation and purification of the reduced POM prior to its reaction with metal salts provides NPs solutions that contain fewer salt or other by-products.[75–79]

The different UV-visible absorption spectra of the reduced polyoxometalates and many nanoparticles (nanostructures) make it possible to simultaneously monitor oxidation of the POMs by metal salts, and the formation of the nanoparticles (nanostructures) (Fig. 2).

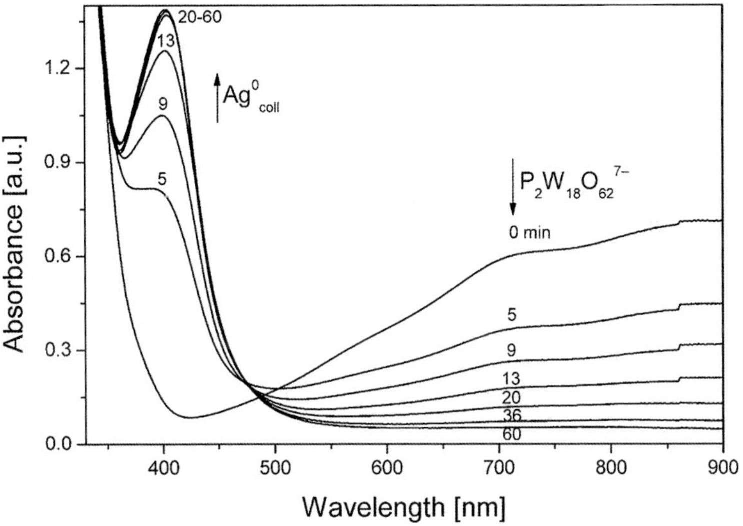

Fig. 2. Absorption spectra demonstrating the consumption of a 1e⁻-reduced POM ($P_2W_{18}O_{62}^{7-}$; band at 714 nm), and the concomitant formation of silver nanoparticles (band at 420 nm), upon mixing $P_2W_{18}O_{62}^{7-}$ (2×10^{-4} *M*) with Ag^+ (10^{-4} *M*). (Reprinted from Ref. 80 with permission from John Wiley & Sons, Inc.)

In solutions of suitable ionic strength, nanoparticle size can be controlled by varying: (1) the initial concentration of reduced POM, (2) POM structure, (3) the extent of reduction of the POM, and (4) the concentration of metal cation salt. Recent findings[81] indicate that the nature of the POM counter-cation can be critically important.

Papaconstantinou controlled the sizes and polydispersities of POM-stabilized Ag(0) NPs[80] by changing the rate of silver reduction. Faster reduction of Ag^+ led to smaller and more uniform nanoparticles. More rapid rates of reduction were achieved by either increasing the concentration of the reduced POM or, interestingly, by using reduced POMs with more negative reduction potentials, in some cases by more extensively reducing a single POM. Rates of Ag^+ reduction paralleled increasingly negative POM reduction potentials (Fig. 3; plotted as functions of the last one-electron redox couple). Curiously, increasing the concentration of Ag^+ resulted in larger nanoparticles.[80]

To further analyze these findings, the concentrations of Ag NPs can be calculated. For a colloidal solution of monodisperse Ag NPs, the concentration of the particles is calculated according to Eq. 5.

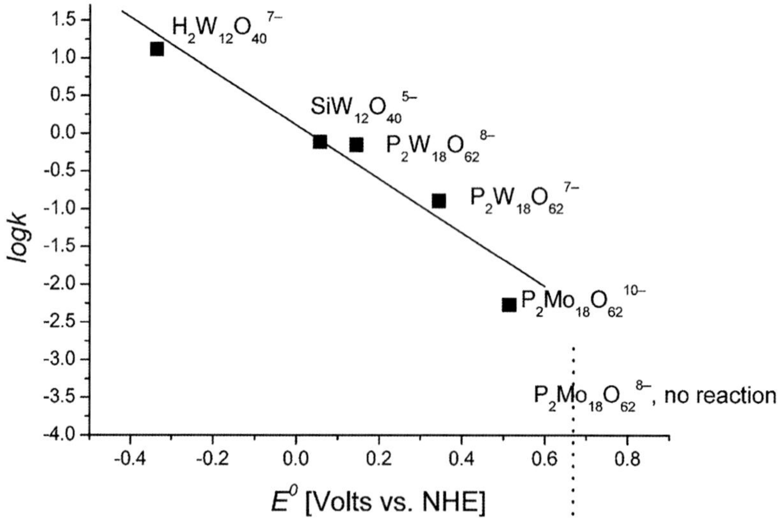

Fig. 3. Linear function of log k vs. the reduction potential of various reduced POMs for the reduction of Ag^+ (0.5×10^{-4} M) by reduced POMs (1.0×10^{-4} M). (Reprinted from Ref. 80 with permission from John Wiley & Sons, Inc.)

$$\left[\text{Ag NP}\right] = \frac{\left[\text{Ag}^{+}\right]}{\frac{4}{3}\pi r^{3} d/(M.W./N_{A})}. \qquad (5)$$

In Eq. 5, r is the radius of the Ag NP, *M.W.* is the molecular weight of Ag, d is the density of metallic Ag (1.05×10^7 g m^{-3}), and N_A is Avogadro's number. Provided the average diameter is used, Eq. 5 can also be used to calculate concentrations of particles with narrow size distributions (< 5%) and to estimate concentrations of particles with somewhat wider size distributions (< 20%), such as those discussed immediately above.[80]

It can be seen from Table 1 and Fig. 4 that, as the concentration of $SiW_{12}O_{40}^{5-}$ increased 2.6 fold (from 2.3 to 8.3×10^{-4} M), the concentration

Table 1. Average sizes and calculated concentrations of silver nanoparticles.

[$SiW_{12}O_{40}^{5-}$] (M)	Ag^+ (M)	Diameter[a] (nm)	Relative standard deviation (%)	Calculated [Ag NP] (M)
8.3×10^{-4}	8.4×10^{-5}	28.2	19.0	1.2×10^{-10}
5.3×10^{-4}	8.4×10^{-5}	38.0	21.4	5.0×10^{-11}
2.3×10^{-4}	8.4×10^{-5}	44.0	10.6	3.2×10^{-11}
7.0×10^{-4}	5.7×10^{-4}	62.3	17.3	7.7×10^{-11}
7.0×10^{-4}	2.0×10^{-4}	42.9	17.4	8.3×10^{-11}
7.0×10^{-4}	1.0×10^{-4}	33.3	20.6	8.8×10^{-11}

[a]Average diameter based on the counting of ca. 150 particles from the corresponding TEM images.

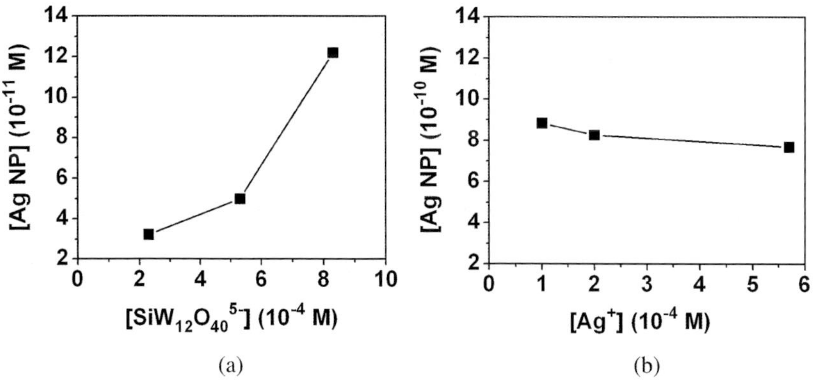

Fig. 4. Concentrations of Ag nanoparticles as functions of (a) initial polyoxometalate and (b) Ag^+ concentrations.

of nuclei increased by a similar factor of 2.8 (from 3.2×10^{-11} to 1.2×10^{-10} M). A less significant influence on nucleation was observed by varying [Ag^+] from 1.0×10^{-4} to 5.7×10^{-4} M at constant [$SiW_{12}O_{40}^{5-}$].

Interestingly, for synthesis of non-metallic Se(0) nanoparticles by reaction of $SiW_{12}O_{40}^{5-}$ with a Se(IV) salt, the concentration of one-electron reduced POM, [POM_{1e}], also plays a dominant role in controlling nucleation.[82] Upon increase of [POM_{1e}] from 2 to 4 to 10 mM, concentrations of Se(0) NPs increased from 0.39 to 1.0 to 2.4×10^{-10} M.

The reduction potential of $HAuCl_4$ is more positive than that of Ag^+. Under standard conditions, the reduction potentials of the $HAuCl_4$/Au and Ag^+/Ag couples are 1.002 and 0.799 V vs. NHE, respectively. Although actual rates depend on the mechanism(s) of reduction, reaction rates between $HAuCl_4$ and reduced POMs are generally faster and less well controlled. As the concentration of $HAuCl_4$ increases, rates of surface-growth increase as well, leading to fewer nuclei and smaller particle concentrations. At the same time, more strongly reducing POMs lead to more rapid nucleation and narrower size distributions. From Table 2, it is clear that Au NP size is strongly affected by both [POM] and [$HAuCl_4$].

However, an analysis of the nucleation (from Eq. 5) reveals that the nuclei concentrations are not increased (and in fact, are even suppressed) by larger $HAuCl_4$ concentrations (this is seen in entries 1–3 and 6–8). Therefore, it appears that [$HAuCl_4$] values control rates of surface growth processes, while [POM] values have a larger impact on nucleation rates.

Table 2. Average sizes and calculated concentrations of gold nanoparticles.

Entry	POM used	[POM]/ mM	[$HAuCl_4$]/ mM	d/nm	Calculated [Au NP]/M
1[77]	$Na_2[Mo_3(\mu_3\text{-}S)(\mu\text{-}S)_3(Hnta)_3]$	0.5	0.25	6 ± 0.5	3.8×10^{-8}
2[77]	$Na_2[Mo_3(\mu_3\text{-}S)(\mu\text{-}S)_3(Hnta)_3]$	0.5	0.5	9.5 ± 0.5	1.9×10^{-8}
3[77]	$Na_2[Mo_3(\mu_3\text{-}S)(\mu\text{-}S)_3(Hnta)_3]$	0.5	1.0	90	4.4×10^{-11}
4[75]	$\beta\text{-}H_3[H_4PMo_{12}O_{40}]$	0.1	0.1	10	3.2×10^{-9}
5[75]	$\beta\text{-}H_3[H_4PMo_{12}O_{40}]$	0.5	0.5	60	7.5×10^{-11}
6[75]	$\beta\text{-}H_3[H_4PMo_{12}O_{40}]$	1.0	0.1	~27	1.6×10^{-10}
7[75]	$\beta\text{-}H_3[H_4PMo_{12}O_{40}]$	1.0	0.4	~68	4.1×10^{-11}
8[75]	$\beta\text{-}H_3[H_4PMo_{12}O_{40}]$	1.0	1.0	110	2.4×10^{-11}

2.3.2 *Synthesis of complex nanostructures*

Photocatalytic reductions of POMs have been used as the basis for the syntheses of more complex nanostructures, such as Au-Ag core-shell bimetallic nanoparticles. These were prepared by utilizing Keggin anions, bound to the surfaces of gold nanoparticles, as "UV-switchable" reducing agents. Exposure of photochemically reduced Keggin-ion-capped gold nanoparticles to aqueous solutions containing salts of Ag^+, Pd^{2+} and Pt^{4+} ions (Scheme 1) resulted in selective reduction of the metal ions onto the surfaces of the Au NPs, and thus, to phase-pure core–shell nanostructures (Fig. 5).[66–67,83]

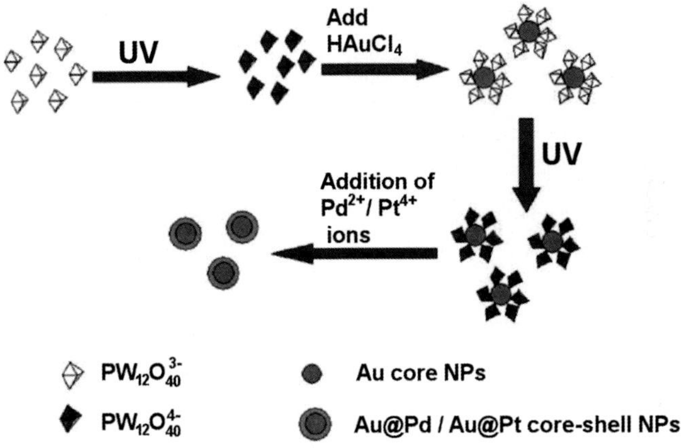

Scheme 1. Illustration of the Keggin ion-mediated synthesis of Au core–Pd/Pt shell nanoparticles. For simplicity, the Keggin ions are shown as octahedrons. (Reprinted from Ref. 67 with permission from Royal Society of Chemistry.)

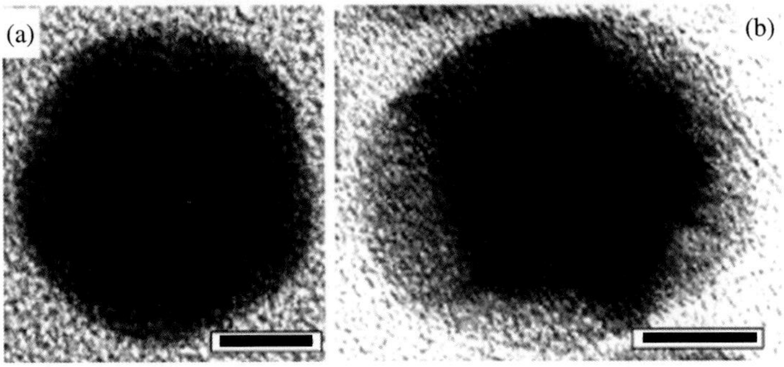

Fig. 5. High-resolution TEM images of (a) a gold nanoparticle, and (b) a Au core–Ag shell nanoparticle. Scale bars in (a) and (b) correspond to 5 and 10 nm, respectively. (Reprinted from Ref. 66 with permission from the American Chemical Society.)

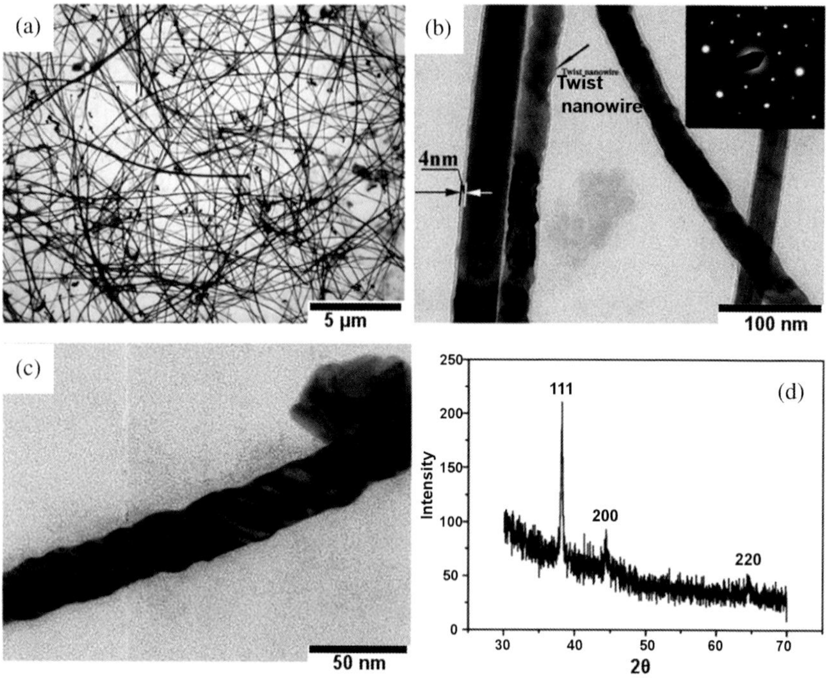

Fig. 6. (a) TEM image of Ag nanowires, (b) higher-magnification image of the nanowires, (c) a magnified twisted helical nanowire, and (d) the XRD pattern of a Ag nanowire. (Reprinted from Ref. 76 with permission from the American Chemical Society.)

Reaction of a more reducing POM, $(NH_4)_{10}[(Mo^V)_4(Mo^{VI})_2O_{14}(O_3PCH_2PO_3)_2(H_2OPCH_2PO_3)_2]$, with Ag_2SO_4, gives spherical Ag NPs, while reaction of a weaker reductant, such as $H_7[\beta\text{-}P(Mo^V)_4(Mo^{VI})_8O_{40}]$, gives irregularly shaped Ag NPs at initial stages of the reaction, and Ag nanowires (~40 nm diameter and tens of millimeters in length; Fig. 6) with prolonged reaction time.[76] A controlling parameter, defined as γ = [metallic salt]/[POM], was optimized for each synthesis.[84]

Analogous to the formation of Ag nanostructures, reactions of $\beta\text{-}[PMo_{12}O_{40}]^{7-}$ with $HAuCl_4$ are controlled by the same parameter, γ, defined immediately above, as well as by the absolute POM concentration. In pH 2–3 solutions of 1 mM POM, $\gamma \geq 1$ gave quasi-spherical Au nanoparticles of ~110 nm, while for $\gamma < 1$, one obtains quasi-monodisperse spherical nanoparticles accompanied by triangular and hexagonal nanoplates.

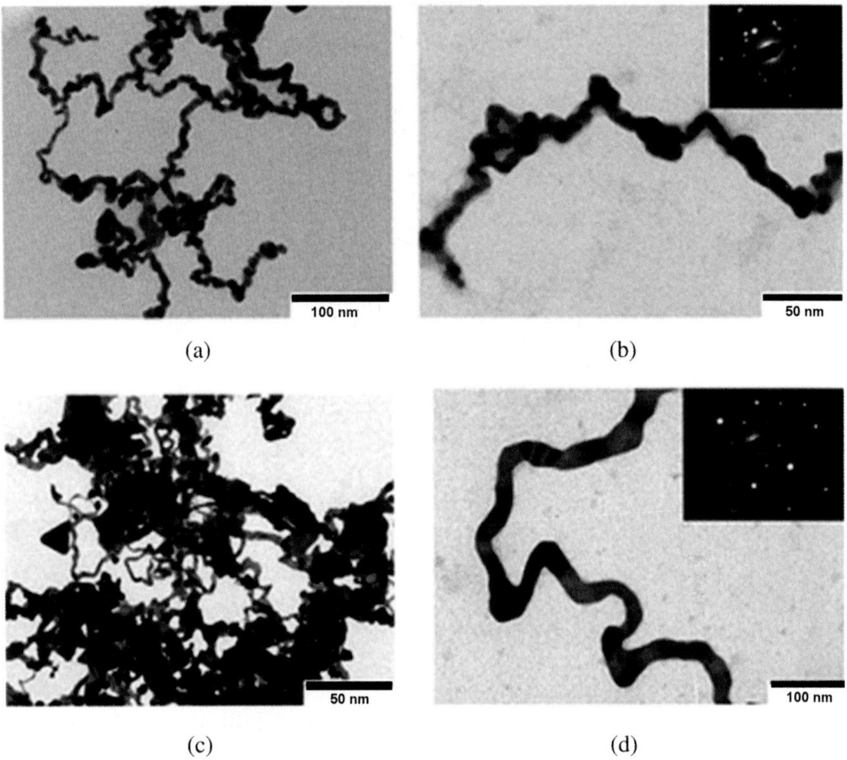

Fig. 7. Worm-like Au nanostructures prepared with C_{POM} = 0.1 mM and γ = 4 for (a) and (b), and γ = 0.2 for (c) and (d). (Reprinted from Ref. 75 with permission from Royal Society of Chemistry.)

At 0.1 mM POM, a striking γ dependence was observed. In addition to the quasi-spherical NPs, worm-like structures also formed (Fig. 7). The average diameter of the nanowires was approximately 10 nm. These twisted nanowires were interconnected to form a network structure.[75]

In situ fabrication of flower-like gold nanoparticles was observed using surfactant-polyoxometalate hybrids.[85] These were prepared using photochemically-reduced $PMo_{12}O_{40}^{3-}$ anions that were integrated in spherical assemblies (hydrodynamic radius ≈ 213 nm) of a surfactant complex, didodecylamine hydrobromide, to reduce $HAuCl_4$. The organic matrix serves as a "soft" template for the formation of the flower-like Au-NP structures (Fig. 8).

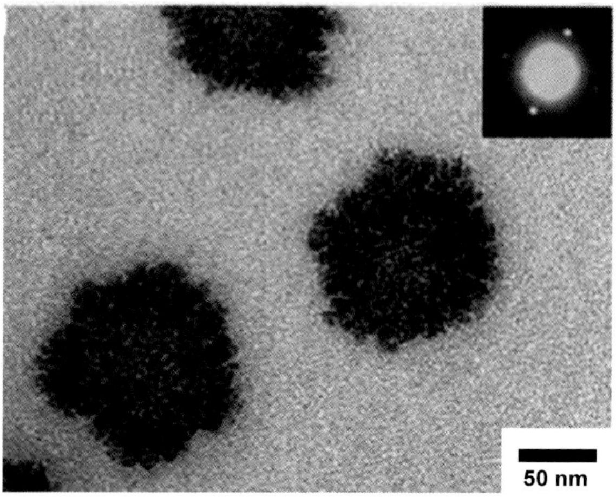

Fig. 8. TEM image of flower-like Au-NPs. The inset displays the electron diffraction pattern. (Reprinted from Ref. 85 with permission from Royal Society of Chemistry.)

2.4. *Ligand-exchange reactions*

Place-exchange of thiols on the surfaces of metal and semiconductor NPs is a versatile method for preparing functionalized supramolecular assemblies.[86] And, despite numerous publications concerning the size- and shape-controlled syntheses of nanostructures, it remains quite difficult to control these parameters by varying concentrations of precursors and reactants. Hence, ligand-exchange provides a valuable option for engineering specific sized and/or shaped structures with functionalized surface structures. This general class of reactions can also be used to prepare POM-stabilized metal(0) nanoparticles by exchanging POMs for the ligands used in initial nanoparticle preparations.[51,52,61,87,88]

In a notable example by Kulesza and Cox, $PMo_{12}O_{40}^{3-}$-stabilized Au NPs were prepared by reacting acidic aqueous solutions of $PMo_{12}O_{40}^{3-}$ with hexane solutions of 3-nm core hexanethiol-protected Au NPs.[88] The liquid-liquid extraction into aqueous solution occurred without change in the size of the Au-NP core.

Tong[87] reacted citrate-stabilized colloidal Ag NPs with mono-defect ("lacunary") derivatives of the Keggin structure, α-$PW_{11}O_{39}^{7-}$ and α-$SiW_{11}O_{39}^{8-}$. UV-vis spectroscopic analysis of the surface plasmon

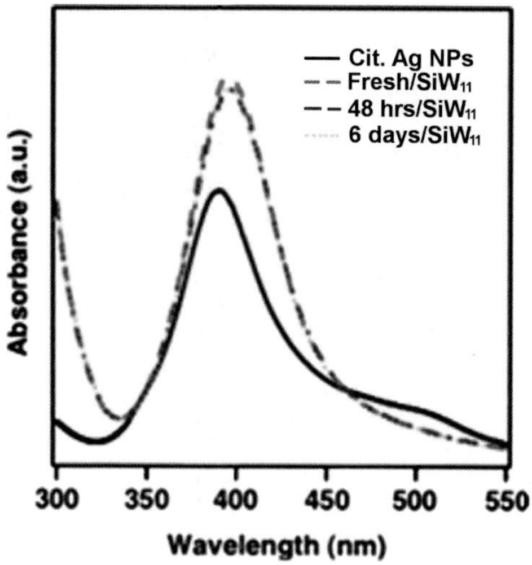

Fig. 9. UV-vis spectroscopic analysis of the displacement of citrate by $\alpha\text{-}SiW_{11}O_{39}^{8-}$ on Ag NPs. After addition of $\alpha\text{-}SiW_{11}O_{39}^{8-}$ to a solution of citrate-stabilized Ag NPs (solid black curve), absorbance due to the surface plasmon resonance (SPR) increased to new values (dashed red curve) and then remained constant (blue and green curves). (Reprinted from Ref. 87 with permission from Springer Science & Business Media.) **See Color Plate 1**.

resonance (SPR) was used to monitor the displacement of citrate by POM anions (Fig. 9), and surface-enhanced Raman spectroscopy (SERS) data were consistent with ligand exchange at the metal-NP surface.

Recently, we used ligand exchange to synthesize POM-stabilized Au and Ag NPs of different sizes.[51,52, 61] These were prepared by reacting POMs with aqueous solutions of "pre-formed" citrate-stabilized metal nanoparticles. UV-vis analysis, combined with a linear relationship between SPR absorbance and POM coverage predicted by Mie theory, was used to quantify extents of surface coverage, Θ, of the metal particles by the POMs, Eq. 6. (This linear relationship and other structural features of these assemblies, was confirmed by cryo-transmission electron microscopy, cryo-TEM, which is discussed in Sec. 3.2.)

The absorption maximum in UV-visible spectra of citrate-protected gold nanoparticles is located at 520 nm. Upon incremental additions of $\alpha\text{-}K_9AlW_{11}O_{39}$, absorbance values increase until reaching a plateau, while the position of the absorbance maximum shifts from 520 to 526 nm (Fig. 10b). A corresponding color change is visible by eye (Fig. 10a).

(a)

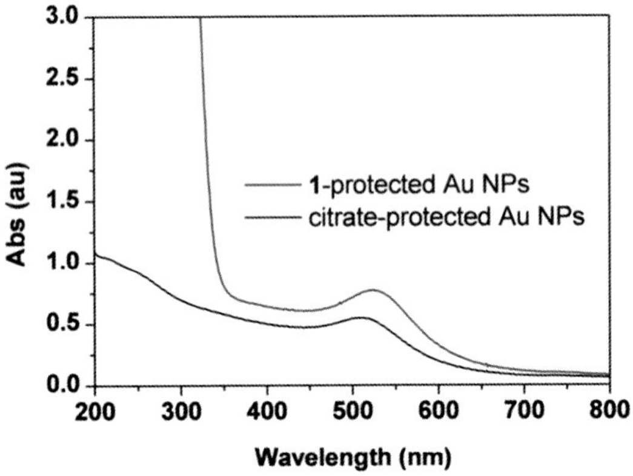

(b)

Fig. 10. (a) Photographs and (b) spectra of citrate-protected Au NPs and α-$K_9AlW_{11}O_{39}$-protected Au NPs. The cuvette at the left in panel (a) contains citrate-protected Au NPs. Addition of α-$K_9AlW_{11}O_{39}$ (**1**) to a final concentration of 2.5 mM, causes a visible color change (cuvette at right). Corresponding UV-visible spectra are shown in panel (b). Reprinted from Ref. 61 with permission from the Elsevier. **See Color Plate 2**.

Based on Mie theory, change in absorbance, A, at a certain wavelength λ, is related to surface coverage (Θ) by Eq. 6.

$$\Theta = \frac{A - A_0}{A_1 - A_0} \tag{6}$$

In Eq. 6, A_0 and A_1 correspond to $\Theta = 0$ and 1, respectively, where $\Theta = 1$ refers to full monolayer coverage. This relationship can be used to describe the Langmuir isotherm (Eq. 7).

$$A = \frac{A_0 + A_1 KC}{1 + KC} \tag{7}$$

In Eq. 7, C is the concentration of added POM, which is close to the equilibrium concentration as amounts adsorbed are small; K is the Langmuir constant which is correlated with the curvature of the isotherms as shown in Fig. 11, and indicates the affinity of the POMs for the Au-NP surface.

Type I Langmuir isotherms were obtained for a series Keggin ions, with charges of 5- and 6-, and mono-defect Keggin-ion derivatives, with charges of 7-, 8-, and 9- (Fig. 11). Fitting of the data to Eq. 7 provided values for K, as well as for A_1, the absorbance associated with complete surface coverage.

For selected POMs adsorbed onto 13.8 nm citrate-stabilized Au NPs, K values (in units of M^{-1}) increased from 900 to 4,300 to 8,000 to 12, 300 to 16, 300, respectively, for α-$Na_5AlW_{12}O_{40}$, α-$Na_6H_2W_{12}O_{40}$, α-$K_7PW_{11}O_{39}$, α-$K_8SiW_{11}O_{39}$, and α-$K_9AlW_{11}O_{39}$. Cryo-TEM images directly confirmed

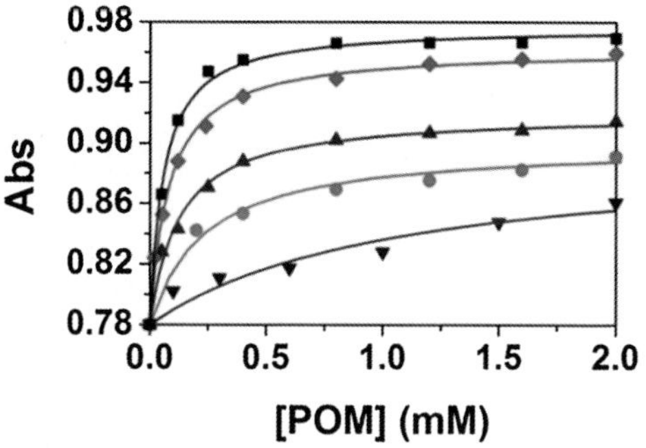

Fig. 11. Adsorption isotherms for reactions of citrate-protected gold nanoparticles with a series of POM anions. The plots are absorbance at 526 nm versus POM concentration for α-$K_9AlW_{11}O_{39}$ (■), α-$K_8SiW_{11}O_{39}$ (◆), α-$K_7PW_{11}O_{39}$ (▲), α-$Na_6H_2W_{12}O_{40}$ (●) and α-$Na_5AlW_{12}O_{40}$ (▼). The curves are fits to Eq. 7. (Reprinted from Ref. 51 with permission from the American Chemical Society.)

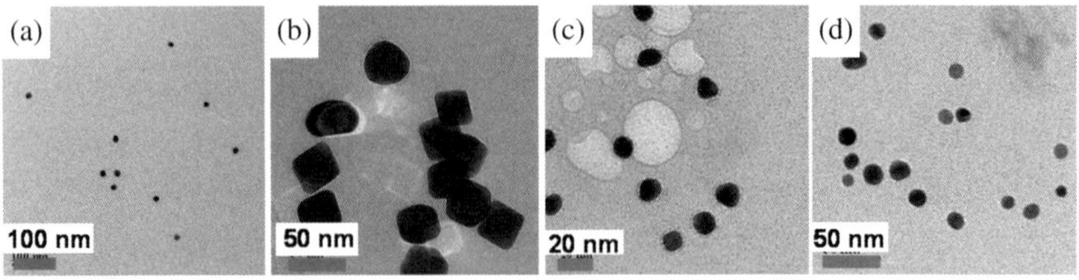

Fig. 12. Cryo-TEM images of POM stabilized Au and Ag nanoparticles synthesized by place exchange. (a) Monodisperse 13.8 nm Au NPs stabilized by α-$K_9AlW_{11}O_{39}$.[51] (b) α-$K_9AlW_{11}O_{39}$-stabilized Au octahedrons synthesized by reacting CTA^+-stabilized Au octahedrons with α-$K_9AlW_{11}O_{39}$. (c) Preyssler-type POM, $K_{14}P_5W_{30}O_{110}$-stabilized 13.8 nm Au NPs using the same method as that of Ref 51. (d) α-$K_9AlW_{11}O_{39}$-stabilized Ag nanoparticles by reaction of citrate-stabilized Ag NPs with α-$K_9AlW_{11}O_{39}$. In panel (c), the black dots surrounding and beyond the 13.8 nm Au NPs are POM molecules. At the lower magnifications used to obtain the images in panels (a), (b) and (d), individual POM molecules are not visible.

that K values reflect affinities of the POM anions to the gold nanoparticles.[51] Hence, the combined UV-spectroscopic and cryo-TEM analysis showed that POM affinities increase in the order: $C_6H_5O_7^{3-}$ < α-$AlW_{12}O_{40}^{5-}$ < α-$H_2W_{12}O_{40}^{6-}$ < α-$PW_{11}O_{39}^{7-}$ < α-$SiW_{11}O_{39}^{8-}$ < α-$AlW_{11}O_{39}^{9-}$.

These Langmuir constants, K, are "relative" to citrate, such that absolute affinity values should be much larger than the ones listed above. Therefore, the POM anions possess very large affinities for the nanoparticle surface, which facilities the use of POM anions for the displacement/exchange of other stabilizing ligands or surfactants such as cetyltrimethylammonium bromide (CTAB), polyelectrolytes, and polyols. An example of the use of ligand exchange for the synthesis of POM-stabilized nanoparticles of specific sizes and shapes is shown in Fig. 12 (cyro-TEM imaging is discussed in Sec. 3.2).

3. Structures of Inorganic Cluster Anions Stabilized Nanoparticles

Monolayers of POM anions on a variety of planar surfaces have been characterized in detail at the sub-nanometer scale. Much less is known, however, about POMs on the curved surfaces of metal(0) nanoparticles (NPs).

The deliberate synthesis of POM-stabilized metal(0) NPs traces its history back to 1952, when colloidal Ag(0) was prepared by reacting photochemically reduced $H_3PW_{12}O_{40}$ with Ag^+ in water.[89,90] Since then, POMs of many compositions and structures have been shown to stabilize a variety of metal NPs in organic solvents and water. Meanwhile, detailed structural information about the nature of the metal/protecting-anion interface has only recently been obtained. In many respects, this situation is similar to that associated with alkanethiol ligand shells on gold nanoparticles. Although spectroscopic and chemical methods provide important information, direct images of organic ligands on gold NPs are extremely difficult to obtain. Substantially less is known about the structural organizations of protective monolayers comprised of inorganic ligands. This is beginning to change, however, and new information about the structural organization of POMs on metal nanoparticles is the focus of this section.

3.1. *Evidence for the stabilization of metal*(**0**) *nanoparticles by polyoxometalates*

Following is a brief summary of evidence that polyoxometalates stabilize metal nanoparticles.

(i) Elementary analyses of isolated POM-stabilized Ir nanoparticles confirmed the presence of the polyoxometalate anions.[46,49,54] For the $Ir_{\sim300}$ NPs (ca. 300 Ir atoms), 33 POM anions were found by elemental analysis, while the maximum number that could bind to the metal surface is ~17. For $Ir_{\sim900}$ NPs (ca. 900 Ir atoms), 60 POM anions were found, while no more than 39 could bind to the metal surface. Apparently, a significant fraction of the POM anions associated with each NP were bound to the Ir(0) surface. In work by Sastry, isolated nanoparticles,[63] air dried after purification by dialysis,[66] contained significant amounts of W atoms as indicated by X-ray photoelectron spectroscopy (XPS).

(ii) POM-stabilized Rh(0) and Ir(0) nanoparticles are negatively charged. Analysis of electrophoresis data (see Sec. 2.2) showed that each Ir nanoparticle carried an average negative charge of 3.6e.

(iii) In many preparations of POM-stabilized metal NPs, no plausible protecting ligands other than POMs are present.[62] The photo-catalytic synthesis of POM-protected metal(0) nanoparticles by Papaconstantinou is an example: neither the alkali-metal cations nor NO_3^- anions present are effective stabilizers, so the polyoxometalate anions present must be serving in that capacity. This is reasonable given that polyoxometalates bind strongly to planar surfaces of noble metals such as Ag(100) and Au(111).

(iv) Surface enhanced Raman spectroscopy (SERS) is an important *in situ* technique for studying interactions of polyoxometalate anions with surfaces of Ag(0) and Au(0) nanoparticles. Compared to pure solutions of $Mo_7O_{24}^{6-}$, association with Ag(0) NPs causes the frequency of the asymmetric Mo=O stretch of this POM anion (at 886 cm^{-1}) to become more intense, while that for the symmetrical Mo=O stretch (936 cm^{-1}) becomes less intense. This has been attributed to a preferred orientations between $Mo_7O_{24}^{6-}$ and the Ag(0)-NP surface. Changes in the Raman intensities of α-$PMo_{12}O_{40}$ were also observed.[91] The displacement[87] of citrate anions by α-$SiW_{11}O_{39}^{8-}$ or α-$PW_{11}O_{39}^{7-}$ was revealed by Raman spectra, which not only shows that the POMs serve as protecting ligands, but also indicates they bind more strongly than does citrate. One shortcoming of SERS studies of ligand binding is that the surface-enhancement can occur over space, i.e., over ca. 10 Å. Hence, this technique alone doesn't prove the existence of direct contact (bonding) between a ligand and Ag or Au surfaces. The Raman spectrum of pure α-$K_9AlW_{11}O_{39}$ (gray) and the SER spectrum of a solution of α-$K_9AlW_{11}O_{39}$-protected Ag(0) NPs (red) are compared in Fig. 13. Based on cryo-TEM observation (see Sec. 3.2), the SER spectrum provided here can be reliably assigned to POM anions directly bound to the metal NP surface.[87,91] The SER bands are red-shifted and selectively enhanced. The terminal W=O bonds give rise to two bands in the SER spectrum, at 945 and 930 cm^{-1}, indicative of the C_s symmetry mono-defect Keggin anion.

3.2. *Direct imaging of POM-ligand monolayers by cryo-TEM*

Studies summarized above provide multiple lines of evidence that POM anions bind to metal(0) nanoparticles. Until very recently, however,

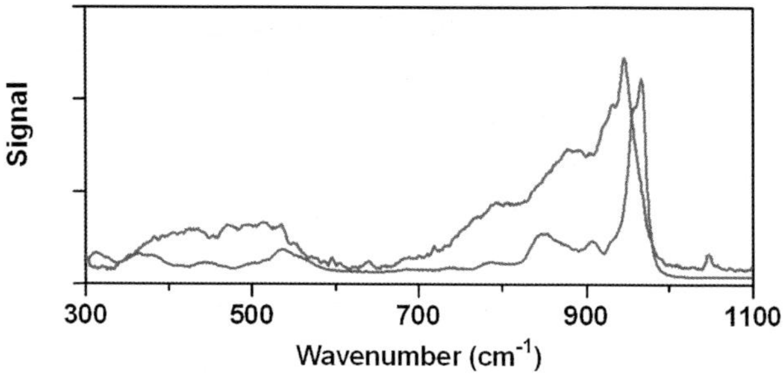

Fig. 13. Raman spectrum of solid α-$K_9AlW_{11}O_{39}$ (gray line) and the SER spectrum of an aqueous solution of α-$K_9AlW_{11}O_{39}$-protected Ag(0) NPs (red line). (Reprinted from Ref. 50 with permission from the American Chemical Society.) **See Color Plate 3**.

there was no detailed information about the structure of the protecting-ligand shell, or about the metal/POM interface. In TEM images of dried samples of POM-stabilized metal(0) NPs, the POMs are either not clearly observed, and/or the metal particles are found to be embedded within a poorly defined "sea" of aggregated POM salts. It was therefore speculated that as solutions are dried on coated Cu grids for TEM analysis, surface tension from the evaporating water might "shear" the highly solvated POM salts from the metal(0) particles. To address this, cryo-TEM was used to "trap" the actual solution-state structures in a matrix of vitrified water.[50]

Cryogenic sample preparation for transmission electron microscopy (cryo-TEM) was developed as a tool for imaging highly hydrated biological structures and macromolecules. These structures are modified extensively during drying, and cryogenic methods make it possible to rapidly "trap" and preserve them in their native hydrated states. For analysis by TEM, crystallization of the water or solvent matrix is avoided by rapid vitrification of the sample (i.e., rapid conversion from a liquid to its glassy state). This is accomplished by rapid cooling — at rates greater than 10^5 Ks^{-1} — to temperatures below the boiling point of liquid ethane (boiling point 184.5 K).

More recently, scientists have been using this technique to obtain images of non-biological samples. For example, cryo-TEM has been used to better understand nucleation and growth phenomena in solution.[92–97] Kumar *et al*. used cryo-TEM to study the early stages of pure-silica MFI zeolite growth, Pouget *et al.*[93] studied initial stages of template-controlled

$CaCO_3$ formation, and Talmon and co-workers[95,96] studied the formation of silica-based mesoporous materials. Cryo-TEM has also been used to characterize bulk dispersions and nanometrically thin films.[97] POM-protected Ag NPs for analysis by cryo-TEM were prepared by $NaBH_4$ reduction of Ag^+ in the presence of α-$K_9AlW_{11}O_{39}$. After routine analytical methods were used to ensure the integrity of the POM, samples were examined by electron microscopy. TEM images of dried samples show the Ag(0) particles embedded in amorphous α-$K_9AlW_{11}O_{39}$ [Fig. 14(a)]. At high resolution, energy dispersive spectroscopy (EDS) was used to identify α-$K_9AlW_{11}O_{39}$ and Ag(0) [(arrows in Fig. 14(b)]. Electron diffraction showed the Ag(0) to be polycrystalline, and no organized arrangements of α-$K_9AlW_{11}O_{39}$ are observed on or near the Ag(0) particle surfaces. When trapped in their native solution states by cryogenic

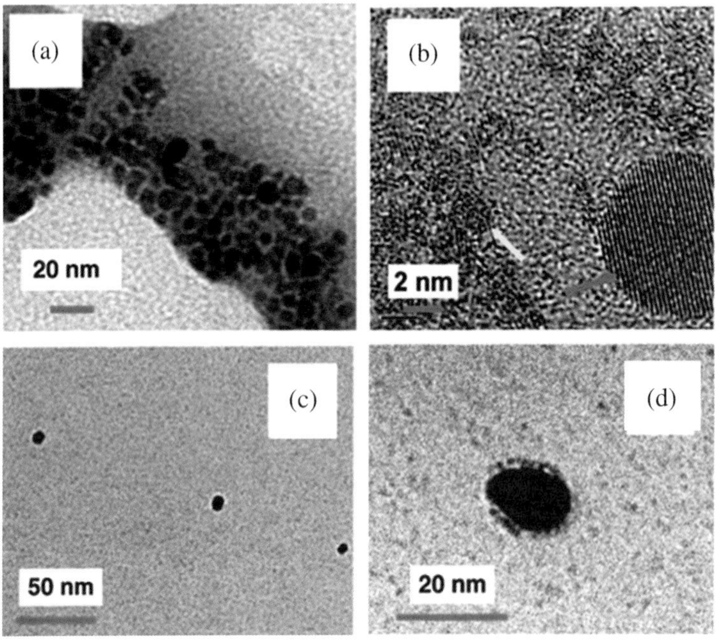

Fig. 14. (a) TEM image of Ag(0) (dark) embedded in amorphous $K_9AlW_{11}O_{39}$ (gray). (b) HRTEM image of the dry sample showing α-$K_9AlW_{11}O_{39}$ (yellow arrow) and a Ag(0) particle (red arrow). (c) Cryo-TEM image of individual α-$K_9AlW_{11}O_{39}$— protected NPs with no evidence of aggregation. (d) α-$AlW_{11}O_{39}^{9-}$ — protected Ag(0) particles "trapped" in cryogenically frozen water–glass. Freely solvated POMs appear as small dark objects in the background. (Reprinted from Ref. 50 with permission from the American Chemical Society.) **See Color Plate 3**.

cooling, however, the particles are dispersed throughout the vitreous matrix [(Fig. 14(c)]. At higher magnification, [(Fig. 14(d)], an apparent self-assembled monolayer (SAM) of individual cluster anions, α-$K_9AlW_{11}O_{39}$, is clearly observed around the periphery of the Ag(0) particles. The image in Fig. 15 is a 2-dimensional projection of the particle and its POM-protecting monolayer. Scattering by the silver particle precludes observation of POM ligands on the superior and anterior surfaces of the metal particle, while the spacing between molecules of α-$K_9AlW_{11}O_{39}$ on the silver surface makes it possible to image individual cluster anions. The dark ~1-nm-diameter objects randomly distributed throughout the remainder of the visual field in Fig. 14d are freely solvated POM anions.

Using the ligand exchange method introduced by Tong,[87] Keggin anions were reacted with 13.8 nm citrate-protected Au NPs. For example, after adding α-$K_9AlW_{11}O_{39}$ (2 mM) to ca. 10^{-9} M solutions of the Au NPs, cryo-TEM images again revealed a "ring" of individual molecules of α-$K_9AlW_{11}O_{39}$ around the periphery of the metal particle (Fig. 15), indicative of monolayer formation. The informative illustrations in Fig. 15 show the POM surface coverage implied by the structures observed in the cryo-TEM image.

Around the perimeter of 13.8-nm gold-core particle in Fig. 16(a) is a ring of 30 ± 1 molecules of α-$K_9AlW_{11}O_{39}$. Measurements of center-to-center distances between >100 bound POMs give a center-to-center distance of 1.57 ± 0.04 nm. For hexagonal packing — observed for center-to-center distances of ~1.6 nm between Keggin ions on planar

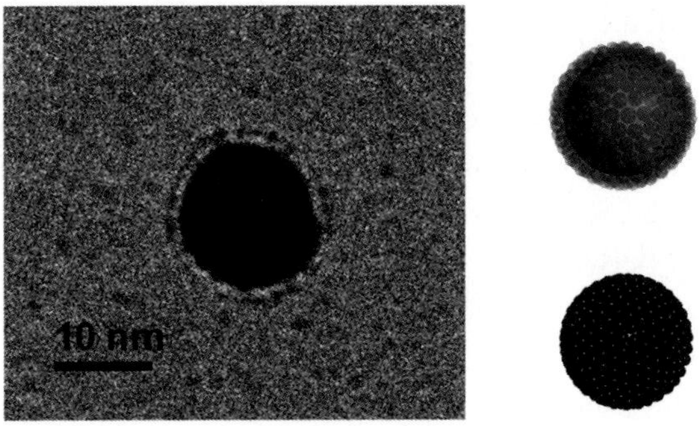

Fig. 15. Cryo-TEM image of α-$K_9AlW_{11}O_{39}$— protected gold nanoparticles. The drawings illustrate why coverage of the particle surface by a self-assembled monolayer of POM anions is seen as a peripheral ring of POM anions in cryo-TEM images. (Reprinted from Ref. 51 with permission from the American Chemical Society.)

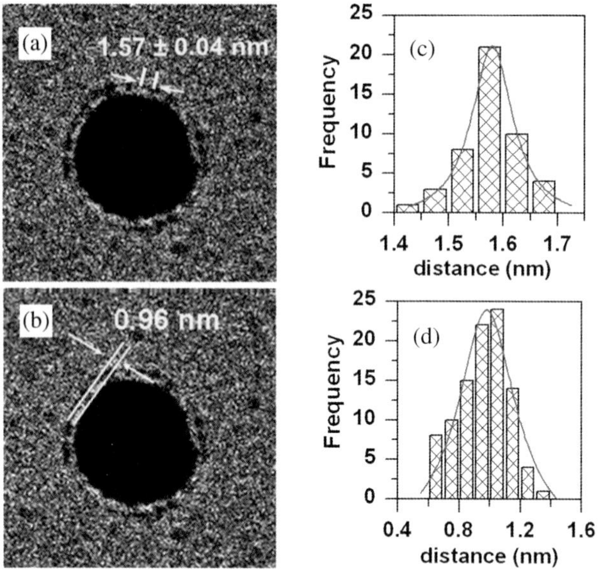

Fig. 16. Geometrical analysis of gold-surface covered with α-$K_9AlW_{11}O_{39}$: (a) the center-to-center distance between two adjacent molecules of α-$K_9AlW_{11}O_{39}$ and (b) the radial distance between the center of the monolayer shell and the surface of the gold particle. Panels (c) and (d) show the range of values obtained for both measurements over statistically meaningful numbers of particles and α-$K_9AlW_{11}O_{39}$ anions. (Reprinted from Ref. 51 with permission from the American Chemical Society.)

surfaces — geometrical extrapolation from the perimeter (2πr) to the entire surface ($4\pi r^2$) gives a total of 330 ± 33 anions in the spherical monolayer shell, and from that, a calculated ligand density of $7.4 \pm 0.7 \times 10^{-11}$ mol cm^{-2}.

In cryo-TEM images, the radial distance between the center of the monolayer shell and the surface of the gold particle is ~9.6 Å [(Fig. 16(b)]. Subtraction of the Keggin ion's crystallographic radius of 5.6 Å gives a distance of 4 Å between the terminal oxide ligands and the gold surface. This is larger than the 2.0 to 2.4 Å distances determined by specular reflectivity for α-$SiW_{12}O_{40}^{4-}$ ions on planar silver.[41] However, the phase contrast imaging used in cryo-TEM makes it difficult to know precisely where the metal surface ends and the POM-ligand begins. The zeta potential, ζ, of the citrate-protected 13.8-nm core Au NPs is −31.5 mV, while that of the α-$K_9AlW_{11}O_{39}$-protected Au NP is −62.1 mV. The negative shift in ζ values is consistent with displacement of citrate anions by α-$K_9AlW_{11}O_{39}$. Calculations based on the ζ values[51,52] assigned net charges of 15.0e to the citrate-protected particles and 29.5e to those possessing a monolayer of α-$K_9AlW_{11}O_{39}$.

α-$AlW_{11}O_{39}^{9-}$-stabilized 5.4-nm and 3-nm diameter core Au NPs were also prepared.[61] The peripheral ring around the 5.4 nm Au NP in Fig. 17b is comprised of 12 POM anions. Geometrical analysis, assuming hexagonal packing of the POMs on the Au NP surfaces, indicates that complete monolayer coverage of the surface involves 51 POM anions. The surface density of α-$AlW_{11}O_{39}^{9-}$ on these particles is calculated to be 2.4×10^{-11} mol cm^{-2}, slightly smaller than that on a 13.8 nm Au core (see above). This might indicate that the distances between adjacent POM anions increases with surface curvature. The zeta potential value was -43.2 ± 4.0 mV, from which a net charge of 10*e* was calculated. Langmuir isotherm data (from UV-visible spectroscopy) were fitted to a function of the binding constant, *K*, for α-$AlW_{11}O_{39}^{9-}$ and α-$AlW_{12}O_{40}^{5-}$. The binding constant, *K*, for the more negatively charged POM (α-$K_9AlW_{11}O_{39}$) was ca. 20,000 M^{-1}, while that for α-$AlW_{12}O_{40}^{5-}$ was 149 M^{-1}. These values are consistent with those obtained for binding of these two POM anions to 13.8 nm Au NPs.

A 3-nm diameter core seems to define the small-size limit for the observation of 1 nm POM anions on the nanoparticle surface. At this high-magnification limit, focusing is more difficult for the cryogenic samples, and the smaller metal NPs are no longer distinct from the POM clusters on their surfaces. Also, because the size of the Au core approaches that of the Keggin-anion ligand, "monolayer" coverage is no longer observed, nor is it expected. Many fewer POM ligands can occupy sites on the particle surface, and the POM/Au nanostructure now appears more polygonal, as indicated in the cartoon in Fig. 17 (right-most image).

At 40% surface coverage by α-$K_9AlW_{11}O_{39}$ (Fig. 18a), discrete groupings of α-$K_9AlW_{11}O_{39}$ are seen at the perimeter of the gold particle. These discrete groupings are due to "islands" or "patches" of POM anions. This is highlighted in Fig. 18b, an illustration of a surface structure implied by the image in panel a. These islands are observed during time-resolved cryogenic cooling during monolayer assembly, and therefore indicate the course of monolayer growth on the particle's surface.

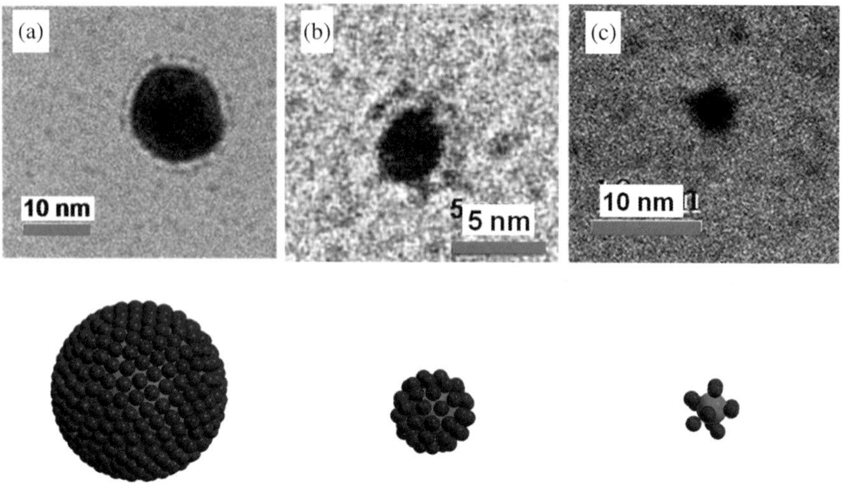

Fig. 17. Cryo-TEM images and cartoon illustration of polyoxometalate stabilized Au nanoparticles of 14 nm, 5.4 nm and 3 nm. (Panel a is reprinted from Ref. 51 with permission from the American Chemical Society.)

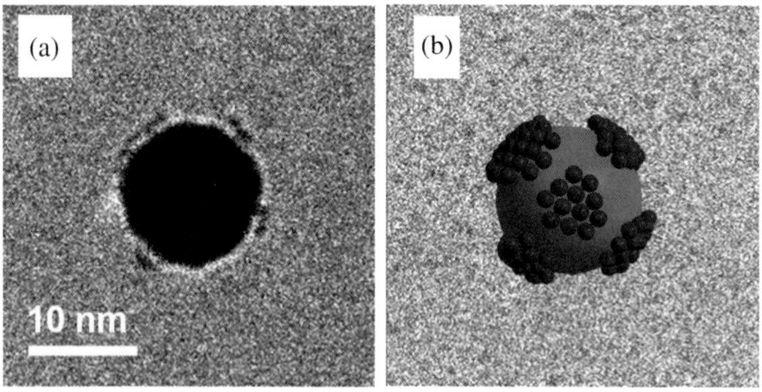

Fig. 18. (a) Cryo-TEM image of gold nanoparticle with 40% surface coverage of α-$K_9AlW_{11}O_{39}$ ($\Theta = 0.4$). (b): An illustration showing how the image in panel a implies the presence of "islands" (discrete regions) of POMs on the gold surface. The "empty" areas in the illustration are protected by citrate anions. (Reprinted from Ref. 51 with permission from the American Chemical Society.)

This "island" on a Au nanoparticle is analogous to that observed by Gewirth[40] for the partially coverage of planar Ag(111) by silicotungstate (Fig. 19).

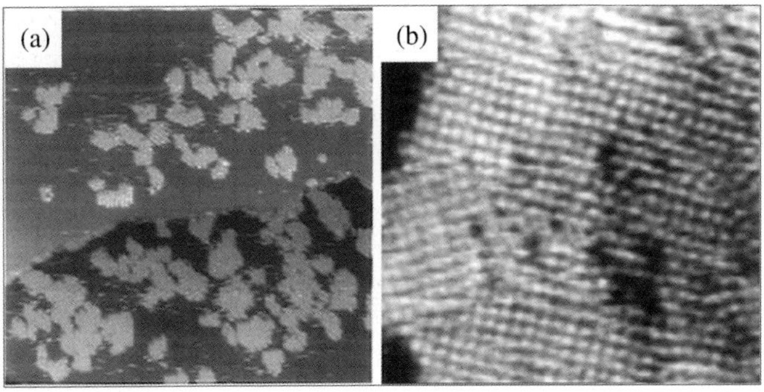

Fig. 19. (a) Partial and (b) full monolayer α-$SiW_{12}O_{40}^{4-}$ anions on Ag(111) surfaces. (Reprinted from Ref. 40 with permission from the American Chemical Society.)

The observation of "islands" points to strong attractions between the POM anions: once a polyanion is adsorbed, adjacent sites become more favored for binding of the next POM anion. This is most likely due to electrostatic forces between POMs and their counter cations, and/or to hydrogen bonds between POMs and cation-bound water molecules. In retrospect, the islands of α-$SiW_{12}O_{40}^{4-}$ observed on Ag(111) surfaces[40] imply that analogous net attractions, mediated by counter cations and/or hydrogen bonding, regulate POM-monolayer self-assembly on both planar and curved (nanoparticle) surfaces.

3.3. *Electric double layer of a polyoxometalate-stabilized nanoparticle*

The Gouy-Chapman-Stern model with modification by Valleau and Torrie was used to describe the gold nanoparticles and their double layer structures. In this model, the Stern layer is comprised of counter cations tightly associated with the negatively charged "surface" of the POM-ligand shell. The distance from the outer surface of the Stern layer to the slipping plane is the Debye length, κ^{-1} (Fig. 20).

Debye lengths, which are functions of ionic strength, are 4.8 nm for the citrate-protected 13.8-nm core Au NPs, 1.0 nm for the α-$K_9AlW_{11}O_{39}$-protected Au NPs, 1.1 nm for $K_8SiW_{11}O_{39}$-protected Au NPs and 1.3 nm for $K_7PW_{11}O_{39}$-protected Au NPs (see Table 3).

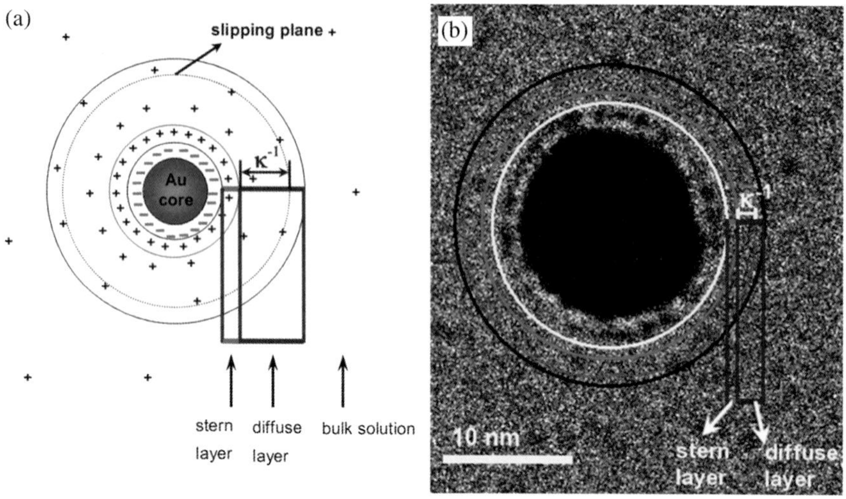

Fig. 20. The electrical double layers of POM-protected gold/Ag nanoparticles. The Stern layer and Debye length (see panel (a) and below) are highlighted on the cryo-TEM image in panel (b). Note that the dimensions in panel a are not to scale. In this model, the tightly associated counter cation layer next to the POM monolayer is regarded as the Stern layer. (Reprinted from Ref. 51 with permission from the American Chemical Society.)

Table 3. Calculated Debye-Hückel parameters and Debye lengths of 13.8-nm Au nanoparticles.[51]

Solution content	Ionic strength/mM	κ/nm^{-1}	κ^{-1}/nm
0.25 mM Au (citrate)	4.1	0.2	4.8
2 mM α-$K_9AlW_{11}O_{39}$ + 0.25 mM Au	94	1.0	1.0
2 mM $K_8SiW_{11}O_{39}$ + 0.25 mM Au	76	0.9	1.1
2 mM $K_7PW_{11}O_{39}$ + 0.25 mM Au	60	0.8	1.3

The *Debye-Hückel parameter*, κ (nm^{-1}; and Debye length, κ^{-1}) is calculated from

$$\kappa^2 = \frac{1000F^2}{\varepsilon RT}\Sigma c_i z_i^2 \tag{8}$$

where F is Faraday's constant (96480 C mol^{-1}), ε is the permittivity of water (7.12×10^{-10} F m^{-1} at 298 K), R is the universal gas constant

(R = 8.314 J K^{-1} mol^{-1}), T is the temperature (298 K), and c_i and z_i are concentrations and charges of all significant ions in solution, respectively.

Calculated Debye-Hückel parameters and Debye lengths (1/κ) for different solutions are listed in Table 3. The images in Fig. 20 are perhaps the first to reveal structural features of the electric double layers of an anion-protected metal(0) nanoparticle.

Further analysis showed that the counter cations (mainly K^+) are extensively integrated into the POM monolayer itself (i.e., between the metal surface and the beginning of stern layer).[51,61] Figure 21 shows five molecules of α-$K_9AlW_{11}O_{39}$ at actual center-to-center distances (drawn to scale) on the gold nanoparticle surface, and spheres that represent the relative crystallographic (grey) and hydrated (blue) sizes of K^+. Subtraction of the diameter of the Keggin anion (1.12 nm) from the center-to-center distance of 1.57 nm leaves 4.5 Å between molecules of α-$K_9AlW_{11}O_{39}$ on the gold surface. Notably, the fully solvated K^+ cation, with its ~17 waters of hydration, drawn to scale near the outer region of the monolayer, is

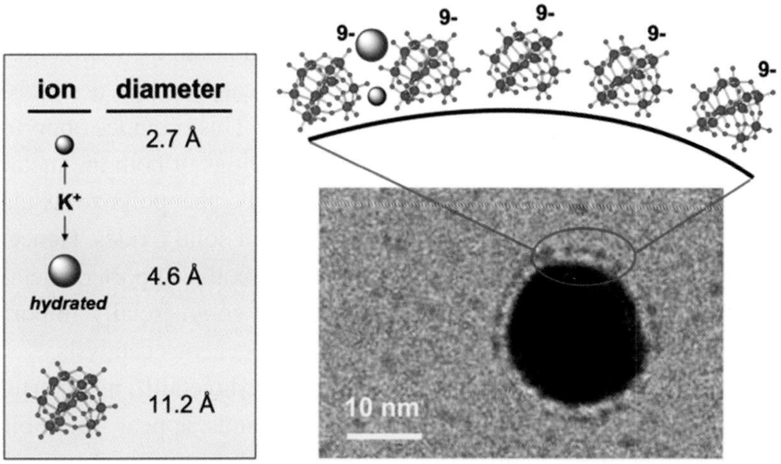

Fig. 21. Scale model of α-$AlW_{11}O_{39}^{9-}$ and K^+ counter cations on the surface of a gold nanoparticle. The key at left shows the crystallographic and hydrated sizes of K^+ relative to a scale model of α-$AlW_{11}O_{39}^{9-}$. The structures of α-$AlW_{11}O_{39}^{9-}$ on the gold surface are drawn with their defect sites facing the gold surface, although this may not be the only viable orientation. (Reprinted from Ref. 51 with permission from the American Chemical Society.)

effectively the same size as the space between molecules of α-$K_9AlW_{11}O_{39}$. A fully dehydrated cation is shown closer to the metal surface. There, the hydrophobic character of the gold surface might facilitate partial dehydration of K^+, a process that would be compensated for by direct bonds between K^+ and the bridging and terminal O^{2-} ligands of the POM. Whether hydrated or partially dehydrated, aqua ligands on K^+ could form H-bonds to O^{2-} ligands of α-$K_9AlW_{11}O_{39}$.

4. Application of POM-stabilized Au NPs in Catalysis

An intriguing application of POM-stabilized metal(0) nanoparticles is in catalysis. Compared to organic ligands such as PPh_3 and alkanethiols, which usually render the nanoparticles inert in catalytic reactions,[46,58] inorganic POM ligands are not only more resistant to oxidizing conditions, but are also more labile. While effectively stabilizing metal(0) nanoparticles, the POMs form inorganic ligand shells whose labilities are controllable through choice of metal-particles size and POM structure, charge, and counter-cations.[61] The option for engineering reversible POM binding thus provides for greater control over access of incoming substrates to the catalytically active metal(0) nanoparticle surface. In addition, the POMs are not simply inert ligands, but are themselves active catalysts for a variety of selective processes in water and organic solvents. This provides opportunities for cooperative behavior involving the reactivities of both the metal(0) surface atoms and nearby POM ligands. Finally, the polyoxometalates themselves may be viewed as soluble fragments of solid oxides. Hence, as first articulated by Finke, "soluble polyoxoanion 'oxide-supported' metal(0) nanoclusters can be regarded as analogs of the commercially important solid-oxide-supported metal(0) particles in heterogeneous catalysts."[46]

This section concerns the use of POM-stabilized metal(0) nanoparticles in catalysis. In many instances the POM ligands appear to stabilize metal(0) nanoparticles, with activity closely matching that of the metal(0) nanoparticles themselves. In other cases, intriguing new activities and/or selectivities are observed. The detailed mechanisms responsible may become more clear as additional information is obtained concerning the structures of POM cluster-anion ligand shells, and the nature of bonding at the POM/metal(0) interface.

4.1. *Hydrogenation reactions using POM-stabilized metal NPs*

Finke used polyoxometalate-stabilized Ir,[46,98] Ru[49,54] and Pt[99] nanoparticles as catalysts for the hydrogenation of cyclo-olefins (Eq. 8). For example, in acetonitrile or acetone at 22°C, ~1 mM of the precatalyst (Ir^I or Rh^I salts, i.e., $(Bu_4N)_5Na_3[(1,5\text{-cyclooctadiene})IrP_2W_{15}Nb_3O_{62}]$ or $(Bu_4N)_5Na_3[(1,5\text{-cyclooctadiene})RhP_2W_{15}Nb_3O_{62}]$), was combined with 1.6 M of purified cyclohexene, and pressurized to 40 psig H_2. Under these conditions, Ir(0) or Ru(0) nanoparticles were formed *in situ*, and subsequently served as catalysts for hydrogenation of the cyclic olefin (see also Sect. 2.2).[46] An induction period was always observed (Fig. 22, curve A), its length depending on the reaction conditions, such as acidity, indicating that metal(0) nanoparticle formation was autocatalytic.[46,48,54] Olefin hydrogenation also proceeded using the isolated POM stabilized nanoparticles, but without an induction period (Fig. 22, curve B).[54]

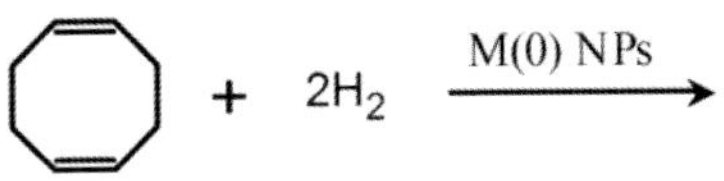

(8)

In Eq. 9, M = Ir,[46,98] Rh[49,54] and Pt[99].

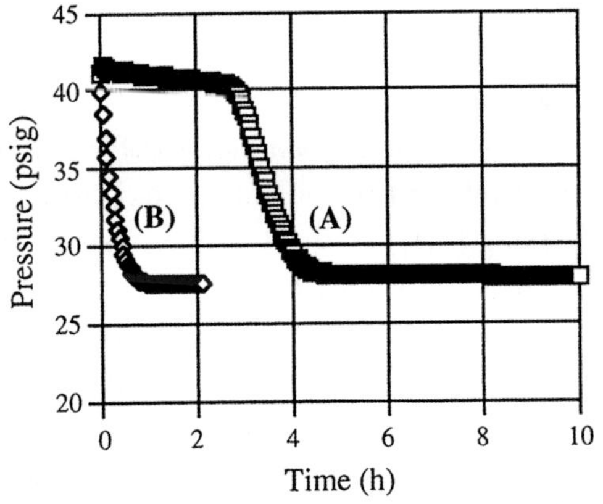

Fig. 22. Cyclohexene hydrogenation beginning with (A) the precatalyst, $[(n\text{-}C_4H_9)_4N]_5Na_3[(1,5\text{-cyclooctadiene})Rh{\cdot}P_2W_{15}Nb_3O_{62}]$, and (B) the isolated Rh nanoparticles resulted from experiment (A). Reaction conditions were indicated in the description above. Curve B shows that when the isolated nanoclusters were used, the reaction proceeded with no induction period. (Reprinted from Ref. 54 with permission from the American Chemical Society.)

In these studies, POMs served to modify the sizes and poly-dispersities of the nanoparticles they then stabilized. The metal(0) NPs are the active hydrogenation catalysts, as demonstrated by (i) catalyst isolation and re-use, (ii) dependence of reaction rates on the M(0) NP concentration, (iii) activity inhibition by added POMs, and (iv) reversible catalyst poisoning and recovery, (v) poisoning of the metal(0) NPs with PPh_3 or Hg leading to inactivity of the catalysts.[46]

The *in situ* formed metal nanoparticles were highly active. For $Ir_{\sim190\text{–}450}$ NPs, an average TON was ca. 3000 turnovers/h, comparable to commercial 7.9% dispersed Ir/γ-Al_2O_3 (~3900 turnovers/h). The *in situ* generated, $P_2W_{15}Nb_3O_{62}^{9-}$-stabilized 40 ± 6 Å Rh(0) NPs, exhibited catalytic lifetimes for olefin hydrogenation of greater than 193, 000 turnovers,[49] — approaching that of a solid-oxide-supported heterogeneous Rh catalyst, and an order of magnitude larger than for previously reported unsupported NPs in solution.

4.2. *Molecular oxygen activation reactions for organic synthesis*

The selective aerobic epoxidation of alkenes using only molecular oxygen and no sacrificial reducing agents is an immensely important and largely unsolved research objective with significant commercial implications. Ronny Neumann and co-workers have addressed this goal using POM-stabilized metal(0) nanoparticles.

Much of that work was carried out using $H_5PV_2Mo_{10}O_{40}$, which like many POMs, tends to inhibit free radical autooxidation reactions. At the same time, $H_5PV_2Mo_{10}O_{40}$ is itself an active and versatile catalyst for oxidations by O_2. In Ronny Neumann's work, Ag(0),[71] Ru(0)[71] and Pt(0)[100] NPs were prepared by the reaction of metal salts with POMs that had been pre-reduced by Zn(0) (see Sec. 2.3.1) to give $ZnH_5PV^{IV}_2Mo_{10}O_{40}$. The POM-protected nanoparticles were supported on metal-oxides (such as α-Al_2O_3, γ-Al_2O_3 and silica) by impregnation. The supported POM-protected nanoparticles were effective catalysts for a variety of reactions in the liquid phase, such as aerobic epoxidation, dehydrogenation, aerobic selective oxidation of secondary alcohols and coupling reactions.[71]

Aerobic epoxidation of cyclohexene and 1-methylcyclohexene. A major obstacle to the direct and aerobic epoxidation of alkenes is the relative weakness of allylic C–H bonds, which often leads to very low epoxidation selectivities. For this reason, well-known heterogeneous silver catalysts are only effective for direct aerobic epoxidations of ethylene, which has no allylic C–H bonds. To address this, the combination of metal(0) nanoparticles with a polyoxometalate capable of activating molecule oxygen was expected to generate an active and possibly highly selective catalyst for direct aerobic epoxidations of a wider range of substrates.[71] $PV_2Mo_{10}O_{40}^{5-}$-stabilized Ag and Ru nanoparticles were thus used to catalyze the aerobic epoxidation of cyclohexene and 1-methylcyclohexene, which are sensitive to autoxidaton and therefore give rise to low selectivities during epoxidation by molecular oxygen. The presence of the nanoparticles did not change the conversion rate of the cyclohexenes. However, the use of $PV_2Mo_{10}O_{40}^{5-}$ to stabilized the metal(0) NPs inhibited autooxidation, and led to enhanced selectivity (see Scheme 2).[71] Using 1-octene as a model substrate, initial selectivity for the formation of 1-octene oxide was high (17% conversion, 88% selectivity). As conversion progressed, however, the selectivity decreased, and more of the substrate was

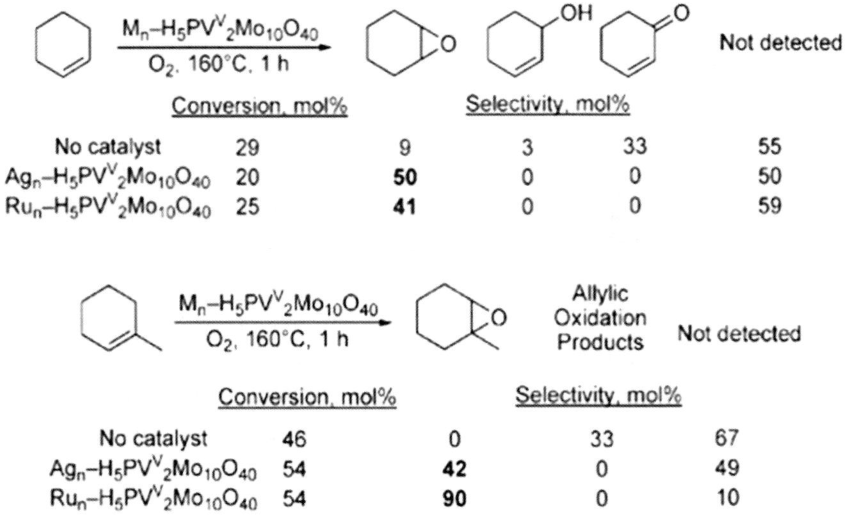

Scheme 2. Aerobic epoxidation of cyclohexene and 1-methylcyclohexene. (Reprinted from Ref. 71 with permission from the Royal Society of Chemistry.)

converted to autoxidation products that were not detected by gas chromatography.

The authors tentatively attributed this to the gradual loss of the POM structure. However, the addition of pure $H_5PV_2Mo_{10}O_{40}$ did not restore selectivity.

Aerobic oxidation of secondary alcohol. Oxidations of primary and secondary alcohols to the corresponding carbonyl compounds are synthetically important. However, the selective aerobic oxidation of primary alcohols to aldehydes, which is very difficult to achieve, and the oxidation of secondary alcohols, which are much less reactive, are rarely reported.

In this context, 2.6-nm diameter Pt NPs stabilized by $PV_2Mo_{10}O_{40}^{5-}$, and supported on α-alumina, were used for the aerobic oxidation of secondary alcohols to corresponding ketones.[100] Oxidations of the secondary alcohols were typically performed using 0.5 mmol alcohol and 0.05 mmol catalyst for 14 h under 2 bar O_2 at 125°C. A 5% catalyst loading on α-alumina was found to be optimal. Under the optimized conditions, selected secondary alcohols, such as 2-decanol, 2-octanol, 2-hexanol, 2-pentanol, cyclohexanol and cyclooctanol, were oxidized to ketones with 100% selectivity, For the acid form of $PV_2Mo_{10}O_{40}^{5-}$, turnover numbers (TONs) were typically ~10 under the optimized conditions. By replacing the POM's H^+ counter-cations by rhodium cations, TONs increased to 50–90 under the otherwise identical conditions. The mechanism was not reported. The intriguing cooperative effect between the POM, its counter cation and the Pt(0) nanoparticles remains to be better understood and perhaps further exploited.[100]

Aerobic dehydrogenation reactions. Conversion of alkanes to alkenes, and of alkylarenes to vinylarenes, can be accomplished by dehydrogenation and oxy-dehydrogenation. Neumann reported that 2–6 nm Pd NPs stabilized by $[(C_6H_{13})_4N]_4[SiW_{11}O_{40}(SiCH_2CH_2CH_2SH)_2]$ are effective binary catalysts for aerobic oxy-dehydrogenation of substrates with reactive allylic tertiary C–H bonds (see Scheme 3 for an example).[101] For the oxy-dehydrogenation of 4-vinylcyclohexene at elevated temperatures (140°C) for 14 h, the conversion was high (80%), and yielded styrene (77%) and ethylbenzene (23%) as the only products with 100% mass balance. No aggregation of Pd or formation of Pd black was observed during the reaction.

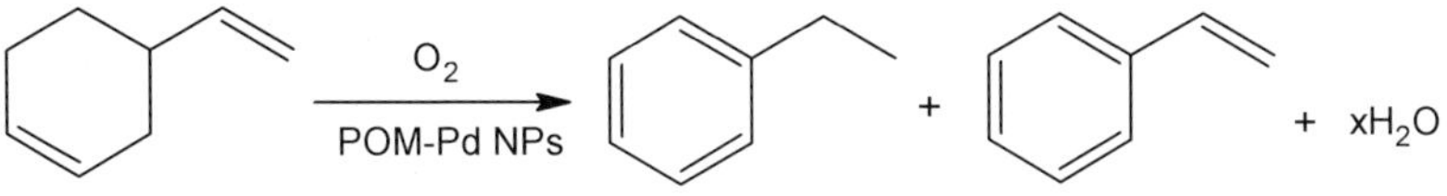

Scheme 3. Aerobic oxydehydrogenation of 4-vinylcyclohexene catalyzed by a heteropolytungstate-stabilized Pd nanoparticle.

4.3. *Other organic reactions catalyzed by polyoxometalate-stabilized metal nanoparticles*

Coupling reactions. Spherical, 15–20 nm palladium particles stabilized by α-$PW_{11}O_{39}^{7-}$ were synthesized by reduction of α-$K_5P(Pd)W_{11}O_{39}$ with H_2.[64] The nanoparticles were effective catalysts for Suzuki-, Heck-, and Stille-type carbon–carbon coupling and carbon–nitrogen coupling reactions of bromoarenes in aqueous media (Scheme 4).[64] The reactions were typically carried out in a mixture of water and an organic solvent, i.e., 4:1 (v/v) of water:EtOH for the Suzuki and Heck coupling, and 1:1 water:DMF for Stille-type and carbon-nitrogen coupling reactions at elevated temperatures of 80–140°C. Yields of these reactions were high (see Scheme 4). Mechanistic studies indicated that the coupling reaction was initiated by reaction of the haloarene with the Pd catalyst at the surface of the nanoparticle rather than with dissolved Pd species.

In work by Mandal *et al.*,[63] Pd(0) nanoparticles (9 ± 2 nm) were first synthesized by reduction of $Pd(NO_3)_2$ with photochemically reduced $H_3PW_{12}O_{40}$, then extracted into a toluene solution containing octadecylamine (~95% yield). The nanoparticles were then isolated by rotary evaporation. The obtained material was soluble in organic solvents such as toluene and chloroform, a function of the octadecylammonium counter cations. The octadecylammonium — and α-$PW_{12}O_{40}^{3-}$— stabilized Pd nanoparticles were used for catalyzing the Heck reaction (Scheme 4, panel d), the ring-opening reaction of epoxides, and an acylation reaction (Scheme 4). The Heck reaction was typically carried out at 120°C for 3–24 h under N_2 in a mixed solvent of DMF and chloroform. For X = I and Br, and R = H, the conversion was 97.5% (selectivity 99.1% trans, turnover frequency, TOF = 1710 h^{-1}) and 96% (selectivity 97.5% trans,

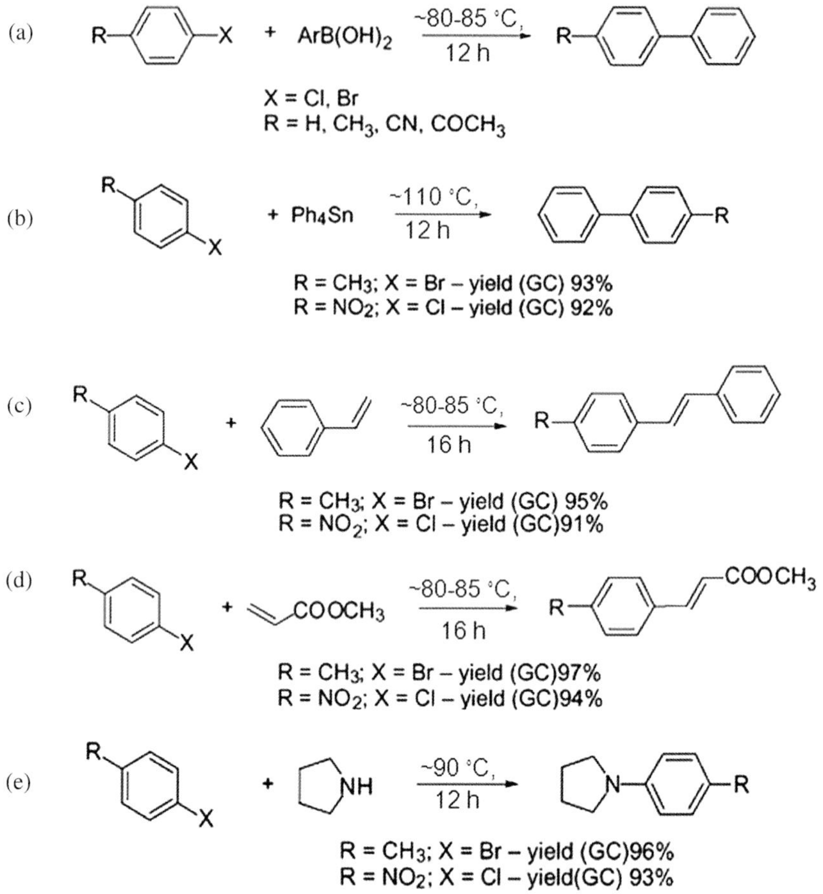

Scheme 4. (a) Suzuki coupling, (b) Stille coupling, (c) and (d) Heck coupling and (e) carbon-nitrogen coupling catalyzed by α-$PW_{11}O_{39}^{7-}$ — protected Pd NPs. (Reprinted from Ref. 64 with permission from the American Chemical Society.)

TOF 44 h^{-1}), respectively, superior to commercial Pd/C and Pd/Al_2O_3 catalysts. The activity of the recycled catalyst was comparable to that of the fresh catalyst. TEM images indicated that although a small increase in the average size was observed, no aggregation or appreciable sintering of Pd nanoparticles had occurred. In contrast, octadecylammonium-stabilized Pd nanoparticles, with no POM present, showed a higher TOF and similar selectivity in the Heck reaction, while slight aggregation occurred after three cycles. This indicates that the polyoxometalate anion serves to

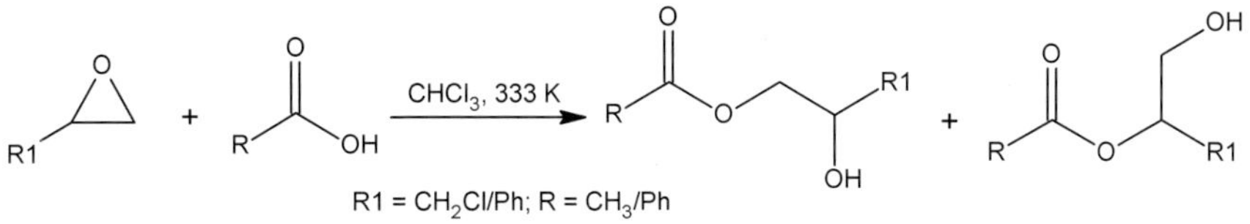

Scheme 5. The ring-opening reaction of epoxy compounds catalyzed by octadecylammonium — and α-$PW_{12}O_{40}^{3-}$— stabilized Pd nanoparticles.

stabilize the Pd nanoparticles. For the ring-open reaction of epoxides, the Pd nanoparticles stabilized by a combination of octadecylammonium and α-$PW_{12}O_{40}^{3-}$ exhibited relatively high activity, i.e., 82.5% conversion with a TOF of 144 h^{-1}. However, the octadecylamine-stabilized Pd nanoparticles was inactive, indicating an important role for the POM in the catalytic reaction.

4.4. *Electrocatalysis*

On electrode surfaces, POM-modified metal(0) NPs often have the synergistic effect of both positively shifting redox potentials, and enhancing electrode currents. A positive shift is not surprising, as the electrode is modified with noble metals, and is no longer simply glassy carbon (GC). Current enhancement is possibly caused by an increase in the surface area by loading of nanoparticles, or a favorable change in the activation barriers associated with heterogeneous electron-transfer.

Assuming that a planar electrode surface with an area of a^2 is entirely covered by square packing of spherical particles of radius, *r*, there will be $(a/2r)^2$ particles, each with surface area of $4\pi r^2$. Ignoring the gaps between the particles as the size of the particles approaches nanometer scale, only ca. half the surface of each particle is exposed to the environment. Therefore, the effective nanoparticle-monolayer surface area is $(a/2r)^2 \times 4\pi r^2$, or πa^2. Therefore, if a planar surface is modified with nanoparticles, the surface area might increase by approximately 3 fold. This could explain the approximately 2-fold increases in current often observed for these systems.[85,88,102–104] At the same time, the activation barriers to electron transfer between the nanoparticle assemblies and redox-active substrates in solution can be modified by the metal nanoparticles, leading to true electrocatalysis.

Oxygen reduction reactions (ORR). The ORR is a multi-step reaction that leads to the 4e$^-$-reduction of O_2 to H_2O (Eqs. 10 and 11). In electrochemical reductions of O_2 in fuel cells, where the facile conversion of O_2 to H_2O is critically important, H_2O_2 sometimes accumulates as an undesired product of partial reduction.

$$O_2 + 2H^+ + 2e^- \rightarrow H_2O_2 \tag{9}$$

$$H_2O_2 + 2H^+ + 2e^- \rightarrow 2H_2O \tag{10}$$

$$O_2 + 4H^+ + 4e^- \rightarrow 2H_2O \tag{11}$$

A typical cyclic voltammogram (CV) of an electrocatalytic ORR is shown in Fig. 23. This was obtained using a Au-NP-modified glassy carbon (GC) electrode in 0.4 M phosphate buffered saline solution (pH = 7).[75] To prepare the Au-NP-modified GC electrodes, Nadjo and coworkers deposited an aliquot of a solution of β-$PMo_{12}O_{40}^{7-}$-stabilized Au NPs onto a polished glassy carbon surface, which was then air-dried at room temperature, covered with a small amount of Nafion solution, and

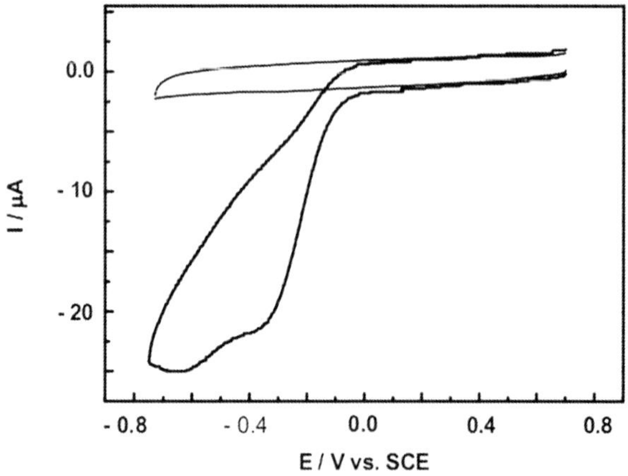

Fig. 23. CV with the glassy carbon electrode modified with Au NPs. The background current is in the absence of O_2. (Reprinted from Ref. 75 with permission from Royal Society of Chemistry.)

air dried again. The CV showed a typical two-wave reduction of oxygen, indicating two closely spaced two-electron steps (Eqs. 10 and 11). In addition, the CV obtained in 0.5 M H_2SO_4 showed simultaneously the characteristic redox features of Au(0) and of β-$PMo_{12}O_{40}^{7-}$, indicating that the polyoxometalate anions were intact on the Au NP surfaces.

In studies by Kulesza *et al.*,[102,105] POM-stabilized 8-nm-diameter Pt NPs were obtained by suspending commercial Pt black in solutions containing $H_3PMo_{12}O_{40}$ or $H_3PW_{12}O_{40}$, which were then sonicated for 24 h and centrifuged 3 or 4 times. Nafion was added to aliquots of the colloidal NP solutions, which were then deposited on GC electrodes, air dried, and heated at 75°C for 1 h to firmly immobilize the NPs on the electrode surface. The Pt-NP-modified GC electrodes were used to study the ORR. The electrode with α-$PW_{12}O_{40}^{3-}$-protected Pt NPs shifted the potential of the ORR by 30 mV in the positive direction, and the rate constant at the electrode surface (as well as the current intensity) was enhanced by 2-fold relative to unmodified Pt NPs. α-$PMo_{12}O_{40}^{3-}$ and α-$PW_{12}O_{40}^{3-}$ on Pt NPs adsorbed on electrode surfaces might both suppress the oxidation of Pt(0), as indicated by inhibition of the current for Pt(0) oxidation. The ORR at potentials < +0.8 V vs. NHE on the POM-stabilized Pt NPs proceeded by a $4e^-$-reduction (Eq. 12), the same as on unmodified Pt surfaces. H_2O_2 yield was usually low, typically < 3% at potentials > 0 V vs. NHE, for both POM-modified and unmodified Pt NPs, consistent with the lower barrier for activation of H_2O_2. Moreover, the yield of H_2O_2 was actually suppressed by 50–100% by POM modification. Overall, α-$PW_{12}O_{40}^{3-}$-protected Pt NPs provided a more active catalyst than unmodified Pt NPs, and α-$PW_{12}O_{40}^{3-}$ was more effective than α-$PMo_{12}O_{40}^{3-}$. Other POMs, such as α-$SiW_{12}O_{40}^{4-}$ and α-$SiMo_{12}O_{40}^{4-}$ were found to slightly inhibit the ORR reaction.[105]

Bromate reduction. $PMo_{12}O_{40}^{3-}$-stabilized Au NPs were prepared by ligand exchange (see Sec. 2.4) and used to modify a GC electrode.[88] The POM, $PMo_{12}O_{40}^{3-}$, retained its redox behavior on the modified electrode, an indication of its integrity after electrode fabrication. For the reduction of bromate, the current was enhanced 2.2 fold and the reduction potential was shifted by +25 mV, comparing to a GC electrode modified with $PMo_{12}O_{40}^{3-}$ alone. Also, bromate reduction was mediated by the $PMo_{12}O_{40}^{3-}$/$PMo_{12}O_{40}^{4-}$ couple, i.e., it involved the first one-electron reduction of $PMo_{12}O_{40}^{3-}$. Overall, the combination of gold nanoparticles and $PMo_{12}O_{40}^{3-}$ provided a synergistic effect, and the limit for bromate detection was ~3 μM by pulse voltammetry.[88]

Alcohol oxidation for fuel cell applications. Pt and Pd NPs stabilized with Mo-based POMs (for synthesis, see Sec. 2.2), such as β-$H_7PMo_{12}O_{40}$, $(NH_4)_{10}[Mo_6O_{14}(O_3PCH_2PO_3)_2(H_2OPCH_2PO_3)_2]$, and ε-$PMo_{12}O_{36}(OH)_4[La(H_2O)_{2.5}Cl_{1.25}]_4$, were studied in the electrocatalytic oxidation of methanol and ethanol, respectively, and found to efficiently catalyze these processes.[79] The NP-modified GC electrodes were prepared by Nadjo and his coworkers, following a protocol analogous to that for preparation of the Au-NP-modified electrodes described above for catalysis of the ORR. As shown in Fig. 24 (the solid curve), the catalytic oxidation of methanol using a Pt-NP-modified GC electrode occurred at peak potentials between +0.600 and +0.620 V vs. SCE for scan to positive potentials, and between +0.475 and +0.465 V vs. SCE for the reverse scan. CV data (Fig. 24,

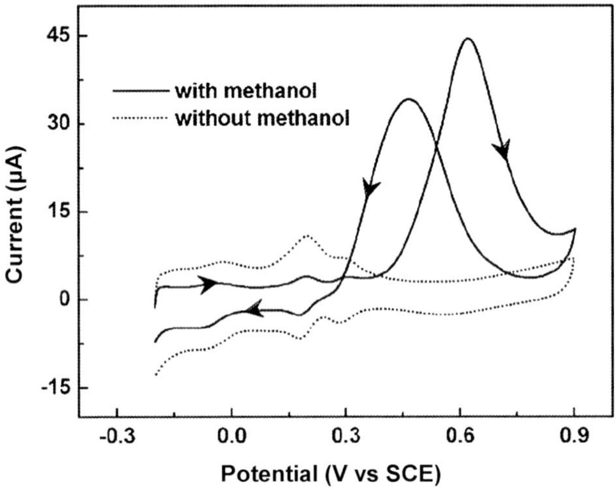

Fig. 24. CVs of 0.5 M methanol (solid curve) and the pure supporting electrolyte (dotted curve) using the Pt-NP-modified GC electrode. Electrolyte: 0.5 M H_2SO_4; scan rate: 50 mV s^{-1}. The superimposed feature of the dotted curve was attributed to adsorbed polyoxometalate anions on the electrode surface. The CV response of the solid curve was considered to correspond to the overall oxidation reaction of methanol as: $CH_3OH + H_2O \rightarrow CO_2 + 6H^+ + 6e^-$. (Reprinted from Ref. 79 with permission from the American Chemical Society.)

dotted curve) demonstrated that the POM redox features on Pt NPs remained unchanged after three washing and centrifugation cycles, indicating that the POM super-structures (possibly monolayers) were remarkably stable.[51] No current decrease was observed after more than 100 cycles between –0.2 and +0.9 V, an indication of the stability and resistance to poisoning of the β-$PMo_{12}O_{40}^{7-}$-protected Pt NPs in the electrocatalytic oxidation of methanol to CO_2.[79]

Hydrogen production. POM-stabilized 6-nm-diameter Pd NPs[106] were synthesized by reaction of K_2PdCl_4 with the reducing V(IV), d^1, ions of the wells-Dawson $[H_4PVW_{17}O_{62}]^{9-}$ anion. Previous study[106] showed that these POM-stabilized Pd NPs tended to spontaneously self-assemble into hollow, spherical, 30–50-nm-radius supramolecular structures in solution. In a recent report, the isolated (by centrifugation) POM-stabilized Pd NPs were used to modify a GC electrode for use in the electrocatalytic production of H_2 in 0.5 M H_2SO_4.[103] CV data demonstrated that the POM anions were firmly attached to the Pd NP surface. Hydrogen production began to occur at ~–0.25 V vs. SCE with an overpotential of near 0 at low current density (Fig. 25). As the current density increased, the overpotential also

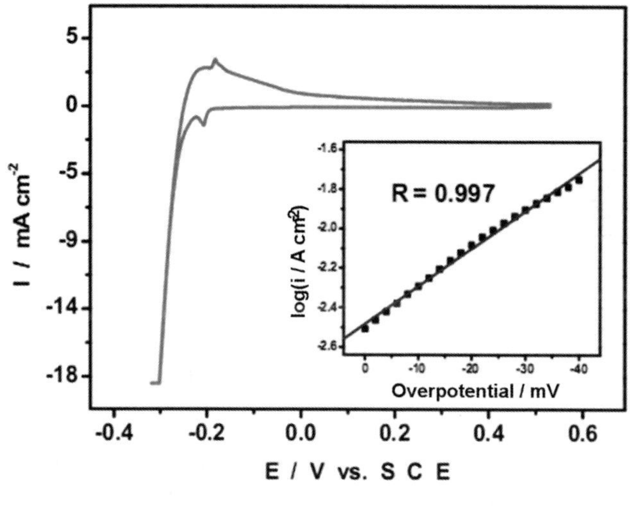

Fig. 25. CV of 0.5 M H_2SO_4 using a GC electrode modified with $H_4PV^{IV}W_{17}O_{62}$—stabilized Pd NPs. Inset shows that the overpotential of H_2 formation increases linearly with current density. (Reprinted from Ref. 103 with permission from MDPI Publishing, Basel, Switzerland.)

increased. It was shown that the presence of the POM had a beneficial effect on electrode activation, and the exchange current density measured using the POM-Pd-NP-modified electrode was 2–2.9 times larger than the corresponding value obtained for unmodified bulk Pd.

In another report by Yuan *et al.*,[104] highly faceted Au NPs (90% icosahedral, 30–50 nm) were synthesized by reaction of $HAuCl_4$ with reduced α-$H_3PW_{12}O_{40}$ at 100°C. The α-$PW_{12}O_{40}^{3-}$-stabilized Au NPs adsorbed on a GC electrode were highly active towards electrocatalytic hydrogen production. At potentials between –0.25 and –0.8 V, the current for hydrogen production from 5 mM H_2SO_4 was larger than that for electrodes modified by the deposition of polycrystalline Au in the absence of POMs. By contrast, the unmodified GC electrode was inactive under the same conditions.

5. Closing Comments

Unlike most ligands used to provide protective shells for metal nanoparticles, polyoxometalates (or POMs) constitute a large and rapidly growing class of discrete molecular structures with numerous applications in catalysis. Many POMs possess extensive and reversible redox chemistries, which are central to their use in many applications. Their reduction potentials, acidities, and other key properties relevant to catalysis and materials science can be extensively yet readily altered by the elemental composition of the POM cluster itself. And, as molecular anions, it is not surprising that polyoxometalates can stabilize colloidal metal nanoparticles in solution. In this regard POMs have been shown to form remarkably robust ligands shells, and while more cases remain to be investigated, cryo-TEM images have revealed well-defined POM monolayers on the surfaces of Ag(0) and Au(0) nanoparticles. The close distances between POMs on the metal(0) surfaces suggested that POM counter-cations must be extensively incorporated as critical structural components of the POM monolayer. Hence, it is now clear that the stability of POM-protected metal(0) nanoparticles likely derives from several phenomena, namely, electrostatic forces typical of anion-stabilized colloids, direct bonding between POMs and the metal(0) surfaces, and the formation of numerous ionic and hydrogen bonds between POMs bound closely to one another on the metal(0) surface.

Numerous POMs are redox active. Therefore, in addition to traditional methods for nanoparticle synthesis, such as the reduction of metal salts in the presence of protecting ligands, and ligand exchange, POM-protected metal(0) nanoparticles can be prepared by direct reductions of metal salts by reduced POM anions, whose oxidized forms then form protecting-ligand shells. The few systematic studies of these processes reveal clear, and practically useful, correlations between particle sizes, and the reduction potentials of the POM reducing agents. This is an important control option beyond that traditionally afforded by varying ratios of reducing agents to metal-salt nanoparticle precursors.

Another unique feature of POM-stabilized metal(0) nanoparticles is that their protecting shells — the POM anions — are themselves catalytically active. Hence, two catalytically active components, the metal(0) atoms at the nanoparticle surface, and the POM ligands, can act together, in or in rapid sequence, to provide for new reactions or product selectivities. Moreover, POM counter-cations play an important structural role in the formation and stability of POM-ligand monolayers, and in fact, are integral components of the POM-ligand shells. This provides an additional opportunity for the design of catalytically active POM-stabilized nanoparticle assemblies wherein the POM counter-cations are now reactive transition-metal cations of cationic complexes.

For example, Anson[35] has shown that $Ru(bpy)^{3+}$ can be deployed as a structural counter-cation in POM monolayers on the surfaces of hanging drop Hg electrodes, and more directly, Neumann[100] has observed that Rh(I) counter cations can considerably enhance the catalytic O_2-oxidation of alcohols by POM-protected Pt(0) nanoparticles. With new and more detailed information about the presence and structures of POM monolayers on metal(0) nanoparticles, and the recognition that POM counter-cations are an integral component of these monolayers, it may soon be possible to rationally design metal(0)/POM/counter-cation nano-reactors for an increasingly broad range of catalytic applications.

References

1. Pope M. T., Polyoxo Anions: Synthesis and Structure. In *Comprehensive Coordination Chemistry II: From Biology to Nanotechnology*, Wedd, A. G. (Ed.) Elsevier, Oxford, UK; Vol. 4, pp. 635–678, (2004).

2. Hill C. L., Polyoxometalates: Reactivity. In *Comprehensive Coordination Chemistry-II: From Biology to Nanotechnology*, Wedd, A. G. (Ed.) Elsevier, Oxford, Vol. 4, pp. 679–759, (2004).
3. Weinstock I. A., *et al.*, *Nature* **414** (2001) 191–195.
4. Neumann R., *et al.*, *Inorg Chem* **49** (2010) 3594–3601.
5. Fleming C., *et al.*, *Nat Nano* **3** (2008) 289–233.
6. Müller A. and Roy S., Oxomolybdates: From Structures to Function in a New Era of Nanochemistry. In *The Chemistry of Nanomaterials*, Rao C. N. R., Müller A. Cheetham A. K. (Eds.) Wiley-VCH, Weinheim, Vol. 1, pp. 452–475, (2004).
7. Yamase T. and Prokop P. V., *Angew Chem Int Ed* **41** (2002) 466–469.
8. Müller A., *et al.*, *Angew Chem Int Ed* **37** (1998) 1220–1223.
9. Lyon D. K., *et al.*, *J Am Chem Soc* **113** (1991) 7209–7221.
10. Anderson T. M., *et al.*, *Science* **306** (2004) 2074–2077.
11. Ozeki T. and Yamase T., *Acta Cryst B* **50** (Part 2) (1994) 128–134.
12. Luo Q., *et al.*, *Inorg Chem* **41** (2002) 6112–6117.
13. Belai N., *et al.*, *Inorg Chem* **44** (2005) 169–171.
14. Zhang C., *et al.*, *Inorg Chem* **44** (2005) 3569–3578.
15. Clemente-Juan J. M., *et al., Inorg Chem* **44** (2005) 3389–3395.
16. Mizuno N., *et al.*, *J Am Chem Soc* **120** (1998) 9267–9272.
17. Weiner H. and Finke R. G., *J Am Chem Soc* **121** (1999) 9831–9842.
18. Bar-Nahum I., Khenkin A. M. and Neumann R., *J Am Chem Soc* **126** (2004) 10236–10237.
19. Neumann R., *Prog Inorg Chem* **47** (1998) 317–370.
20. Hill C. L., *Chem Rev* **98** (1998) 1–389.
21. Kim W. B., *et al.*, *Science* **305** (2004) 1280–1283.
22. Geletii Y. V., Gueletii A. and Weinstock I. A., *J Mol Catal A Chem* (1–2) (2007) 59–66.
23. Yamase T. and Pope M. T., *Polyoxometalate Chemistry for Nano-Composite Design*. Kluwer Academic/Plenum Publishers, New York, Vol. 2, (2002).
24. Pope M. T., *Heteropoly and Isopoly Oxometalates*. Springer-Verlag, Berlin, (1983).
25. Weinstock I. A., *Chem Rev* **98** (1998) 113–170.
26. Bi L. H., *et al.*, *Inorg Chem* **44** (2005) 896–903.
27. Kozik M., Hammer C. F. and Baker L. C. W., *J Am Chem Soc* **108** (1986) 7627–7630.
28. Toth J. E. and Anson F. C., *J Am Chem Soc* **111** (1989) 2444–2451.

29. Kozik M. and Baker L. C. W., *J Am Chem Soc* **112** (1990) 7604–7611.
30. Grigoriev V. A., *et al.*, *J Am Chem Soc* **123** (2001) 5292–5307.
31. Geletii Y. V., *et al.*, *J Am Chem Soc* **128** (51) (2006), 17033–17042.
32. Geletii Y. V. and Weinstock I. A., *J Mol Catal A Chem* **251** (2006) 255–262.
33. Geletii Y. V., *et al.*, *Inorg Chem* **44** (2005) 8955–8966.
34. Snir O., *et al.*, *J Am Chem Soc* **132** (2010) 11678–11691.
35. Rong C. and Anson F. C., *Anal Chem* **66** (1994) 3124–3130.
36. Keita B., *et al.*, *J Electroanal Chem* **384** (1995) 155–169.
37. Keita B. and Nadjo L, *J Electroanal Chem* **227** (1987) 265–270.
38. Keita B., Nadjo L. and Haeussler J. P., *J Electroanal Chem* **230** (1987) 85–97.
39. Klemperer W. G. and Wall C. G., *Chem Rev* **98** (1999) 297–306.
40. Ge M., *et al.*, *J Am Chem Soc* **118** (1996) 5812–5813.
41. Lee L., *et al.*, *J Am Chem Soc* **123** (2001) 8838–8843.
42. Teague C. M., *et al.*, *J Phys Chem B* **108** (2004) 1974–1985.
43. Kaba M. S., *et al.*, *Inorg Chem* **37** (1998) 398–406.
44. Song I. K., *et al.*, *Chem Mater* **8** (1996) 2352–2358.
45. Tang Z., *et al.*, *Langmuir* **16** (2000) 4946–4952.
46. Lin Y. and Finke R. G., *J Am Chem Soc* **116** (1994) 8335–8353.
47. Watzky M. A. and Finke R. G., *J Am Chem Soc* **119** (1997) 10382–10400.
48. Aiken J. D. and Finke R. G., *J Am Chem Soc* **120** (1998) 9545–9554.
49. Aiken J. D. and Finke R. G., *J Am Chem Soc* **121** (1999) 8803–8810.
50. Neyman A., *et al.*, *J Am Chem Soc* **130** (2008) 16480–16481.
51. Wang Y., *et al.*, *J Am Chem Soc* **131** (2009) 17412–17422.
52. Wang Y. and Weinstock I. A., *Dalton Trans* **39** (2010) 6143–6152.
53. Keita B., Liu T. and Nadjo L., *J Mater Chem* **19** (2009) 19–33.
54. Aiken J. D. and Finke R. G., *Chem Mater* **11** (1999) 1035–1047.
55. Widegren J. A., *et al.*, *Chem Mater* **13** (2001) 312–324.
56. Aiken J. D. and Finke R. G., *J Mol Catal A Chem* **145** (1999) 1–44.
57. Ott L. S. and Finke R. G., *J Nanosci Nanotech* **8** (2008) 1551–1556.
58. Lin Y. and Finke R. G., *Inorg Chem* **33** (1994) 4891–4910.
59. Ott L. S. and Finke R. G., *Coord Chem Rev* **251** (2007) 1075–1100.
60. Chen I., *Langmuir* **12** (1996) 3437–3441.
61. Wang Y., *et al.*, *Inorg Chem Acta* **363** (2010) 4416–4420.
62. Troupis A., Hiskia A. and Papaconstantinou E., *Angew Chem Int Ed* **41** (2002) 1911–1914.
63. Mandal S., *et al.*, *Langmuir* **21** (2005) 2408–2413.

64. Kogan V., *et al.*, *Org Lett* **4** (2002) 3529–3532.
65. Shanmugam S., Viswanathan, B. and Varadarajan T. K., *Nanoscale Res Lett* **2** (2007) 175–183.
66. Mandal S., *et al.*, *J Am Chem Soc* **125** (2003) 8440–8441.
67. Mandal S., Mandale A. B. and Sastry, M., *J Mater Chem* **14** (2004) 2868–2871.
68. Gordeev A. V. and Ershov B. G., *High Energy Chem* **33** (1999) 218–223.
69. Ershov B. G. and Gordeev A. V., *Mendeleev Comm* **11** (2001) 147–148.
70. Maksimov G. M., *et al.*, *Kinetics and Catal* **41** (2000) 844–852.
71. Maayan G. and Neumann R., *Chem Commun* (2005) 4595–4597.
72. Maksimova G. M., *et al.*, *Kinetics Catal* **45** (2004) 870–878.
73. Keita B., *et al.*, *Electrochem Commun* **6** (2004) 978–983.
74. Keita B., *et al.*, *J Phys Chem C* **111** (2007) 8145–8148.
75. Zhang G. J., *et al.*, *J Mater Chem* **19** (2009) 8639–8644.
76. Zhang G., *et al.*, *Chem Mater* **19** (2007) 5821–5823.
77. Keita B., *et al.*, *J Mater Chem* **18** (2008) 3196–3199.
78. Keita B., Liu T. B. and Nadjo L., *J Mater Chem* **19** (2009) 19–33.
79. Keita B., *et al.*, *J Phys Chem C* **111** (2007) 8145–8148.
80. Troupis A., *et al.*, *Eur J Inorg Chem* (2008) 5579–5586.
81. Wang Y., *et al.*, *Inorg chem* **51** (2012) 7436–7438.
82. Triantis T., *et al.*, *Catal Today* **144** (2009) 2–6.
83. Sastry M., *et al.*, *J Mater Chem* **15** (2005) 3161–3174.
84. Dolbecq A., *et al.*, *Chem Eur J* **15** (2009) 733–741.
85. Li H., *et al.*, *Chem Commum* **46** (2010) 3750–3752.
86. Dahl J. A., Maddux B. L. S. and Hutchison J. E., *Chem Rev* **107** (2007) 2228–2269.
87. Lica G., Browne K. and Tong Y., *J Clust Sci* **17** (2006) 349–359.
88. Ernst A. Z., *et al.*, *Electroanal* **19** (2007) 2103–2109.
89. Chalkley L., *J Phys Chem* **56** (1952) 1084–1086.
90. Graham C. R., Ott L. S. and Finke R. G., *Langmuir* **25** (2009) 1327–1336.
91. Siiman O. and Feilchenfeld H., *J Phys Chem* **92** (1988) 453–464.
92. Yuwono V. M., *et al.*, *J Am Chem Soc* **132** (2010) 2163–2165.
93. Kumar S., *et al.*, *J Am Chem Soc* **130** (2008) 17284–17286.
94. Pouget E. M., *et al.*, *Science* **323** (2009) 1455–1458.
95. Omer L., *et al.*, *J Am Chem Soc* **131** (2009) 12466–12473.
96. Baytekin H. T., *et al.*, *Chem Mater* **21** (2009) 2980–2992.

97. Nativ-Roth E., *et al.*, *Small* **4** (2008) 1459–1467.
98. Mondloch J. E., *et al.*, *J Am Chem Soc* **132** (2010) 9701–9714.
99. Mondloch J. E., Yan X and Finke R G, *J Am Chem Soc* **131** (2009) 6389–6396.
100. Maayan G. and Neumann R., *Catal Lett* **123** (2008) 41–45.
101. De bruyn M. and Neumann R., *Adv Synth Catal* **349** (2007) 1624–1628.
102. Chojak M., *et al.*, *Electrochim Acta* **52** (2007) 5574–5581.
103. Biboum R. N., *et al.*, *Materials* **3** (2010) 741–754.
104. Yuan J., *et al.*, *Nanotechnology* **17** (2006) 4689.
105. Wlodarczyk R., *et al.*, *J Power Sources* **159** (2006) 802–809.
106. Zhang J., *et al.*, *Langmuir* **24** (2008) 5277–5283.

Chapter 2

WHEN GIANTS MEET DWARVES IN THE SAME POND — UNIQUE SOLUTION PHYSICAL CHEMISTRY OPPORTUNITIES OFFERED BY POLYOXOMETALATE MACROIONS

DONG LI, PANCHAO YIN and TIANBO LIU
Department of Chemistry, Lehigh University, Bethlehem, Pennsylvania, USA, 18015

1. Introduction and Retrospection

Polyoxometalates (POMs) macroions represent a large group of structurally well-defined, giant macroions assembled from multiply transition metal oxides, which are linked through shared oxygen atoms. Owing to the tunable valence and coordination geometry of the central metal ions, various POM macroions with different size, shape and charge have been synthesized. POM macroions also demonstrate interesting catalytic, photochemical, electrical, and biomedical properties, which attracts intensive attention from different scientific regions.[1] Most of the large POMs are quite soluble in aqueous solutions or polar solvents owing to their surface water ligands and their charges. Therefore, these POM clusters can also be treated as macroions in solution.

Macroions (such as proteins, DNA molecules) play important roles in biological systems. However, it is not easy to understand the complex solution behaviors of macroions. The primary reason for this is likely to be the long range Coulombic interaction between ions. The intermolecular and intramolecular electrostatic interactions play a dominating role in many systems. Historically, the most famous method to describe a simple dilute ionic solution is the Debye-Hückel theory[2] which was first formulated by Peter Debye and Erich Hückel in 1923. The general physical conclusions of this theory reveal that (a) overall the solution is electrically neutral; (b) oppositely charged ions attract each other and lower the overall free energy of the solution; (c) an ionic atmosphere is formed for any given ion in solution, and this spherical ionic atmosphere has a net charge opposite in sign to that of the central ion but equal in magnitude. However, in general, the Debye-Hückel theory is only valid for very dilute ionic solutions because it treats ions as point charges, which means that the ionic size is negligible. For ionic polymers and colloidal particles, this theory is no longer valid. On the other hand, charged colloids with size dimension ranging from tens of nanometers to several microns are fundamentally different from simple ions. Colloidal particles are thermodynamically unstable, therefore, in reality, two main strategies are used to increase the stability of colloidal particles: bring charges to the colloidal particle surface and/or increase the steric effect of colloidal particles. To elucidate the solution behaviors of these charged colloidal particles, a theoretical framework of the Derjaguin-Landau-Verwey-Overbeek (DLVO) theory developed in the 1940s is widely applied.[3] The DLVO potentials consist two major parts, one is the attractive van der Waals interaction and the other is the purely repulsive Coulomb interaction. Small counter-ions are believed to take a distribution of thermal equilibrium in a stationary configuration of colloids, and the effective interaction $U = U_R + U_A$ controls the dynamical behavior of the macroions. In other words, the stability of colloidal particles in solution stems from the competition between these two interactions. In D-H theory, the simple spherical ionic atmosphere model gives people a pathway to solve the fundamentally inconsistent Poisson-Boltzmann function by making the first D-H approximation. Subsequently, DLVO theory inherited this basic idea and made further expansions. Although the DLVO theory has successfully applied to some colloidal systems, recent experimental findings and simulation work indicates the idea that the van der Waals forces are the only attraction source may not be accurate.[4]

1.1. *Derivation of the Debye-Hückel's limiting theory*[5]

The central idea of the D-H theory is the concept of ionic atmosphere. It states that because there is an electrostatic attraction between opposite

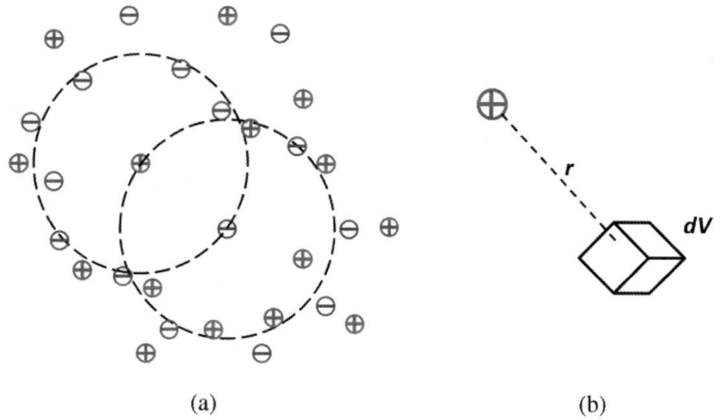

Fig. 1. Illustration of the concept of ionic atmosphere.

charged ions and an electrostatic repulsion between the same charged ions, for any given ion in the solution, there is an excess of its counter ions, and those counter ions have a net opposite charge equal to that on the central ion. This time-averaged, spherical cloud is called the ionic atmosphere, as illustrated in Fig. 1(a). Since the electrostatic interactions between ions are responsible for the non-ideal behaviors of an electrolyte solution, the chemical potential of an *i*th ion can be expressed as

$$\mu_i = \mu_i^{\text{ideal}} + \Delta\mu_i^{\text{et}} \tag{1}$$

For an ideal solution

$$\mu_i^{\text{ideal}} = \mu_i^{\circ} + RTlnx_i \tag{2}$$

For an electrolyte solution

$$\mu_i = \mu_i^{\circ} + RTlnx_i + RT\ln\gamma_i \tag{3}$$

where γ_i is the activity coefficient of the *i*th ion. Compare Eq. 1 to 3, it is clear that the contribution from electrostatic interaction is

$$\Delta\mu_i^{\text{el}} = RTIn\gamma_i \tag{4}$$

To calculate γ_i, assumption must be made so that in the electrolyte solution there is 1, 2 … *i* kinds of ions with bulk concentration being n_1, n_2 … n_i per unit volume and their charge number is z_1e, z_2e … z_ie (z is the valence and e is the unit charge). If we choose one cation as our target (as shown in Fig. 1(b)), then in a small volume *dV* which is separated from the central cation by a distance *r*, the local concentration for the *i*th ion n_i' can be written as Eq. 5 according to the Boltzmann distribution equation, in which φ is the central cation mean electrical potential averaged by time, and *k* is the Boltzmann constant

$$n_i' = n_i \exp\left(-\frac{z_i e\varphi}{kT}\right) \tag{5}$$

Since the mean charge density ρ in *dV* is actually a sum of charge density all ions $\Sigma\, n_i' z_i e$, we have

$$\rho = \sum_i n_i e z_i \exp\left(-\frac{z_i e\varphi}{kT}\right) \tag{6}$$

Now, if we consider the solution is homogeneous and isotropic at thermal equilibrium and the counter-ion distribution in the ionic atmosphere is highly symmetric, according to Poisson's equation,

$$\nabla^2\varphi = -\frac{\rho}{\varepsilon} \tag{7}$$

a simpler form of this equation can be derived:

$$\frac{1}{r}\frac{d^2(r\varphi)}{dr^2} = -\frac{1}{\varepsilon}\sum_i n_i e z_i \exp\left(-\frac{z_i e\varphi}{kT}\right) \tag{8}$$

This equation does not have an explicit general solution because there is a fundamental inconsistency and it can be solved only for certain limiting cases. Therefore, if we assume that this electrolyte solution is so dilute that the electrostatic energy between ions is much smaller that the thermal energy, $z_i e\varphi << k\text{T}$, we have

$$\frac{1}{r}\frac{d^2(r\varphi)}{dr^2} = \frac{e^2}{\varepsilon kT}\sum_i n_i z_i^2 \varphi \tag{9}$$

If we define

$$r_d^2 = \frac{e^2}{\varepsilon kT} \sum_i n_i z_i^2 \tag{10}$$

Then Eq. 9 becomes

$$\frac{1}{r} \frac{d^2(r\varphi)}{dr^2} = r_d^2 \varphi \tag{11}$$

where r_d^{-1} is called the Debye length. By applying two boundary conditions: a) when $r \to \infty$, $\varphi \to 0$, and b) $r \to 0$, $\varphi \to \varphi_0$ when no counter-ions present around the central ion; we can solve this Poisson-Boltzmann equation as

$$\varphi = \frac{z_i e}{4\pi \varepsilon r} e^{(-r_d r)} \tag{12}$$

By using the charge-up strategy, the work done by the electrostatic interactions between ions is equal to

$$w_{el} = w_A - w_o = \Delta G_{el} = -\sum_i \frac{N i z_i^2 e^2 r_d}{12\pi \varepsilon} \tag{13}$$

where ΔG_{el} is the Gibbs free energy change raised by the electrostatic interactions. Since

$$\Delta \mu_i^{el} = \frac{\partial \Delta G_{el}}{\partial N_i} = RT ln\gamma_i$$

We have

$$ln\gamma_i = -\frac{z_i^2 e^2 N_A}{8\pi \varepsilon RT} \left(\frac{2e^2 N_A \rho_{slu}}{\varepsilon kT} \right)^{\frac{1}{2}} \sqrt{I} \tag{14}$$

or

$$ln\gamma_i = -A z_i^2 \sqrt{I} \tag{15}$$

where I is the ionic strength of the solution. From the derivation of the D-H theory we can notice that several important assumptions are used and we should keep in mind that without these limiting conditions, the D-H theory will not hold.

1. The ion distribution follows the Boltzmann distribution form and Poisson equation holds.
2. Ions are point charges with no polarization ability
3. The Coulomb interaction is the only interaction that dominants the ion behaviors. The attraction between ions is relatively small, which is lower than 1 k_BT.
4. The solvent is treated as a continuous dielectric medium.

1.2. *General features of the DLVO theory*

A colloidal system is a thermodynamically unstable system, and it is no longer a homogeneous system as compared with simple ionic solutions. A colloidal particle (diameter larger than 100 nm) has a well defined surface and volume. To increase the stability, modifying the surface to include steric interaction, and modifying the surface to tune charges are two main strategies. The surface charges of a colloidal particle have two possible origins. An initially uncharged surface may preferentially adsorb ions from the solution. Second is the ionization of functional groups on the surface. At the interphase of colloidal particle surface and the solvent, counter-ions tend to accumulate near the colloidal particle surface to form an electrical double layer. To describe the electrical potential of this double layer, Helmholtz first came up with a capacitor model, which simplify the situation of the interphase of charged colloidal particle surface and the accumulated counterions in solvent as a parallel plate capacitor. These two charged surfaces are separated by the dielectric solvent medium. The potential difference between these two plates is $\Delta\varphi$, which decays in a linear fashion inside the double layer. This model is oversimplified in a sense that it totally ignored the thermal motion of the small counterions which tends to diffuse them all over the solutions. Gouy (1910) and Chapman (1913) corrected this parallel capacitor model by replacing the static counter-ion layer into a diffuse layer.[5] This diffuse layer contains two parts: one part is closely associated with the colloidal particle surface while the other part is loosely interact with the colloidal particle surface and gradually diffuse into the bulk solutions. In this case, the counter-ion distribution can be described by Boltzmann distribution. Recall that in D-H theory the ionic atmosphere around a central ion is also in Boltzmann distribution, therefore, the approximation that applied in D-H theory to describe the electrical potential of a point charge can also be applied here. The only difference is that in the colloidal systems, we have to consider the size of the colloidal particles. Solving the linear Poisson-Boltzmann equation for spherical electrical double layer under the D-H approximation, the mean electrical potential can be written as

$$\varphi = \varphi_0 \frac{R_s}{r} e^{[-\kappa(r-R_s)]} \tag{16}$$

where κ^{-1} is the Debye screening length, r is the distance from the particle center to any point in the double layer and R_s is the radius of the spherical particle. Since this relationship between φ and r is derived based upon the D-H approximation, Eq. 16 is only valid at low surface potential. The parameter κ is similar to r_d in the deviation of the D-H theory for simple ionic solutions, and κ^{-1} represents the thickness of the diffused double layer.

The next step in developing the DLVO theory is to consider the interactions between two parallel plates with the same charges immersed in an electrolyte. In this case, as the distance between the two plates decreases, the potential of each of the plates starts to overlap and the counterion concentration in the middle of the plates starts to increase. At equilibrium, the electrostatic forces and the osmotic pressure operate on any volume element between the two plates must balance with each other. A general expression for the repulsive potential between two parallel plates with the same charges immersed in a dielectric medium can be written as

$$\Phi_R = 64kTn_\infty \gamma_0{}^2 \kappa^{-1} \exp(-l\kappa) \tag{17}$$

Eq. 17 is only valid when the distance between the two plates is relatively large and the electric potential is relatively low. To describe the repulsive interaction between two spherical particles, an easy way is to use the Derjaguin approximation. And the repulsive potential can be written as

$$\Phi_R = 64\pi R_s kTn_\infty \gamma_0^2 \kappa^{-2} \exp(-s\kappa) \tag{18}$$

where R_s is the radius of the two spherical particle and s is the closest distance from the surface of the two spheres.

1.2.1. *DLVO potential and the primary and secondary minima*

In the DLVO theory, the only possible source of attraction between two colloidal particles is believed to be the van der Waals forces, which can be written as

$$\Phi_A = -\tfrac{1}{12} A \tfrac{R_s}{s} \tag{19}$$

For two spheres with s is the closest distance from the surface of the two spheres, R_s is the radius of the colloidal particles and A is the *Hamaker* constant.

Now, with the knowledge of repulsive potential and attractive potential, we can write the DLVO potential as

$$\Phi_{\text{net}} = \Phi_R + \Phi_A = 64\pi R_s \kappa Tn_\infty \gamma_0^2 \kappa^{-2} \exp(-sk) - \tfrac{1}{12} A \tfrac{R_s}{S} \tag{20}$$

for two large spherical colloidal particles immersed in a dielectric medium. From Eq. 20, we can see that the repulsive potential ϕ_R depends sensitively on the concentrations of small counter-ions and large particles through κ, however, for the attractive potential ϕ_A, it is nearly insensible to the change of salt concentrations. Therefore, it gives us a great opportunity to study the solution behaviors of colloidal particles simply by varying the salt concentration, *e.g.* the ionic strength.

As shown in Fig. 2,[6] we can plot the DLVO potential ϕ_{net} versus the distance between two particles. And we notice that when the repulsive potential energy barrier becomes lower than the thermal energy of k_BT under high salt concentration, the colloidal particles fall into the primary

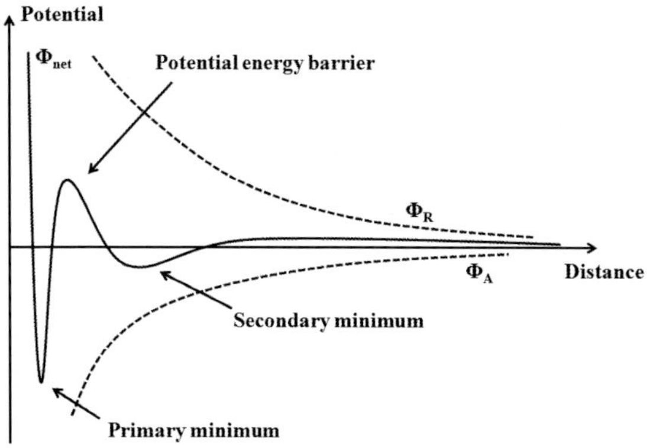

Fig. 2. Plot the DLVO potential φ_{net} versus the distance between two particles.[6]

minimum because of the strong van der Waals' attraction. This is an irreversible process called coagulation. However, under certain circumstances the superposition of ϕ_R and ϕ_A may produce a relatively shallow minimum called the secondary minimum. The depth of the secondary minimum is quit sensitive to the salt concentration and if it is larger than the thermal energy k_BT, the colloidal particles tend to flocculate and form colloidal crystallites. This flocculation is a reversible process.

1.2.2. *Limitations of the DLVO theory*

The DLVO theory has been very successful in providing a general tool to explain and predict important behaviors of colloidal systems. However, in the last 40 years, people have observed new experimental phenomena which could not be explained by the DLVO theory. For example, Ise and co-workers noticed large void regions in equilibrium with densely packed region for highly charged silica latex particles.[7] Hachisu in 1973 found that an increase of the salt concentration results in melting of colloidal crystals, which according to DLVO theory, the increase of salt concentration should deepen the secondary minimum therefore stabilize the colloidal crystals.[8] Moreover, Schurr and co-workers observed a strange "ordinary-extraordinary phase transition" when they study the effect of adding salt to the dynamics of poly-L-lysine $(lys)_{955}$ solutions. Over a narrow ionic strength range, they found the apparent diffusion coefficient

D_{app} — monitored by dynamic light scattering (DLS) — dropped 20 times; indicating there is some sort of clustering is formed in solution.[9] From these experimental results, it seems the cause of failures of applying DLVO theory to these systems is from the basic assumption that the short-ranged van der Waals is the only attraction source in the colloidal systems. In fact, in order to explain these experimental phenomena another long-ranged attraction source is needed, which is the counterion mediated electrostatic attraction between colloidal particles. Nevertheless, this attraction is weak compared to the dominant van der Waals forces for colloids due to their large sizes. Therefore, the existence of the counterion mediated attraction and their importance are not accepted by some scientists. However, as we shown below, when the size of solutes become much smaller, this effect becomes significantly important.

2. Characterization of Macroion Solution Behaviors

Previous discussions reveal the complexity in describing the properties of the ionic solutions. Simple ions and large colloids represent two types of completely different systems. The interesting question is that how would the macroions behave with their size and charge density between simple ions and colloids? Will they behave more like simple ions or colloids, or neither? Macroions to the related exploration will help to fill the gap between D-H theory and DLVO theory, as we summarized below. The most appropriate models for our study are polyoxometalate (POM) macroanions and metal-organic cationic nanocages.

2.1. *Polyoxometalates (POMs) type macroanions*

POMs represent a large family of inorganic molecular clusters which generally formed by covalently connecting various early transition metal (Mo, V, Cr, Fe, W, Mn, *etc.*) oxide polyhedrals. Recent achievements in synthetic inorganic chemistry have even created heavy metallic (U, Nb) POMs[10] and inert metallic (Au, Pd) POMs.[11] These POM clusters exhibit a large diversity of size and shape. Because of their delicate structures and components (different metal centers), they show a great variety of magnetic, electronic and catalytic properties.

$\{Mo_{154}\}$	$\{Mo_{132}\}$	$\{Mo_{72}Fe_{30}\}$	$\{Mo_{72}V_{30}\}$	$\{Mo_{72}Cr_{30}\}$	$\{W_{48}P_8Cu_{20}\}$	$\{Mo_{368}\}$	$\{P_4Y_8W_{43}\}$
3.5 nm	2.9 nm	2.5 nm	2.5 nm	2.5 nm	2.0 nm	6.0 nm	2.0x4.0 nm

Fig. 3. Some typical giant polyoxometalate molecular clusters (and their sizes) described in the chapter. Their surface charge densities are listed in Table 1.

Table 1. Some general features of giant, hydrophilic POM macroanions covered in this chapter.

POM abbreviation[a]	Counterion number	Total charge Number	Size (nm)	Charge density (c/m^2)
$\{Mo_{154}\}$[12b, 17]	15	−15	3.6	0.08
$\{Mo_{132}\}$[18]	42	−42	2.9	0.25
$\{Mo_{72}V_{30}\}$[19]	34	−34	2.5	0.28
$\{Mo_{72}Fe_{30}\}$[20]	0[b]	−4 ~ −8	2.5	0.04 ~ 0.07
$\{Mo_{72}Cr_{30}\}$[21]	0[b]	−4 ~ −8	2.5	0.04 ~ 0.07
$\{W_{48}P_8Cu_{20}\}$[22]	25	−25	2.0	0.21
$\{Mo_{368}\}$[23]	48	−48	6.0 × 4.0	0.38
$\{P_4Y_8W_{43}\}$[24]	21	−21 ~ −30[c]	2.0 × 4.0	0.11 ~ 0.16

[a] Detailed formulae can be found in references listed next to each of the POM abbreviation.

[b] For these two molecular clusters, they are neutral in the crystal state but undergo deprotonation when dissolved in aqueous solutions. Therefore, their total surface charge number depends on the degree of deprotonation.

[c] For this cluster, besides 15 K^+ and 6 Na^+ counter-ions, it also contains 9 H_3O^+ counter-ions. Therefore its surface charge density is also affected by solution pH.

Polyoxomolybdates, polyoxotungstates and their heteropolyoxometalate derivatives are the most commonly seen POMs and quite a few of them are very large in size, as shown in Fig. 3 and Table 1. All of these POMs are hydrophilic in nature and can be readily dissolved in polar solvents, owing to their surface charge and terminal $-OH_2$ groups. According to the nature of their charges, POMs (including but not limited to those list in Table 1) can be categorized into two subclasses: strong electrolytes and weak electrolytes. The POM clusters in the first group carry inherent charges in crystals which are balanced by multiple small cations to keep the whole structure neutral. When this kind of POMs dissolves in solution, small counterions are released into the solvent while the POM part usually stays as macroanions. In the second group, the POMs are almost neutral clusters in crystals but in aqueous solution their water ligands attached to non-Mo (or non-W) metals can partially deprotonate. As the result, the clusters become negatively charged by carrying certain amount of localized charges. Overall, this kind of clusters behaves as weak acids in solution, and their charge density can be controlled by solution pH.

2.1.1. *Characterization of the self-assembly of POM macroanions in dilute solutions*

The pioneering work of synthesizing giant polyoxomolybdate clusters was initiated by Achim Müller and his colleagues, who discovered various large polyoxomolybdate clusters from molybdenum blue solutions. The first identified crystalline structure of molybdenum blue was $(NH_4)_{28}[Mo_{154}(NO)_{14}O_{448}H_{14}(H_2O)_{70}]\cdot nH_2O$ containing the $\{Mo_{154}\}$ giant wheel anions.[12] Interestingly, unexpected Tyndall effect was observed from those solutions, as shown in Fig. 4(a). Tyndall effect is the result of scattered light from suspended particles of colloidal size and is not likely to appear in the current $\{Mo_{154}\}$ inorganic macroanionic solutions. The $\{Mo_{154}\}$ inorganic macroanions do not exist as discrete ions; instead, they tend to slowly further aggregate into large, spherical assemblies (TEM images shown in Fig. 4(b)). The "aggregates" do not look like regular aggregates formed by less soluble species, which usually have broad size distributions and tend to continue to grow until precipitating from the solution. This self-assembly process is contradictory to our common sense that hydrophilic macroanions should not attract to each other but homogeneously distribute in solution to decrease the total Gibbs free energy.[13] Interestingly, these large assemblies may be hollow. An energy dispersive X-ray spectroscopy analysis (EDX) showed considerable quantities of solvent existing inside the aggregates. And the aggregates suddenly burst when switching the atomic force microscopy (AFM) from conventional mode to the high vacuum mode.[14]

To thoroughly explore the unique self-assembly of $\{Mo_{154}\}$ macroanions, a combination of dynamic light scattering (DLS) and static light scattering (SLS) techniques are employed. In general, SLS gives information such as

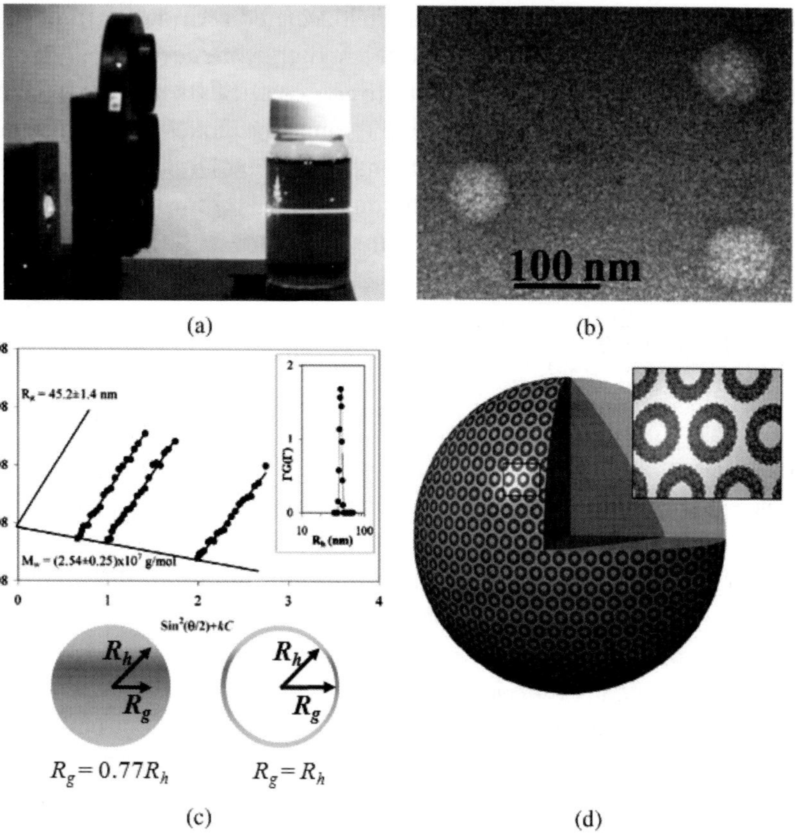

Fig. 4. (a) Tyndall effect of a $\{Mo_{154}\}$ aqueous solution when a green laser (532 nm) passes through the solution. (b) TEM image of the blackberry structure formed by $\{Mo_{154}\}$ macroanions. (c) Zimm plot of a series of $\{Mo_{154}\}$ aqueous solutions reveals the radius of gyration R_g and weight averaged molecular weight (M_w) of the large blackberry structure. (d) A cartoon illustrates the blackberry structure. (Reprinted with permission from Ref 13. Copyright 2003 Nature Publishing Group.) **See Color Plate 4.**

weight-average mass (M_w), radius of gyration (R_g), as well as the nature of interparticle interactions of the studied solute (the second virial coefficient A_2) through measuring the absolute scattered intensity from sample solutions at different scattering angles and concentrations. On the other hand, DLS measurements can determine the hydrodynamic radius (R_h) and the size distribution of particles (e.g., polymers, colloids, or biomacromolecules) in solution by measuring the intensity-intensity time correlation function (then converted to electric field time correlation function). Using constrained regularized (CONTIN) method, the time correlation function is analyzed to provide the apparent diffusion coefficients of the solute which can be further used to calculate the R_h through the Stokes–Einstein equation. As shown in Fig. 4(c), a CONTIN analysis of the DLS measurement on a 0.01 mg/mL $\{Mo_{154}\}$ aqueous solution at pH=3.0 shows that large assemblies have an average R_h of 45 ± 1 nm with a narrow size distribution. SLS study show that their average R_g is 45.2 ± 1.4 nm, which reflects that $R_g = R_h$. This is interesting because if the assemblies are solid spheres, $R_g = 0.77\ R_h$ holds. Only when a spherical object distributes all its mass on the surface, $R_g / R_h = 1$. Therefore, $\{Mo_{154}\}$ macroanions form spherical, vesicle-like hollow structures in solution. Meanwhile, the apparent molecular weight (M_w) of the large assemblies can be estimated as $(2.54 \pm 0.3) \times 10^7$ g/mol, equivalent to ~1165 single $\{Mo_{154}\}$ macroanions, which barely covers a single layer of a 45-nm-radius sphere. The accurate inter-$\{Mo_{154}\}$ distance is estimated to be ~1 nm. The above experimental results leads to a model shown in Fig. 4(d), and the new vesicle-like structure was given a nickname "blackberry structure" due to the similarities between these two species. The blackberry structure is fundamentally different from the bilayer vesicles because on blackberry surface the $\{Mo_{154}\}$ macroanions have no obvious hydrophobic regions and do not touch with each other.[13]

Similar blackberry structures are also observed in many other POM macroionic polar solutions (*e.g.* $\{Mo_{132}\}$, $\{Mo_{72}V_{30}\}$ and $\{Mo_{368}\}$ in various polar solvents),[14,15] indicating that this type of self-assembly is quite a general process for hydrophilic macroanions in solution, as long as the macroions are large in size (from 2 nm to 6 nm) and carry a moderate number of charges. Currently, the smallest POM cluster shows this self-assembly behavior is $\{W_{48}P_8Cu_{20}\}$.[16]

2.1.2. *The driving forces that responsible for the unexpected self-assembly of macroanions*

In order to uncover this extraordinary self-assembly process in highly charged, fully hydrophilic macroanionic systems, we need to recognize the forces that overcome the electrostatic repulsion among the like charged macroions and bring them together. In general, van der Waals forces and hydrogen bonding are two important attractive interactions that are widely seen in many systems. Therefore, we shall take a close look at them first.

2.1.2.1. Van der Waals attractions

The van der Waals forces, which are generally attractive in nature, are short ranged and decay rapidly to zero away from a surface. The origin of the van der Waals forces lies in the instantaneous dipole generated by the fluctuation of electron cloud surrounding the nucleus of electrically neutral atoms. In colloidal systems the van der Waals forces are significant because of the relatively large size of colloids (Eq. 19). For POM macroanions, the van der Waals forces are much weaker due to their relatively small sizes. Eq. (20) can be used to estimate the effective distance S at which the van der Waals attractive potential has the same magnitude as the repulsive potential. For instance, in 0.01 mg/mL $\{Mo_{154}\}$ aqueous solution at pH = 3.0, this distance is ~ 0.2 nm, much lower than the interparticle distance estimated from the SLS results (~ 1 nm between two adjacent $\{Mo_{154}\}$ macroanions), indicating that the van der Waals forces are not likely the primary or dominant attractions in the blackberry formation process.

Direct experimental evidence is obtained when studying a "Keplerate" $\{Mo_{132}\}$ cluster in water/acetone mixed solvents.[25] Each $\{Mo_{132}\}$ carries ~42 negative charges when it is dissolved in water, and it also shows good solubility in water/acetone mixed solvents even with a lot of acetone (in this case, the clusters exist as almost neutral, discrete solutes). The blackberry formation occurs in solvents containing 3–70 v% acetone (Fig. 5). The single cluster-to-blackberry transition from at 70 v% acetone is interesting. When the solvent contains more water, the $\{Mo_{132}\}$ macroions are expected to carry more negative charges and should repel stronger with each other. However, instead, the blackberry formation is observed, showing obviously stronger attraction among them. During the whole process, the magnitude of attractive van der Waals forces basically does not change. Therefore, these observations confirm that the van der Waals forces are not the major driving forces for the blackberry formation. A similar conclusion can also be obtained when studying the

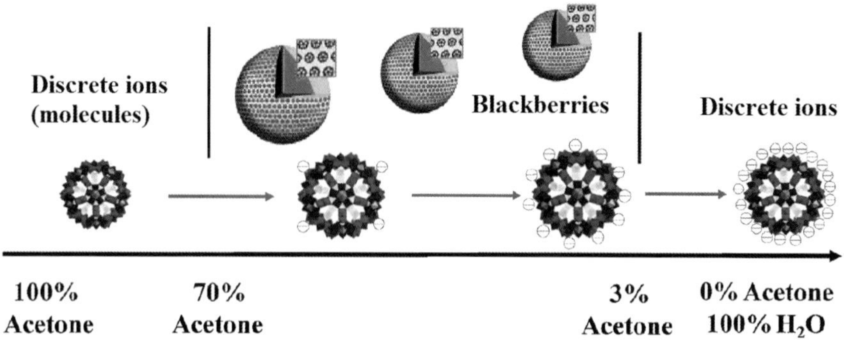

Fig. 5. The reversible transition between discrete macroions (molecules) and blackberries based upon changing solvent content for 1.0 mg/mL $\{Mo_{132}\}$ in water/acetone mixed solvents.[25] (Reprinted with permission from Ref 25. Copyright © 2007 American Chemical Society.)

pH-dependence of the weak acid type POMs in aqueous solution, as discussed in a following section.

2.1.2.2. Hydrogen bonding

Hydrogen bonding (5–20 kJ/mol) in general is stronger than the van der Waals forces and recently experimental results from our group and other groups indicate that the hydrogen bonds formed between adjacent POMs on the blackberry surface may contribute to the blackberry formation. For example, an AFM study indicates that when placing $\{Mo_{72}Fe_{30}\}$ blackberries on silica surface, their height continues to drop with time while their overall morphology on the x-y plane (the surface) does not change. This can be explained as that the blackberries continue to lose their internal water when exposed on a surface. The strong hydrogen bonds prevent them from falling apart. In other words, the soft and robust blackberry surface makes them behave like "water balloons".[26] By using dielectric relaxation measurements, Oleinikova *et al.*,[27] noticed that during the self-assembly of $\{Mo_{154}\}$ wheel-shaped macroions in aqueous solution, "the strength of the hydration extends as cluster aggregation takes place with more water molecules being more strongly bound between the wheels and the presence of relatively fewer less strongly bound water molecules"; that

Fig. 6. Illustration of how the hydrogen bonded water molecules and small counter-ions arrange themselves on the surface of the blackberry structure to create a curved spherical shape. (Reprinted with permission from Ref 28. Copyright © 2009 American Chemical Society.)

is, the water stays between $\{Mo_{154}\}$ macroions and shows higher viscosity (i.e. more hydrogen bonds formed). In other words, the additional hydrogen bonds help to "glue" the hydrophilic surface of POMs together. The special hydrogen bonding formed between macroions on the blackberry surface is reflected by the softness of the blackberry wall. Moreover, Schmitz provides an interesting hypothesis of hydrogen bonded water molecule "fingers" to explain the spherical geometry of blackberries.[28] Fig. 6 illustrates the general picture of how the hydrogen bonds may glue $\{Mo_{154}\}$ macroions together on the blackberry surface.

However, this middle-ranged attraction still could not explain why highly charged macroanions tend to attract each other especially when the electric double layer of the macroanions is large ($\kappa^{-1} \sim 10$ nm for 0.01 mg/mL $\{Mo_{154}\}$ solutions at pH=3.0). There must be some long-ranged attractive interactions which are responsible for the blackberry formation.

2.1.2.3. Counterions mediated attractions

The large size of macroions make the macroions behave fundamentally different from simple electrolyte solutions and the D-H approximation is no longer suitable to describe the macroion solutions (the macroions cannot be treated as point charges anymore). The electrostatic interactions between macroions-macroions, macroions-counterions and counterions-counterions are much more complex than the simple electrolyte solutions. For example, in biomacroionic systems, the self-association of cylinder shaped DNA macroions induced by divalent or multi-valent counterions was observed under many solution conditions.[29] In these systems, high-valence counterions are strongly adsorbed on to the surface of DNAs and may eventually reverse the sign of DNA surface charge, which is known as over-charging. The reversed DNA macroions then attract another DNA macroions with a layer of counterions in between. This counterion mediated attraction is named counterion condensation, a concept developed by Manning and Oosawa in the late 1960s.[30] However, the counterion condensation is normally seen in cylinder macroions with high valence counterions, for spherical macroions with monovalent counterions, this effect is believed to be very small. In the case of POMs, their sizes are much smaller than colloidal particles but large enough to display the size disparity. (The size ratio between the Keplerates and cations is approximately 10:1, which is noticeable but not as significant as in colloidal suspensions with a size ratio on the order of 1,000:1. In the latter case, the counterions can be effectively treated as point charges.) More importantly, during the blackberry formation, no extra or high valent counterions are needed. All these conditions indicate that the blackberry structure formation is indeed quite unique.

a) Interactions between discrete macroions and counterions

For very large POMs such as the Keplerate clusters, the counterion-macroion interaction is expected to be more significant compared with small POMs, such as Keggins. Our first attempt to evaluate the macroion-counterion interaction is done by comparing the apparent solution conductivity values of two POM macroanionic solutions with their theoretic values.[31] At very low POM concentrations ($< 1 \times 10^{-5}$ mol/L), nearly all the counterions are free in solution. However, some cations (mainly NH_4^+ and Na^+) were found to closely associate with those macroanions at higher macroionic concentrations, and thus do not contribute to solution conductivity. The degree of counter-ion association shows obvious concentration dependence before and after dialysis, with larger fraction of counter-ions associated with macroions at higher concentrations.

A feasible way of directly measuring the counterion distribution around the macroions is small-angle X-ray scattering (SAXS). Keplerate $\{Mo_{72}V_{30}\}$ (2.5-nm size and carrying 31 charges with the counterions being 14 K^+, 8 Na^+, 2 VO^{2+}, and 5 H^+) is a good model system due to its stability in dilute solution, simple and highly symmetrical structure and metal counterions which can effectively generate scattering signals. Its SAXS scattering curve can be calculated based on a spherical shell form factor and then compared with experimental data. Moreover, it does not have co-ions and/or excess salts so that the true physical picture of counterions can be revealed.

As demonstrated in Fig. 7, in a diluted 0.26 mM aqueous solution of $\{Mo_{72}V_{30}\}$, the overall SAXS curve can be nicely fit by the form factor of $\{Mo_{72}V_{30}\}$ cluster with a spherical shell model. When the $\{Mo_{72}V_{30}\}$ concentration is less than 0.052 mM, the R_g value obtained from the Guinier

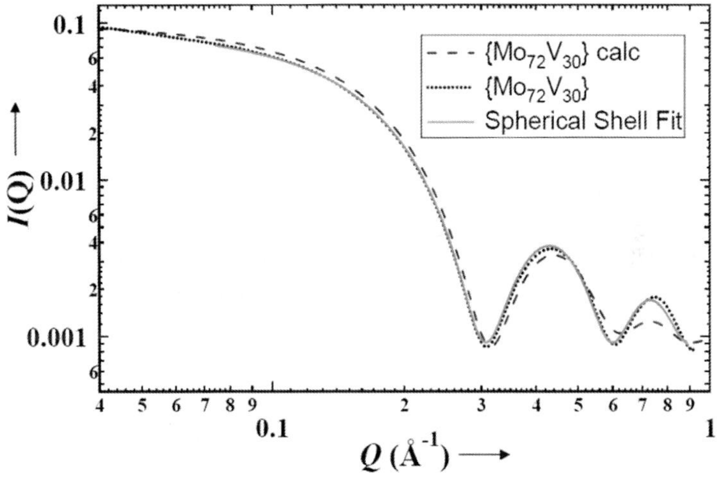

Fig. 7. Log-log plot of the calculated SAXS response for $\{Mo_{72}V_{30}\}$ generated by using CRYSOL from the atomic coordinates from the single crystal data (dashed line), calculated form factor of a 2.5-nm-diameter spherical shell (solid line), and experimental scattering data for 0.26 mM $\{Mo_{72}V_{30}\}$ aqueous solution (dotted line). (Reprinted with permission from Ref 32. Copyright © 2009 Wiley InterScience.)

Table 2. Table of Rg values measured at different $\{Mo_{72}V_{30}\}$ concentrations in aqueous solutions. (Reprinted with permission from Ref 32. Copyright © 2009 Wiley InterScience.)

$\{Mo_{72}V_{30}\}$g/L	R_g(Å)
0.25 (0.013 mM)	10.8 ± 0.5
0.5 (0.026 mM)	10.8 ± 0.5
1 (0.052 mM)	10.9 ± 0.5
5 (0.26 mM)	11.2 ± 0.6

plot (10.8 ± 0.5 Å) suggests that there is no counterion association around the macroions.

Not only does the counterion association show concentration dependency, it is also influenced by the solvent dielectric constant. As shown in Fig. 8, the Gunier plot for a series of $\{Mo_{72}V_{30}\}$ solutions with different amount of acetone reflects that the average R_g increases with the acetone content, which means the counterion association becomes more significant when more acetone is introduced. Meanwhile, the distance pair distribution p(r), which is the probability of finding the vector length r in a molecule that will become zero at the maximum vector length, can be generated from the Moore analysis. For $\{Mo_{72}V_{30}\}$ in dilute aqueous solutions, the p(r) curves shown in Fig. 8 (top) correspond to discrete $\{Mo_{72}V_{30}\}$ clusters (a core shell spherical particle with a maximum linear dimension of ~ 26 Å) without counterion association. When certain amount of acetone is introduced into the solution, another relatively small and isolated peak appears which is centered at ~ 30 Å and extends the effective distribution to ~ 34 Å (Fig. 8, bottom). The original peak distributed from 0 to 26 Å remains unchanged, indicating that the $\{Mo_{72}V_{30}\}$ macroanions still exist as discrete ions without self-assembling into blackberries (blackberry formation is a very slow process). This additional peak suggests that some counterions are closely associated with the macroions and distribute in the range of 2–9 Å to the surface of macroions. The appearance of the peak due to associated counterions is consistent with the appearance of the blackberry structures, indicating the direct connection between these two issues.[32]

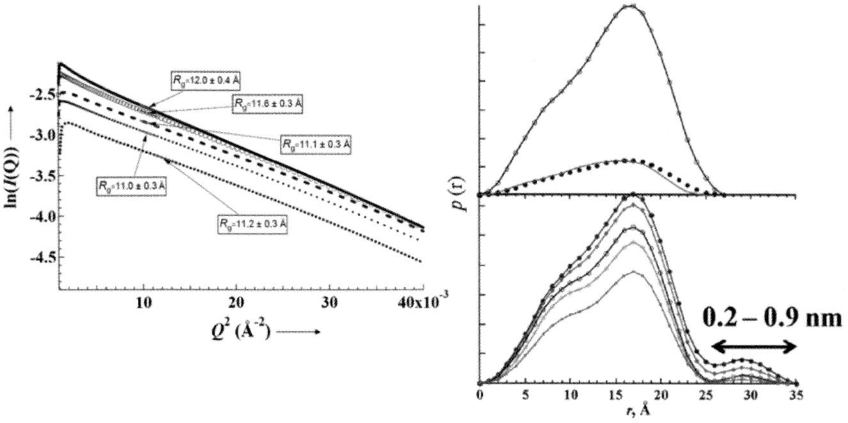

Fig. 8. *Left*: Guinier plots for 0.26 mM {Mo72V30} macroions in water and water/acetone mixed solvents containing various amount of acetone (in vol %) (solid line: 75% acetone/water, open circles: 65% acetone/water, dashed line: 45% acetone/water, dotted line: 10% acetone/water, open squares: in pure water). *Right top*: Distance distribution functions based on calculated and experimental scattering curves for $\{Mo_{72}V_{30}\}$ as obtained by using an indirect Fourier Transform of the primary SAXS data. (closed circles: 0.1 mg/mL, or 0.013 mM $\{Mo_{72}V_{30}\}$, open circles: 0.052 mM $\{Mo_{72}V_{30}\}$, red line: $\{Mo_{72}V_{30}\}$ calculated). *Right bottom*: Experimental distance distributions for 0.26 mM $\{Mo_{72}V_{30}\}$ in water and acetone/water mixed solvents with various acetone content (in vol %). (green: 75% acetone/water, blue: 65% acetone/water, red: 45% acetone/water, light blue: 10%acetone/water, magenta: in pure water).[32] (Reprinted with permission from Ref 32. Copyright © 2009 Wiley InterScience.) **See Color Plate 5.**

b) Effect of surface charge density

The number of charge on the POMs plays an important role during the blackberry formation. POMs can be categorized in two groups based on the origin of their surface charge and the self-assembly mechanisms: weak electrolytes and strong electrolytes. Strong electrolytes contain multiple inherent charges, balanced by counterions. Some strong electrolytes, such as $\{Mo_{72}V_{30}\}$, do not self-assemble in dilute aqueous solution due to its high surface charge density. To decrease the charge, a small amount of water-soluble cationic surfactants, such as cetyltrimethylammonium bromide (CTAB), trimethyltetradecylammonium chloride (CTAT), dodecyltrimethylammonium bromide (DTAB), and octyltrimethylammonium bromide (OTAB) is introduced into the $\{Mo_{72}V_{30}\}$ solutions, such that the charge density on $\{Mo_{72}V_{30}\}$ macroions is expected to decrease and enter the blackberry-formation regime. The long-chain CTAB and CTAT surfactants can interact with $\{Mo_{72}V_{30}\}$ stoichiometrically.

As a result, $\{Mo_{72}V_{30}\}$ blackberries can be observed at the $\{Mo_{72}V_{30}\}$ to surfactant molar ratio of 1:1.2, and the average R_h of the blackberries continues to increase with increasing surfactant concentration, as shown in Fig. 9. The hydrophobic interaction from the surfactant tails might also contribute to the self-assembly, but the effect should be very minor as there are only a few alkyl chains on the relatively large surface area of each macroion.[33]

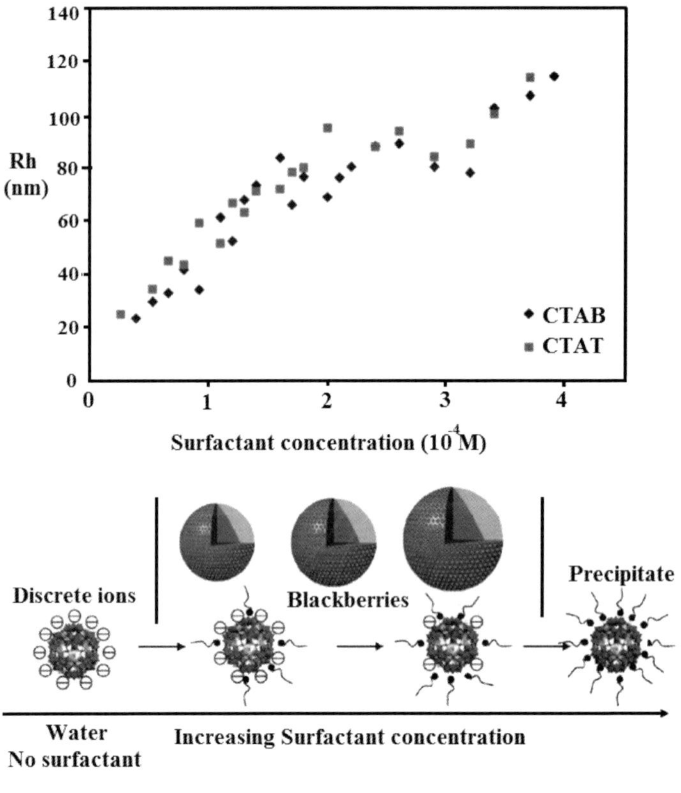

Fig. 9. *Top*: Average hydrodynamic radius (R_h) of the $\{Mo_{72}V_{30}\}$ blackberries in aqueous solution increases with surfactant (CTAB or CTAT) concentration, as measured by DLS. The concentration of $\{Mo_{72}V_{30}\}$ is 0.5 mg/mL, equivalent to a molar concentration of 2.6×10^{-5} M. *Bottom*: The average blackberry size increases with increasing surfactant amount (i.e., decreasing charge density on $\{Mo_{72}V_{30}\}$). (Reprinted with permission from Ref 33. Copyright © 2009 American Chemical Society.)

For weak electrolytes, as in the case of $\{Mo_{72}Fe_{30}\}$, its surface charge comes from the partial deprotonation of water ligands attached to its 30 Fe^{III} centers, so that its surface charge density can be easily tuned by changing the solution pH. For example, when dissolved in pure water to make a 1.0 mg/mL solution, the $\{Mo_{72}Fe_{30}\}$ solution pH is 3.4, that is, 7 or 8 protons are released from each $\{Mo_{72}Fe_{30}\}$ cluster. If we assume that the different deprotonation sites are independent of each other due to the relatively long distance between them, the $\{Mo_{72}Fe_{30}\}$ cluster can be treated as a weak acid with a single pKa value. Therefore, by adding a small amount of NaOH or HCl, the deprotonation degree of $\{Mo_{72}Fe_{30}\}$ and its charge density can be adjusted. Fig. 10 shows the blackberry formation in $\{Mo_{72}Fe_{30}\}$ aqueous

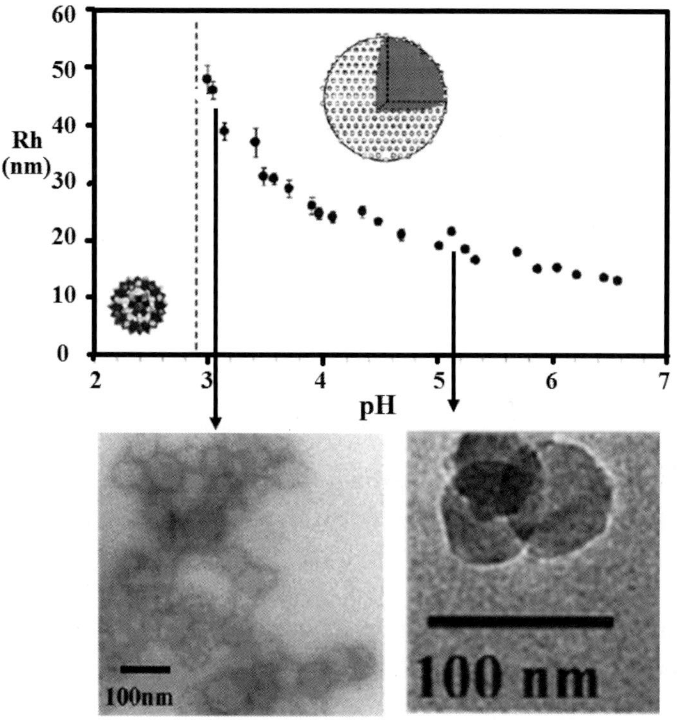

Fig. 10. Average R_h of the blackberries formed in 0.5 mg/mL aqueous solutions of $\{Mo_{72}Fe_{30}\}$ at different pH (adjusted by NaOH or HCl), as measured by DLS at 90° scattering angle. TEM images of aggregates on carbon film formed at pH ~ 3.0 (*left*; conventional TEM) and pH ~ 4.6 (*right*; more appropriate cryo-TEM). (Reprinted with permission from Ref 34. Copyright © 2006 American Chemical Society.)

solution. At pH < 2.9, $\{Mo_{72}Fe_{30}\}$ clusters are almost uncharged and stay as soluble molecules in solution. At pH > 2.9, $\{Mo_{72}Fe_{30}\}$ clusters are more charged. Self-assembly occurs with the blackberry size decreasing with increasing pH, from $R_h \sim 50$ nm at pH = 3.0 to $R_h \sim 15$ nm at pH = 6.0. The transition from single clusters to blackberries with the change in pH again confirms that van der Waals attractions are not the major attractive forces for the self-assembly. More interestingly, the transitions between blackberries at different pH can be achieved by adjusting solution pH.[34]

c) Effect of counterion valence state and counterion hydrated size

The next question is can the hydrophilic macroions selectively attract one (or more) types of counterions when several types of counterions are co-present in the solution? Or whether the counterions with higher affinity to the macroions can replace the original counterions already associated with the macroions? To answer this question, we still use the $\{Mo_{72}Fe_{30}\}$ macroanions, which form blackberries with an average $R_h \sim 25.4 \pm 1.0$ nm when no extra salt presents, as the model system. The original R_h value is used as a reference. As shown in Fig. 11, when small amounts of additional ions are introduced, the average R_h of the $\{Mo_{72}Fe_{30}\}$ blackberries remains steady. After passing a critical salt concentration (CSC), the R_h value starts to increase continuously with increasing salt concentration. The small amounts of Li^+ or Na^+ added to the $\{Mo_{72}Fe_{30}\}$ solution are not directly involved in the self-assembly process. That is, the affinity of the original cations (hydronium ions) to the $\{Mo_{72}Fe_{30}\}$ macroions is stronger than that of Li^+ and Na^+ ions. However, the size increase of the blackberries when K^+ ions are added indicates that the K^+ ions can change the interaction between $\{Mo_{72}Fe_{30}\}$. Therefore, we can speculate that K^+ ions have stronger affinity to the macroions and can replace the H_3O^+ counterions to trigger the blackberry formation more readily than either Li^+ or Na^+. The addition of small amounts of RbCl has the same effect on the $\{Mo_{72}Fe_{30}\}$ blackberry size as K^+. From Li^+, Na^+, K^+ to Rb^+, the ionic radii increase from 0.95, 1.33, 1.48 to 1.69 Å, and the charge density of the cations decreases. Conversely, the hydrated radii and hydration numbers of the cations decrease from 3.40, 2.76, 2.32 to 2.28 Å and from 25, 17, 11 to 10, respectively. In other words, cations like Li^+ and Na^+ are much more

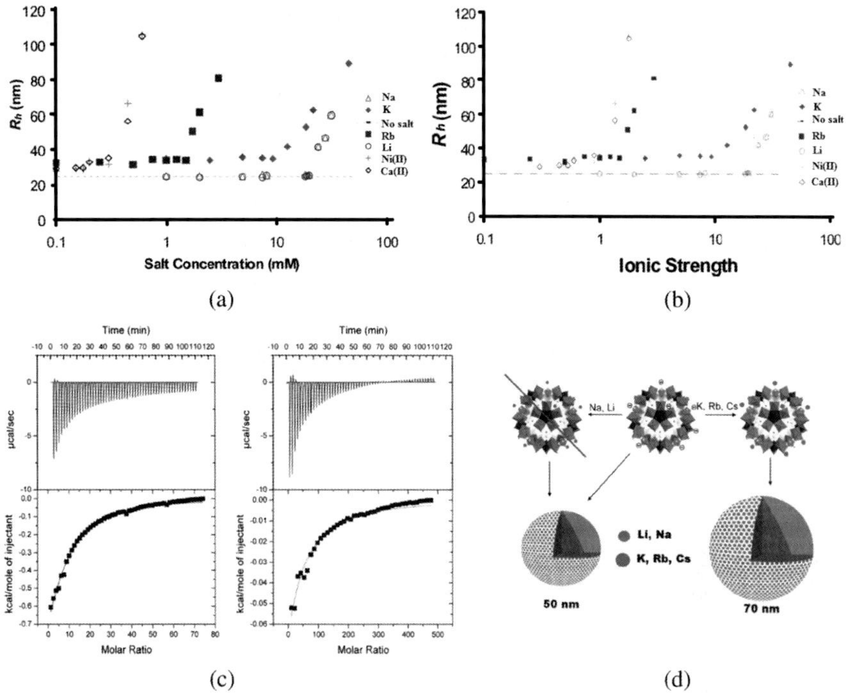

Fig. 11. Change of blackberry size (in R_h) with added chloride salt concentration (a) and total ionic strength (b) for 0.5 mg/mL $\{Mo_{72}Fe_{30}\}$ solutions. For each added cation salt there is a CSC (critical salt concentration), above which the blackberry size increases with increasing salt concentration. (c) ITC results obtained when adding RbCl (*left*) and KCl (*right*) into 2.0 mg/mL $\{Mo_{72}Fe_{30}\}$ solutions. Concentrations of 40 mM for RbCl and 334 mM for KCl were used. The molar ratio needed for saturation of the $\{Mo_{72}Fe_{30}\}$ binding sites is ~ 50–60:1 for K^+ and ~ 10:1 for Rb^+. (d) The countercation replacement on the $\{Mo_{72}Fe_{30}\}$ surface can be achieved by monovalent cations with small hydrated sizes only. (Reprinted with permission from Ref 35. Copyright © 2010 American Chemical Society.)

hydrated, which accounts for their preference to remain as aquated cations rather than binding to large POM macroions. Larger cations like Cs^+ and Rb^+ are much less hydrated than the smaller alkali metals and are more polarizable. Divalent counterions are more effective at screening the interactions between the macroions than monovalent ones as the CSCs are much lower. To further illustrate the difference of counterions affinity towards macroions, isothermal titration calorimetry (ITC) studies are applied (Fig. 11). In addition, the number of K^+ needed for complete saturation of the binding sites is about 50–60 K^+/ $\{Mo_{72}Fe_{30}\}$, with a binding constant of k = 75. Given the fact that each $\{Mo_{72}Fe_{30}\}$ carries about six charges (at 2 mg/mL), this means that about 10 K^+ ions per POM charge are needed to saturate a binding site. Obviously, as K^+ ions are also monovalent, their replacement of H_3O^+ ions is not very selective. For RbCl, less than 10 Rb^+ ions are needed to saturate the 6 $\{Mo_{72}Fe_{30}\}$ binding sites. A k value of 773 was obtained for Rb^+, which is about an order of magnitude larger than for K^+, suggesting that the binding to $\{Mo_{72}Fe_{30}\}$ macroions is much stronger for Rb^+ than for K^+. These differences observed from the ITC measurements may explain the size difference in blackberries formed in the presence of various alkali metals. The ITC studies provide direct confirmation that the binding strength between $\{Mo_{72}Fe_{30}\}$ and monovalent cations follows the order of (Li^+, Na^+) < H_3O^+ < K^+ < Rb^+ < Cs^+, which is completely consistent with the observed corresponding blackberry formation processes. This is also compelling evidence that the counterions cannot be treated as dimensionless point charges in the macroionic solutions. These, in turn, drive the three types of ion association behaviors (i.e., contact-, solvent-shared/-separated, and fully solvated ion-pairing) that are envisioned to lead to the observed variations in blackberry size with alkali metal group countercations.[35]

d) Effect of ionic strength

As mentioned earlier, the Debye screening length κ^{-1} can represent the thickness of the electric double layer of a macroion, and is related to the ionic strength I ($\kappa^{-1} \alpha \sqrt{I}^{-1}$). Therefore, it is important to assess how the ionic strength will affect the blackberry formation and blackberry size. For a solution of $\{Mo_{72}Fe_{30}\}$ with 20 mM KCl added, the blackberries have an average R_h = 61 nm. When diluted with 0.5 mg/ mL $\{Mo_{72}Fe_{30}\}$ aqueous solution to decrease the salt concentration to a point below the CSC (5.0 mM KCl) without changing the macroionic concentration, the average R_h decreased from 100 to 34 nm. These observations again strongly indicate that the blackberry formation process is driven by charge effect, and it is a completely reversible process.[35]

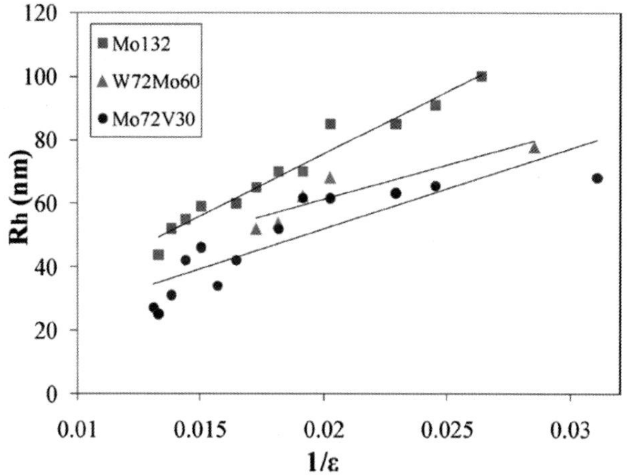

Fig. 12. Plot of the average blackberry radius (in R_h) versus the inversed dielectric constant ($1/\varepsilon$) of the solvent for various POM macroions in water/acetone mixed solvents. Linear relationship roughly follows for these systems.[39] (Reprinted with permission from Ref 39. Copyright © 2009 American Chemical Society.)

2.1.2.4. Solvent effect

Not only the Debye screening length (κ^{-1}) can be influenced by the ionic strength, it is also proportional to the dielectric constant ε_r of the solvent. In other words, when the polarity of a solvent decreases, the Debye length also decreases and the electric double layer is compressed.

Therefore, counterions should be more closely associated with macroions. The average blackberry size increases linearly with increasing $1/\varepsilon_r$ for $\{Mo_{132}\}$[36], $\{Mo_{72}V_{30}\}$[37] and $\{W_{72}Mo_{60}\}$[38] in water/acetone mixed solutions (Fig. 12). Similar trends have also been identified in other POM solutions.

2.1.3. *Kinetics of the blackberry formation*

2.1.3.1. Long equilibrium time

Typically, the blackberry formation is very slow. Fig. 13 demonstrates a representative SLS and DLS study on the formation of $\{Mo_{72}Fe_{30}\}$ blackberries in aqueous solution at different $\{Mo_{72}Fe_{30}\}$ concentrations at room

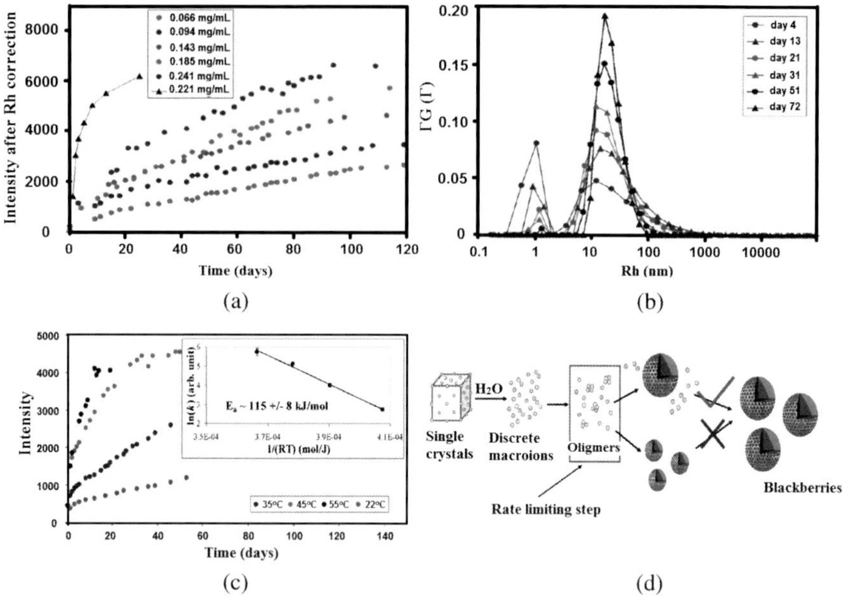

Fig. 13. (a) Change of total scattered intensity of $\{Mo_{72}Fe_{30}\}$ solutions at 90° scattering angle. All solutions were kept at 25°C except one at 45°C (the data shown by Δ). (b) CONTIN analysis of DLS study on $\{Mo_{72}Fe_{30}\}$ aqueous solution at different times. (c) Increment of the scattered intensity (I) from 0.5 mg/mL $\{Mo_{72}Fe_{30}\}/H_2O$ solutions at different temperatures (22, 35, 45, and 55°C) with time indicate the progress of blackberry formation. (inset: Calculation of the activation energy (Ea) for the blackberry formation.) (d) Possible mechanisms of $\{Mo_{72}Fe_{30}\}$ blackberry formation in dilute aqueous solutions. The upper mechanism has been proven to be correct based on SLS and DLS results, while the bottom mechanism can be ruled out.[40(a)] (Reprinted with permission from Ref 40. Copyright © 2003 American Chemical Society. Copyright © 2005 American Chemical Society)

temperature.[40] A continuous increase of the scattered intensity recorded by SLS suggests a continuous formation of large structures in solution. Meanwhile, DLS studies indicate that two modes (corresponding to two different types of particles) co-exist in solution: one with $R_h \sim 1.2$ nm and the other with average $R_h \sim 25$ nm. The first peak can be assigned to discrete $\{Mo_{72}Fe_{30}\}$ macroions whose population becomes smaller and smaller with time; while the second one attributed to blackberries grows larger, indicating that the discrete macroions are continuously forming blackberries. Interestingly, the average blackberry size remains almost unchanged during the whole process, indicating the blackberries are thermodynamically stable in solution. The time needed to achieve equilibrium between blackberries and discrete macroions is strongly influenced by temperature. For instance, at room temperature, the self-assembly process takes months to reach equilibrium while it only takes several weeks at 45°C.

2.1.3.2. High activation energy

The slowness of the blackberry formation is attributed to the high energy barrier between single macroions to blackberries. Time-resolved SLS studies were used to determine the initial "reaction" rates in $\{Mo_{72}Fe_{30}\}$ aqueous solutions at different temperatures. By using the Arrhenius equation

$$Ln(k) = -E_{a/RT} + B \tag{21}$$

the activation energy of the oligomer formation can be calculated as ~115 ± 8 kJ/mol, which is indeed very high (Fig. 13).[40]

2.1.3.3. Slow nucleation and fast aggregation

A closer look at the time-resolved SLS studies on $\{Mo_{72}Fe_{30}\}$ aqueous solutions and other POMs solutions reveals that there is a short lag period at the beginning of blackberry formation (in minutes or days). In a 10.0 mg/mL $\{Mo_{72}Fe_{30}\}$ solution with NaCl (0.9 wt% or 0.17 mol/L) added, this lag phase extended to about 20 days, as shown in Fig. 14 (left).[41] After filtering out the minor precipitates due to the addition of salts, the remaining solution is a stable, saturated $\{Mo_{72}Fe_{30}\}$/NaCl solution (~ 7 mg/mL) at room temperature. The initial scattered intensity from this solution was very low, suggesting that almost all the $\{Mo_{72}Fe_{30}\}$ macroanions exist as discrete ions. However, after a lag period of almost 20 days, the scattered intensity jumped up and continued to increase until it was finally stabilized after several months, indicating a slow formation of large structures. The overall kinetic picture is a sigmoidal shaped curve with a long lag phase. Normally, a sigmoidal curve is considered as a typical feature of a two-step process: slow nucleation and fast aggregation. In the lag period, because of the high activation energy barrier between macroions, oligomer nuclei slowly form in the solutions, and once a critical

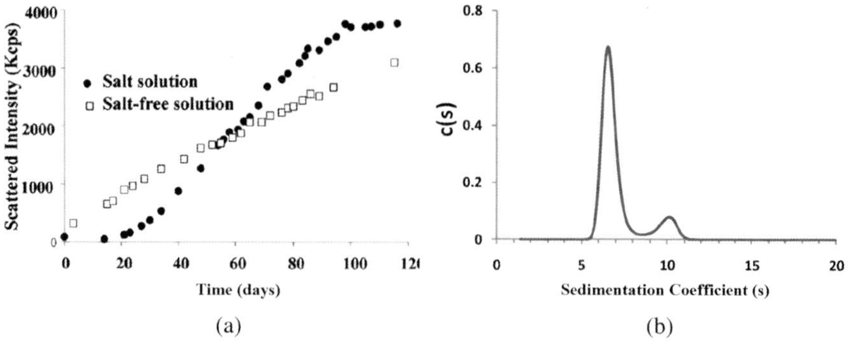

Fig. 14. (a) Comparison of scattered intensity increment of two $\{Mo_{72}Fe_{30}\}$ samples along reaction time in 0.9 wt% NaCl and salt-free solutions. (b) Continuous size distribution c(s) analysis of $\{Mo_{72}Fe_{30}\}$ solution versus sedimentation coefficient, s. Experiments were performed at a $\{Mo_{72}Fe_{30}\}$ concentration of 10 mg/mL in 170 mM NaCl solution at 20°C. (Reprinted with permission from Ref 41. Copyright © 2009 American Chemical Society.)

value of the rate limiting nuclei has reached, subsequent oligomers or monomers are quickly added to the growing assembly structures at a time until it is completed. Sedimentation velocity (SV) experiments performed on the 18th day after sample preparation show direct proof of the existence of oligomers in the lag phase.

The sedimentation coefficients (s) of sedimentating species and their corresponding sedimentation coefficient distributions c(s) show the coexistence of two species: s ~ 6.6 S with the dominant abundance (56%) and s ~ 9.5 S (10%) (Fig.14). The experimental results are fairly close to the theoretical sedimentation coefficients for $\{Mo_{72}Fe_{30}\}$ monomers (6.9 S), dimers (11.0 S) and timers (14.5 S) by solving the Lamm equations. The experimental values are slightly lower than the theoretical ones, possibly due to the fact that the dimers are not spherical. Besides the concentration of additional salts, the length of the lag period also depends on temperature, the valence of the cations and the anions, as well as the solvent content.

2.2. *Self-assembly of macrocations*

All the POMs we have mentioned in the previous sections are macroanions in nature, and their unique self-assembly behaviors in aqueous solutions haven't been reported before and neither the D-H theory nor DLVO theory could give a satisfied explanation. Naturally, we want to ask the following

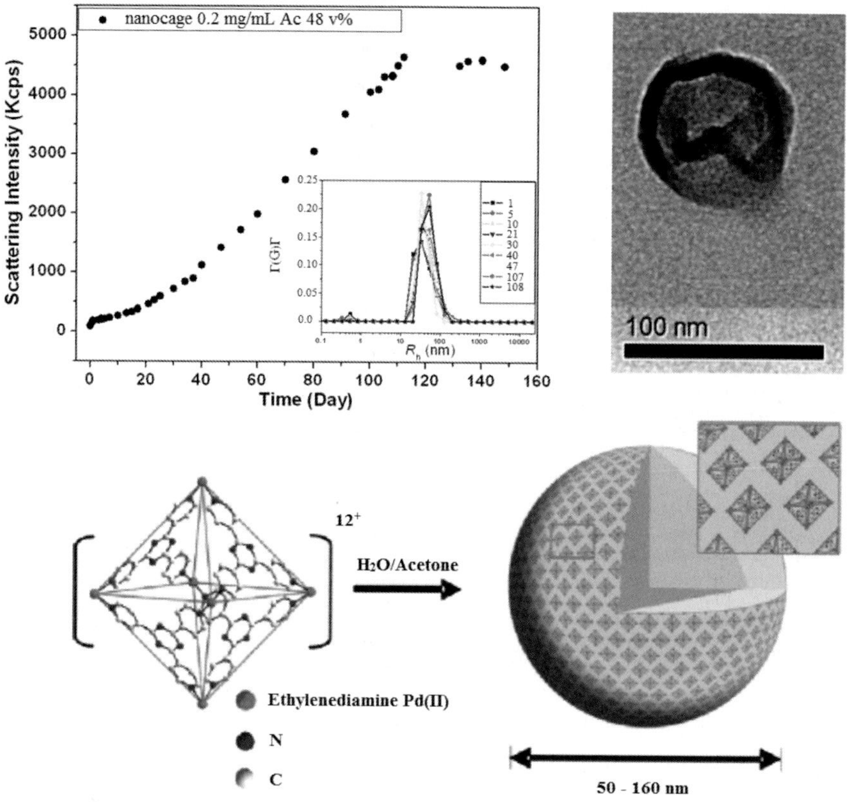

Fig. 15. *Top left*: Scattered intensity increment of a 0.2 mg/mL nanocage solution with 48 vol% acetone at different time. *Top right*: TEM images of vesicle-like structure formed by nanocages in solutions. *Bottom*: An illustration of the self-assembly process of nanocages in water/acetone mixed solvents. (Reprinted with permission from Ref 42. Copyright © 2008 American Chemical Society.)

questions: Does this unique self-assembly only hold true for macroanions or it is a universal phenomenon? Will macrocations also self-assembly into large structures in solution? To answer these interesting and important questions, we first need to find suitable macrocations. As stated before, we will focus our attention to metal-organic macroions with dimension from 2 nm to 6 nm.

The organic-inorganic hybrid macrocation we explored is a Pd_6L_4 type of nanocage {Pd = ethylenediamine palladium (II), L = 2,4,6-tris (4-pyridyl)-triazine}, which has an overall octahedral shape with four open faces, as shown in Fig. 15.[42] Each nanocage carries 12 positive charges owing to 6 Pd^{II} metal centers and the charges are balanced by 12 NO_3^- counterions. Very low scattered intensity was captured by SLS even after several weeks in a diluted aqueous solution of Pd_6L_4 nanocage, indicating that the nanocages exist as discrete macrocations in solution and no self-assembly was noticed. However, after introducing a small amount of acetone (more than 22 v%) into the original solution, the total scattered intensity started to slowly increase, suggesting some large structures are continuously forming. And indeed, as shown in Fig. 15, for a 0.2 mg/mL, Ac 48 v% sample, the scattered light intensity recorded by SLS slowly increased with time until it reached a "platform" around 120 days which indicates that the slow self-assembly process has completed. Also, during the slow-association, the average R_h of this sample keeps around 40 nm, further indicates this supramolecular structure has preferred curvature in solution. Similar trend was also observed for different acetone content samples. The vesicles have very low angle dependence, further implying a spherical shape in solution. The $R_{h,0}$ of 40.5 ± 2.1 nm is quite similar to the average radius of gyration (R_g) of the supramolecular structures of 40.8 ± 0.8 nm. Therefore, just like the blackberry structure formed in the POMs macroanionic solutions; current experimental results indicate that the supramolecular structures we observed in the nanocage solutions have a hollow vesicle-like structure. TEM studies also strongly supported our light scattering results (Fig. 15). It is clearly shown that these hollow spherical vesicles are uniform in size and shape, and some of them are burst because of losing their inside solvent during the evaporation process, which in another way uncovered their soft, biomembrane-like nature.

We also investigated the solvent polarity effect. At a fixed nanocage concentration of 0.20 mg/mL, the average R_h of nanocage formed vesicles increases monotonically from 25 to 80 nm with increasing acetone percentage. By plotting the R_h of the vesicles versus $1/\varepsilon_r$ of the solvent, a linear relationship is demonstrated which implies that the vesicle size, or ultimately, the interaction between single nanocage macrocations is strongly affected by the solvent polarity (Fig. 16). This result also suggests that the nanocage macrocations can spontaneously or reversibly form this vesicle-like structure in aqueous solutions as long as they possess appropriate charge density. This result also indicates that the

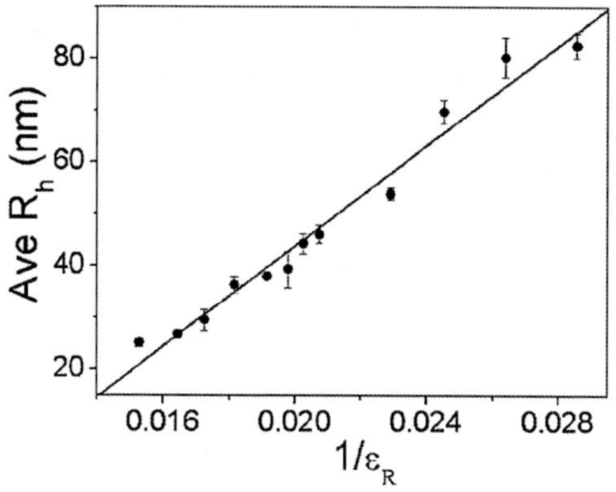

Fig. 16. Average R_h of the vesicle-like structure formed by nanocages demonstrates a linear relationship with the solution polarity. (Reprinted with permission from Ref 42. Copyright © 2009 American Chemical Society.)

vesicle structure formed in solution is not a consequence of the formation of chemical bonds but physical interactions and/or hydrophobic interactions.

3. Connection to Virus Capsid Formation

The blackberry structure self-assembled from POM type macroanions and nanocage type macrocations represent a universal phenomenon. More interestingly, the formation process of blackberry shares a lot of similarities with virus capsids formation. The self-assembly of protein subunits around viral nucleic acids into a highly ordered capsid structure is a crucial step in the virus life cycle.[43] In most instances, sufficient electrostatic interactions between the negatively charged core viral RNA and the positively charged shell proteins are required in order to get properly assembled capsids *in vivo*. However, numerous studies have shown that in many cases even without the viral RNA, empty capsids or similar structures still can be assembled or disassembled.[44] These facts suggest that the self-assembly of viral capsid proteins is mainly driven by the interactions between the capsid proteins themselves. Moreover, a recent study has also demonstrated that for a retrovirus such as the HIV capsid to be properly constructed, electrostatic interactions play an important role during the selective formation of pentamers or hexamers.[45] Considering that POMs, nanocages and capsid proteins are all nanoscale, soluble macroions in solutions, and they all self-assemble into spherical, hollow, single-layered structures, there might exist connections among these assembly behaviors.

3.1. *Trans-membrane transport of counterions over blackberry surface*

One important property of a virus capsid or a living cell obtains is "breathing". When the environmental condition changes (*e.g.* pH, temperature, ionic strength), a virus capsid can feel the change and make relevant responses, such as shrinking or swelling. The blackberry structure, although formed by pure inorganic compounds, are impressively soft and robust. They are different in nature from cell membranes because there is no closely packed hydrophobic region in the middle. For cell membranes, it is difficult for small cations to transport over freely, and special ion channels are needed in order to achieve biological functions.

Blackberry structure formed by macroanions which are not in touch with each other, the membranes might be permeable to cationic species. Fluorescence spectroscopy is sensitive to examine the transport of small ions through the blackberry membrane. The water soluble dyes specifically sensitive for one (or two) type of ions, such as chlorotetracycline (CTC) for Ca^{2+} and Mg^{2+}, 6-methoxyquinoline (6-MQ) for Cl^- and Coumarin 1 for Br^-, were introduced into freshly prepared $\{Mo_{72}Fe_{30}\}$ aqueous solutions not containing any blackberries. The dye molecules were partially incorporated into the blackberries during the self-assembly, as confirmed by an 18 nm shift of their fluorescent signals. This shift also shows that the environment inside the blackberries is different from the bulk solution, as the water inside might have a higher viscosity. After the completion of blackberry formation, additional salts containing the type of ions sensitive to the specific dye were introduced into the solution (here we use CTC and Ca^{2+} as an example). A sudden jump in fluorescence signal suggests that the CTC molecules staying in bulk solution are immediately saturated by the dominant amount of Ca^{2+} ions introduced.

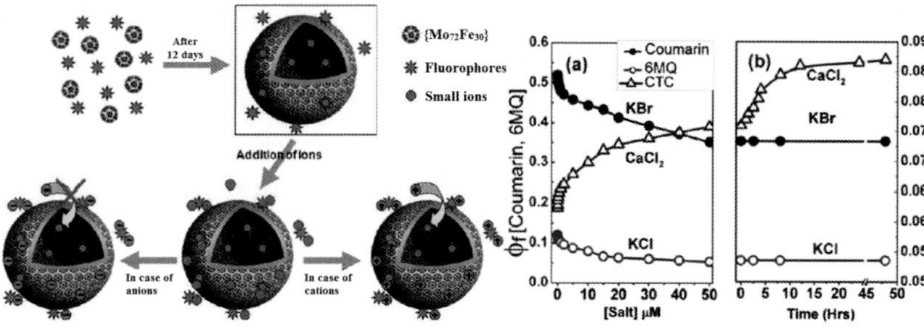

Fig. 17. (*Left*) Formation of fluorophore-containing $\{Mo_{72}Fe_{30}\}$ blackberries in solution. The additional cations, once added into solution, instantly interact with fluorophores in bulk solution and on blackberry surfaces, subsequently enter into the blackberries, and interact with the fluorophores inside. The anions could not cross the membrane. (*Right*) Change in fluorescence quantum yield of Coumarin 1, 6-MQ, and CTC with addition of KBr, KCl, and $CaCl_2$, respectively; (a) instantaneous change occurs with the addition of salts; (b) change in fluorescence quantum yield with time, once the addition of salt is stopped. (Reprinted with permission from Ref 46. Copyright © 2008 American Chemical Society.)

However, after this initial process, a very slow, continuous increment in the fluorescence signal appeared, suggesting a slow, continuous binding between Ca^{2+} ions and CTCs (Fig. 17). This can be explained as that the Ca^{2+} ions slowly transport over the blackberry membranes and are able to bind with the CTC molecules inside. This is strong evidence to show that small cations are able to move across the blackberry membrane. Experiments by using Mg^{2+} cations lead to the same results and conclusions. On the contrary, small anions such as Cl^- and Br^- cannot pass through the blackberry membrane. This is reasonable since the blackberry membrane is negatively charged and should repel with those small anions.[46]

4. Inorganic-Organic Hybrid POMs

4.1. *The motivation of synthesizing new hybrid POMs*

A novel branch of the large POM family, the inorganic-organic hybrid POMs, has been blooming during the last 10 years.[47] The desires for understanding the fundamental interactions between POMs and organic counterions, the demanding for new functional materials which carry properties from both the POMs and organic ligands are the main motivations of scientists to develop these novel hybrids. For instance, polymers that grafted with POMs can be used in photovoltaic cells[48] or in trapping magnetic nanoparticles.[49] Surfaces that covered with POMs (SAM-POM-pyrene) show specific cell adhesion ability in biochemical research.[50] A series of organometallic POM clusters has been tested for norbornene and cyclohexene oxygenation as precatalysts.[51] Moreover, a few hybrid POMs have shown active second-order nonlinear optical properties, indicating their potential applications in optical computing and telecommunication.[52] For our interests, the novel hybrid POMs can be treated as a group of new macroions. Therefore, studying the solution behaviors of these novel hybrid POMs will provide valuable information of the macroion-microion interactions, and serve as an important part of this chapter.

4.2. *POMs based amphiphilic surfactants and their assemblies*

Amphiphilic surfactants have been shown to have important applications in many fields, such as artificial cells, nanoreactors, as well as drug and gene delivery systems.[53] These functions are largely due to the facts that surfactant molecules can interact with two immiscible solvents and lower the interfacial tension between two liquids, and self-organized into micelles or vesicles in solutions. And as mentioned before, exploiting the synergistic effect of the organic component and the POM functionality is one of the motivations to synthesize the inorganic-organic POM hybrids. In this respect, synthesizing new POMs based amphiphilic surfactants becomes a quite interesting research topic. And ever since the first attempt of using POMs as the polar head groups linked to long alkyl chains made by Chambers *et al.*, in 2003,[54] more and more novel POMs based hybrid surfactants have been synthesized. Sulfate, sulfonate, phosphate and carboxylate are the main head groups of normal anionic surfactants known so far, and their sizes are quite small if compared with the POM anions. Therefore, POMs based surfactants with a giant anionic head group could greatly change the surfactant packing parameter N_s, which is widely applied to explain and

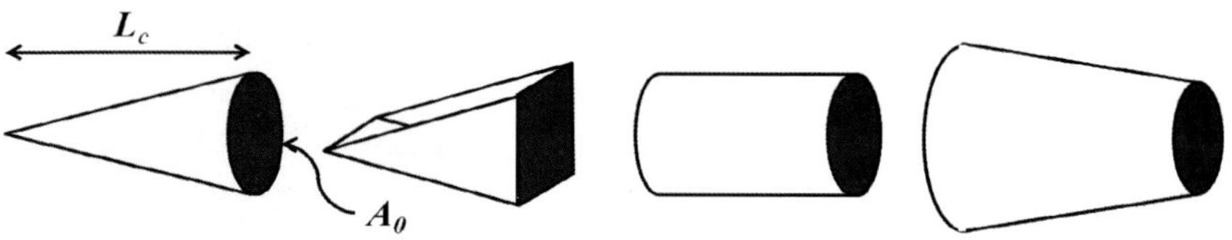

Fig. 18. The different morphologies of self-assembled structures are largely depended on the surfactant packing parameter of the amphiphilic molecules.

predict the molecular self-assembly behavior of a surfactant through the relationship shown in Eq. 22

$$N_S = \frac{V_c}{L_c A_0} \approx \frac{A_c}{A_0} \tag{22}$$

in which V_c is the volume of the hydrocarbon tail, L_c is the length of the hydrocarbon tail, and A_0 is the area per head group. When N_s is small (< 0.37), a spherical micelle structure will form in solution; when N_s is close to 0.5, a cylinder micelle structure will form; and when N_s is close to 1, a bilayer structure or a vesicle will form. If N_s is larger than 1, normally a reversed vesicle or micelle will form, as illustrated in Fig. 18. Moreover, by tethering different organic functional groups to the hydrocarbon tails, the POMs based hybrid surfactants could show better biocompatibility, better catalytic performance and stronger sorption ability in different systems.

4.2.1. *Diblock hybrid surfactants*

Diblock POM hybrid surfactants incorporate one POM polar head and one or two hydrocarbon chains located at only one side of the POM. A typical example is a group of $[PW_{11}O_{39}(SiC_n)_2]^{3-}$ surfactants, reported by Landsmann, S. and co-workers.[55] As shown in Fig. 19, this hybrid surfactant is constructed by covalently linking two organo-alkoxysilane long chains $C_nSi\text{-}(OCH_2CH_3)_3$ (n = 8, 12, 18) with a $[PW_{11}O_{39}]^{3-}$ keggin POM polar head under acidic conditions. The amphiphilic properties of this surfactant largely increased after exchanging the initial counterion TBA^+

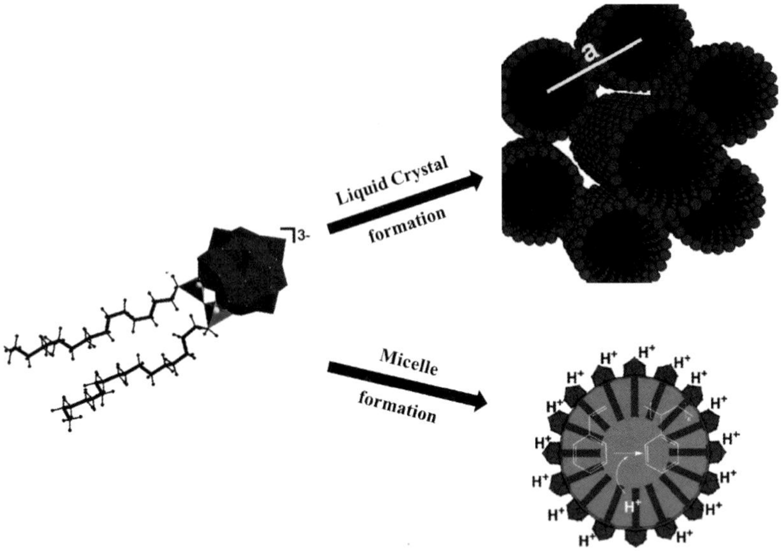

Fig. 19. The hybrid surfactant with a $[PW_{11}O_{39}]^{3-}$ keggin POM polar head and two organo-alkoxysilane $C_nSi\text{-}(OCH_2CH_3)_3$ (n = 8, 12, 18) tails demonstrates interesting self-assembly behaviors in solutions. (Reprinted with permission from Ref 42. Copyright © 2010 American Chemical Society.)

with H^+, Na^+ and K^+. This diblock POM hybrid surfactant exhibits interesting self-assembly behaviors. At high surfactant concentration, the formation of liquid crystalline phases has been observed by using polarization microscopy. Small-angle-X-ray-scattering (SAXS) studies confirmed a hexagonal cylinder packing (space group *P6/mm*) with a core-shell structure. The structure of the liquid crystalline phases is influenced by the type of counterions and the hydrocarbon chain length. Also, the Na^+ balanced surfactant $[PW_{11}O_{39}(SiC_{16})_2]Na_3$ form micelles in diluted aqueous solutions. The size of the micelles also shows dependence on the length of the surfactant tails. The POM surfactants have also been tested as emulsification agents. The cationic polymerization of styrene has been carried out in the POM surfactant ($[PW_{11}O_{39}(SiC_n)_2]H_3$) emulsions. In this scenario, the POM surfactants play two roles at once. They act as amphiphilic to stabilize the emulsion droplets, at the mean time the heteropolyacid Keggin head group is catalytically active.

4.2.2. *Triblock hybrid surfactants*

4.2.2.1. Organic-inorganic-organic type hybrid surfactants

Triblock POMs based hybrid surfactants are different from diblock hybrids in a sense that they normally have three discrete components, which are arranged in either an Organic-Inorganic-Organic (O-I-O) or an Inorganic-Organic-Inorganic (I-O-I) fashion. The different arrangements give the hybrid surfactants different amphiphilic properties. Only recently, an O-I-O type POMs based hybrid surfactants $[n\text{-}Bu_4N]_3[MnMo_6O_{18}\{(OCH_2)_3\text{-}CNHCO\text{-}(CH_2)_{14}CH_3\}_2]$ (Mn-Anderson-C16) has been synthesized by Cronin's group, in which the hydrophilic $\{MnMo_6O_{24}\}$ cluster has a plate-like structure with dimensions of ~ $9 \times 9 \times 3$ Å^3 (Fig. 20).[56] On both sides of the central Anderson $\{MnMo_6O_{24}\}$ structure one hydrophobic alkyl chain ($-COC_{15}H_{31}$) is covalently capped via amide bonds. Compared to conventional surfactants, the hybrid "surfactant" has a much larger inorganic polar head group. A

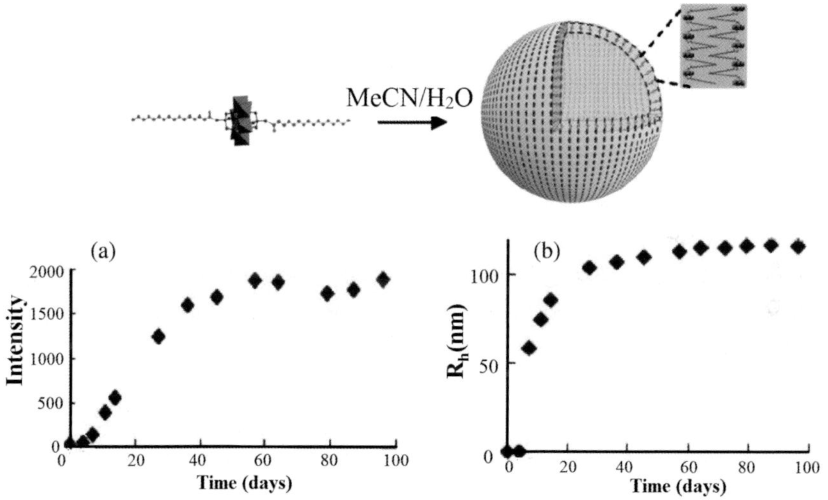

Fig. 20. *Top*: An illustration of the self-assembly process of the hybrid surfactant Mn-Anderson-C16 from monomer into double-layered vesicle structure in water/acetonitrile mixed solvent. *Bottom*: (a) The increment of total scattered intensity versus time of a 0.1 mg/mL Mn-Anderson-C16 sample in water/acetonitrile mixed solvent containing 40 v% MeCN. (b) The average R_h change of the same sample at different time. Reprinted with permission from Ref 57. Copyright © 2008 American Chemical Society.)

similar hybrid surfactant, Mn-Anderson-C6 with shorter chains was also synthesized. The Mn-Anderson-C16 hybrid is insoluble in water probably due to the bulky TBA^+ counterions. But it is quite soluble in acetonitrile (MeCN) and MeCN/water mixtures to form stable and homogeneous solutions (~ 0.1 mg/mL). When water is introduced into the surfactant MeCN solutions, the formation of a supramolecular structure is observed in the mixed solvents of MeCN/water containing 40–65 v% water, as studied by dynamic and static light scattering (DLS and SLS) techniques. The scattered intensity increases with time in a Mn-Anderson-C16 MeCN/water solution containing 60 v% water, as shown in Fig. 20.[57] The scattered intensity slowly and continuously increases with time, indicating a continuous formation of supramolecular structures in the solution. The average hydrodynamic radius (R_h) of the assemblies at the scattering angle of 90° on day 7 is ~ 58.5 ± 1.8 nm. The size of the assemblies continues to grow during the following two months, until it stops at an R_h of 115 nm. The total scattered intensity from the solution also gradually stabilizes, suggesting that the assemblies have reached equilibrium. The large assemblies are spherical and hollow based on DLS and SLS studies, indicating a typical bilayer vesicle structure. Consider the hydrophilic nature of the POM head and the hydrophobic nature of the long alkyl chains in the current MeCN/water mixtures, we proposed a model structure for these vesicles in which the POM part faces outside while the two alkyl chains bend and stay inside to form the solvent-phobic layer. The vesicle formation was also observed in the MeCN/water solutions of Mn-Anderson-C6. The Mn-Anderson-C6 hybrid surfactant has much shorter alkyl chains. The kinetics of the vesicle formation for both hybrid surfactants is similar. It requires nearly two months for the Mn-Anderson-C6 surfactant to reach equilibrium, and vesicles with the radii ranging from 50 to 150 nm are observed in solutions containing 40–66 v% water. However, compared with Mn-Anderson-C16, fewer vesicles are formed for the Mn-Anderson-C6 surfactants in the same type of solvent, probably due to their less amphiphilic nature and the shorter alkyl chains which makes the bending of alkyl chains more difficult. Generally, a lipid surfactant such as dipalmitoylphophatidylcholine (DPPC), which possesses the same double alkyl chains as the Mn-Anderson-C16, tends to form smaller vesicles (~ 10 nm) in aqueous solution. Therefore, compared to the lipid vesicles, the much larger vesicles formed by the Mn-Anderson-C16 surfactant indicate that a giant polar head group will significantly affect the packing of individual surfactant molecules on the vesicle surface. Moreover, since the Anderson cluster has a stiff plate structure, a high bending energy is required to bend two alkyl chains in order to form a vesicular structure.

The solvent is also an important factor which influences the self-assembled structures of the POMs based hybrid surfactants. For instance, we found that the same Mn-Anderson-C6/C16 surfactants can self-assemble into stable reverse vesicles when non-polar solvent such as toluene is introduced. And the formation of reverse vesicles further establishes the amphiphilic nature of the Mn-Anderson-C6/C16 hybrids.[58] As shown in Fig. 21, continuous addition of toluene into the Mn-Anderson-C6/C16/MeCN solution significantly decreases the

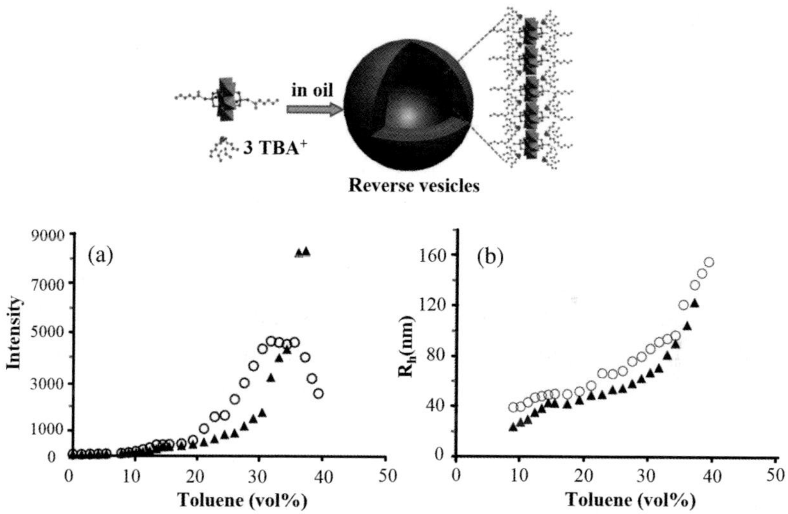

Fig. 21. *Top*: An illustration of the self-assembly process of the hybrid surfactant Mn-Anderson-C16 from monomer into reverse-vesicle structure by titrating toluene into acetonitrile solvent. *Bottom*: (a) The change of total scattered intensity versus toluene content of a 0.1 mg/mL Mn-Anderson-C16 sample (▲) and Mn-Anderson-C6 sample (○) in toluene/acetonitrile solutions. (b) The average R_h change of the same samples at different toluene content. (Reprinted with permission from Ref 58. Copyright © 2010 Wiley InterScience.)

solvent polarity, which is unfavorable for the solubility of the hydrophilic Mn-Anderson clusters but is favored by the alkyl chains. To accommodate the solvent environment change, the solvophobic POM part will have to stay together in the core while the alkyl chains will expose to the non-polar solvent in the corona, which leads to the formation of a single-layer reversed vesicle structure. In the reversed vesicles, the TBA counterions are expected to play a very important role by shielding the repulsive electrostatic interaction between the POMs and by providing additional affinity between hybrid surfactants through their $-C_4H_9$ chains. This is indeed reflected from the fact that the vesicle size is quite sensitive to the solvent quality and changes spontaneously after the addition of more toluene.

We also investigated the alkyl chain length and the temperature effects on the assembly structure. The size difference between the vesicles formed by two types of hybrids is relatively small and we believe that the two types of hybrid surfactants undergo similar self-assembly mechanisms in these mixed solvents. Elevating temperature can accelerate the assembly process, and the length of the alkyl tails on the hybrid surfactants has a minor effect on the vesicle sizes. These results indicate that the vesicle formation is a quite general behavior for the new types of surfactants with large inorganic polar head groups. And we believe the main driving forces are the solvophobic/solvophilic interactions.

A wheel-shaped POM hybrid surfactant with two extremely short hydrophobic components reveals quite different self-assembly behaviors compared with the above two hybrid surfactants.[59] As shown in Fig. 22, four organoruthenium(II) units are covalently grafted on the rim of the central cavity in the wheel-shaped hybrid of $K_{16}Li_{11}[\{K(H_2O)\}_3\{Ru(p\text{-cymene})\text{-}(H_2O)\}_4P_8W_{49}O_{186}(H_2O)_2]$, making the hydrophobic *p*-cymene groups pointing out on both sides of the "wheel".[60] This $\{(p\text{-cymene})_4P_8W_{49}\}$ hybrid surfactant shares the structural similarity to another wheel-shaped $\{Cu_{20}P_8W_{48}\}$ POM derivative ($K_{12}Li_{13}[Cu_{20}Cl(OH)_{24}(H_2O)_{12}(P_8W_{48}O_{184})]\cdot 22H_2O$) reported by Kortz group in 2005,[22] but the four organic ligands grant the hybrid an amphiphilic property. Meanwhile, because the four organic ligands are bound to the center of the wheel, therefore, the hydrophilic and the hydrophobic motifs of the current hybrid are pretty close in space, which makes this hybrid surfactant also quite different from the

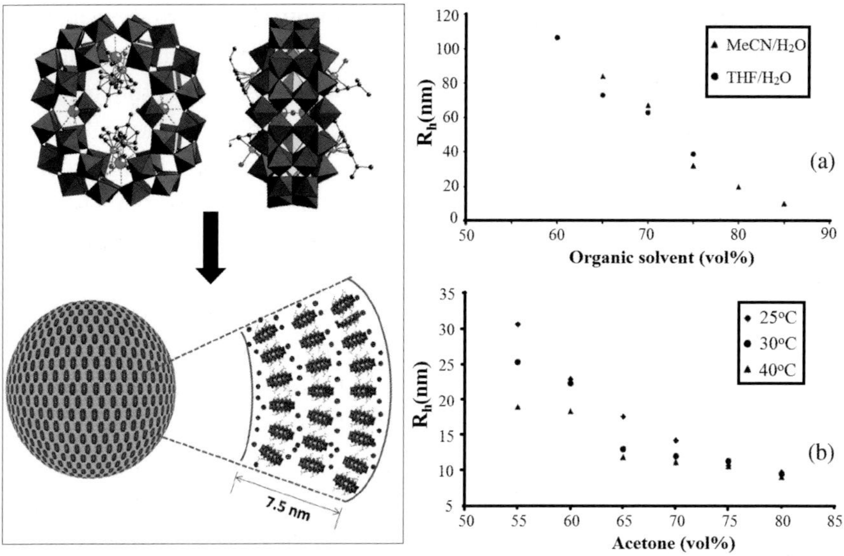

Fig. 22. *Left*: The self-assembly process of the {(p-cymene)$_4$P$_8$W$_{49}$} hybrid wheels in polar solvents. *Right*: (a) Average R_h of {(p-cymene)$_4$P$_8$W$_{49}$}-based vesicles in mixed water/acetonitrile or water/THF solvents. (b) Average R_h of {(p-cymene)$_4$P$_8$W$_{49}$}-based vesicles in mixed water/acetone solvents at different temperature. (Reprinted with permission from Ref 59. Copyright © 2010 Wiley InterScience.)

Mn-Anderson-C$_n$ hybrids, in which the alkyl chains are long and soft. The {(*p*-cymene)$_4$P$_8$W$_{49}$} hybrid can be well dissolved in water and water/acetone mixtures up to 80 v% acetone at 0.10 mg/mL. And a slow self-association process was observed in samples containing 55–80 v% acetone, similar to our previous studies on the self-assembly process of fully hydrophilic POMs. The scattered intensity recorded by SLS remained unchanged for the first few days and then started to increase continuously for about a month, indicating the formation of supramolecular structures. The final R_h deduced from the CONTIN analysis is ~ 23 nm with a narrow size distribution. The DLS and SLS studies indicate that the large assemblies are spherical, hollow, vesicle-like structures. The solutions with different contents of acetone show similar, slow self-association behavior.

However, the size of the large changes in an opposite way as does the size of blackberries when the solvent polarity is changed. Instead of growing into larger assemblies, when the acetone content in the solutions increases (i.e., decreasing dielectric constant), smaller assemblies are formed. For instance, the average size of the vesicle-like structures decreased from ~ 23 to ~ 11 nm when the acetone content increases from 55 to 75 v% (Fig. 22). A similar trend is noticed in other binary solvent mixtures, such as water/acetonitrile and water/THF. Increasing the less polar solvent content always decreases the size of the vesicle-like assemblies. This unusual self-assembly behavior is quite interesting, and we speculate that the four organic ligands on the hybrid surface must play an important role. At low acetone fractions, the hydrophobic interaction between the *p*-cymene groups of neighboring polyanions could be stronger, because acetone is a good solvent for the hydrophobic *p*-cymene part, whereas water is not. Therefore, larger assemblies are formed. In other words, the hydrophobic interaction becomes a more dominant driving force for the self-assembly of the $\{(p\text{-cymene})_4P_8W_{49}\}$ hybrid in relatively polar solvents. However, a critical acetone content is required because in pure water solutions, the electrostatic repulsion between two giant wheels will take over and become the dominant force and indeed, under this condition no vesicle-like structures are observed. Therefore, after passing a critical organic solvent fraction (~55 v% in the case of acetone), the attractive hydrophobic interaction between the *p*-cymene groups will become stronger than the electrostatic repulsion and large assemblies start to form.

The temperature effect on the size of vesicle-like structures is another way of probing the hydrophobic interactions. As shown in Fig. 22, the assembly size shows weak temperature dependence in solutions containing 55–65 v% acetone in the range 25–40°C with larger assemblies formed at lower temperatures. And the temperature effect becomes more significant in lower acetone content. This is reasonable considering the solvophobic groups become more soluble in solvents with low polarity, that is, weaker hydrophobic interactions at higher acetone contents. Overall, the unique solution behavior of amphiphilic 1 might be attributed to a subtle balance between repulsive electrostatic and attractive hydrophobic interactions.

Compared to Mn-Anderson-C_n hybrid surfactants with long and soft alkyl chains, the *p*-cymene groups in the $\{(p\text{-cymene})_4P_8W_{49}\}$ hybrids are relatively small and rigid. Consequently, no bending of the *p*-cymene groups is possible during the self-assembly process. The {(*p*-cymene)$_4$P$_8$W$_{49}$} hybrids must pack in the vesicles in a different way compared to the Mn-Anderson-C$_n$. A possible packing model of the {(*p*-cymene)$_4$P$_8$W$_{49}$} hybrids in the assemblies is proposed in Fig. 22. The molecular main plane of the {(*p*-cymene)$_4$P$_8$W$_{49}$} hybrids might have to stay perpendicular to the vesicle surface in order to bury their hydrophobic and rigid *p*-cymene groups to form an ordered vesicle-like structure.

4.2.2.2. Inorganic-organic-inorganic type hybrid surfactants

A new group of "inorganic–organic–inorganic" POM based hybrid surfactants are synthesized by linking two V3-capped Wells–Dawson-type cluster $TBA_5H_4[P_2V_3W_{15}O_{62}]$ with a linear bis-(Tris) ligands. One of these hybrids is shown in Fig. 23 with the full chemical formula of $TBA_{10}H_2[\{P_2V_3W_{15}O_{59}(OCH_2)_3CCH_2\}_2O]$, which is approximately 3.4 nm long, represent the first example of "dumbbell" hybrid.[61] A series of samples of this dumbbell hybrid at 0.1 mg/mL in water/acetone solvents are monitored by SLS and DLS. The results show that hollow supramolecular assemblies with spherical shape were observed in solutions containing 30–60 v% acetone. The size of the vesicle-like structure increases with increasing acetone content in the solvent, ranging from ~ 61 to ~ 83 nm in solutions containing 30–60 vol% acetone, respectively. A linear relationship is also found between the R_h of the vesicles and the inverse of the solvent dielectric constant. This is the typical feature for a charge-regulated self-assembly process. Moreover, at a fixed solvent content (50:50 v% water/acetone), the size of the vesicle-like structure shows a concentration dependence in the range of 0.05–0.25 mg/mL. DLS studies indicate that the average size of the vesicle-like structures remains constant ~60 nm when the hybrid surfactant concentration is less than 0.15 mg/mL. At higher concentrations, the R_h value increases with increasing concentration to up to 120 nm.

The formation of vesicle-like structures by the current dumbbell hybrids is unique due to their shapes and charges. Consider the geometry of the current hybrid, the packing parameter N_s must be close to 1, suggesting the formation of either the lamella structure or the bilayer vesicle structure. And indeed, because the number of linkers in the solvophobic

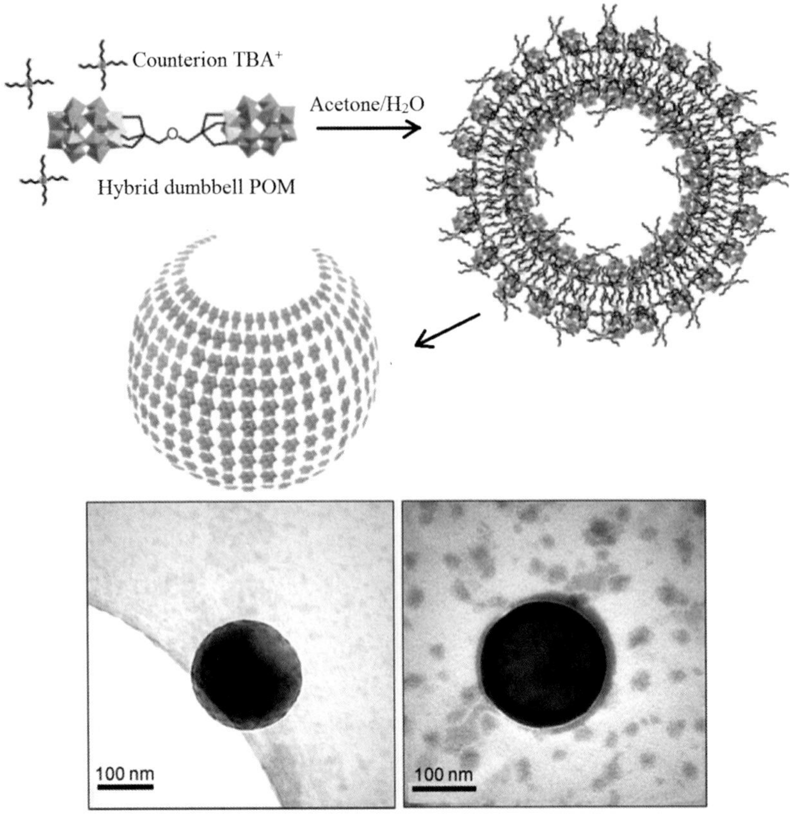

Fig. 23. *Top*: The self-assembly process of the hybrid dumbbell POMs in water/acetone mixed solvents. *Bottom*: TEM images of the vesicle-like structure formed by the dumbbells in mixed solvents. (Reprinted with permission from Ref 61. Copyright © 2009 Wiley InterScience.)

region is limited, the whole molecule is quite rigid, and in the presence of large POM head groups, the closed packing of the organic linker to form micelles becomes impossible. Meanwhile, the high surface charge density on the two POM heads hinders the close contact of individual dumbbells. In order to overcome the electrostatic repulsion between the two POM heads and trigger the vesicle formation, the countercation of TBA play an important role. Each TBA cation possesses four n-C_4H_9 chains whose length is close to that of the hydrophobic linker of the dumbbell, and the middle organic region of each anionic dumbbell can be filled with hydrophobic n-C_4H_9 chains thus provide attractive interactions for the vesicle formation. Furthermore, the electrostatic repulsion among the highly charged dumbbell clusters can be effectively decreased if some TBA counterions are placed between POM polar head groups thus further stabilize the assembled structures.

Based on this, a model is presented in Fig. 23 to demonstrate the possible vesicle structure formed by such novel materials. The vesicles are still negatively charged overall, but most of the remaining counterions are closely associated to the external and internal surfaces of the vesicles, as confirmed by the Zeta potential studies. In addition, the unprecedented formation of such structures from hybrids with large redox-active POM head groups could pave the way for many future materials applications.

4.2.3. *Comparison with blackberries*

The POMs based hybrid surfactants show interesting and distinctive self-assembly behaviors in aqueous solutions, and these novel supramolecular structures may have important applications in many different fields. An important lesson we learned is that the hydrophobic interaction is one of the most essential interactions that control the macroions and microions in solutions. When macroions are fully hydrophilic, the main driving forces for the blackberry type structure formation are counterion mediated attraction and hydrogen bonding, where the hydrophobic interaction is relatively weak. However, when macroions are grafted with organic components, the contribution from hydrophobic interaction can no longer be ignored and under certain conditions it overwhelms the electrostatic interactions. Also, from the thermodynamic point of view, the self-assembly for macroions with/without hydrophobic components is a free energy favored process, which means either the enthalpy term or the entropy term (or maybe both) will contribute to it, as shown in Eq. 23. Since in solutions macroions are always surrounded by small counterions, we can assume that the enthalpy change (ΔH) is relatively small during the association of macroions. Thus, the negative free energy change (ΔG), which indicates a spontaneous process, is largely because a positive entropic gain (ΔS). And the entropic gain is most possible come from the renormalized counterion distribution and the release of water molecules which originally covered the macroions surface and/or the organic components.

$$\Delta G = \Delta H - \Delta S * T \tag{23}$$

5. Conclusions

In this chapter, we reviewed our recent research on the unique solution behaviors of a large group of macroions with size dimension ranging from 2 to 6 nm in polar electrolytes. With the help of both suitable amounts of charge and surface water ligands, these macroions slowly self-assemble into spherical, single-layered, hollow blackberry type supramolecular structures in solution. DLS, SLS, TEM, NMR, zeta-potential and other measurements provide direct and indirect evidences for these interesting structures. We also found that the size of the blackberry structures can be easily tuned by changing pH, solvent polarity, ionic strength, temperature and other parameters, indicating the nature of the macroion-macroion and macroion-microion interactions in current systems are mainly physical based interactions but not chemical bonds. Therefore, we believe the main driving forces for the supramolecular assemblies are counterion mediated attraction and hydrogen bonds. For nanocage type macrocations, the hydrophobic interactions may also contribute to the overall stability of the blackberry structure. Meanwhile, the blackberry structures and their formation process mimic the viral capsid formation, suggesting certain similarities (thermodynamically, kinetically) between these two systems.

By covalently linking the inorganic POM cluster with organic components, a new family of inorganic-organic hybrid surfactants is created. Because of the giant POM head groups which have multiply charges, novel self-association phenomena are observed in solutions containing these hybrid surfactants, ranging from micelles to reversed vesicles. Not only have these hybrids contributed significantly to the knowledge of macroion-microion interactions and the colloidal chemistry, but also they show great potential applications in material science, catalysis, and medicines.

Acknowledgement

Tianbo Liu acknowledges support from National Science Foundation, A. P. Sloan Foundation and Lehigh University.

References

1. (a) Hill C. L. (Ed.), *Chem Rev* **98** (1998) 1–387; (b) Müller A. and Roy S., *Coord Chem Rev* **245** (1–2) (2003) 153–166; (c) Müller A. Kögerler P. and Dress A. W. M., *Coord Chem Rev* **222** (1) (2001) 193–218; (d) Long D. L., Burkholder and E.; Cronin L., *Chem Soc Rev* **36** (1) (2007) 105–121.
2. Debye P. J. and Hückel E., *Phys Z* **24** (1923) 185.
3. (a) Derjaguin B. V.; Landau L., *Acta Physicochim* **89** (1941) 555; (b) Verwey E. J.; Overbeek J. T. G., *Theory of the Stability of Lyophobic Colloids*. Elsevier: Amsterdam, 1948.
4. Sogami I. and Ise N., *J Chem Phys* **81** (1984) 6320.
5. Hiemenz P. C.; Rajagopalan R., *Principles of Colloid and Surface Chemistry*. 3rd ed.; CRC Press: New York, 1997.
6. Verwey E. J. W. and Overbeek J. T. G., *Theory of Stability of Lyophobic Colloids*. Dover Publications: Mineola, New York, 1999.
7. Ise N. Sogami I. S., *Structure Formation in Solution: Ionic Polymers and Colloidal Particles*. Springer Heidelberg, 2005.
8. Kose A., *et al.*, S., *J Colloid Interface Sci* **44** (2) (1973) 330–338.
9. (a) Lee W. I.; Schurr J. M., *Biopolymers* **13** (5) (1974) 903–908; (b) Schmitz K. S.; Lu M.; Singh N., Ramsay D. J., *Biopolymers* **23** (9) (1984) 1637–1646.
10. (a) Burns P. C., *et al.*, *Angew Chem Int Ed* **44** (14) (2005) 2135–2139; (b) Nyman M.; Bonhomme F.; Alam T. M.; Rodriguez M. A.; Cherry B. R.; Krumhansl J. L.; Nenoff T. M.; Sattler A. M., *Science* **297** (5583) (2002) 996–998.
11. (a) Chubarova E. V., *et al.*, *Angew Chem Int Ed* **47** (49) (2008) 9542–9546; (b) Izarova N. V., *et al.*, *Angew Chem Int Ed* **49** (10) (2010) 1886–1889.
12. (a) Müller A., *et al.*, *Angew Chem Int Ed* **34** (19) (1995) 2122–2124; (b) Müller A.; Serain C., *Acc Chem Res* **33** (1) (1999) 2–10.
13. Liu T. *et al.*, *Nature* **426** (2003) 59–62.
14. Muller A., *et al.*, *Chem Commun* (19) (2001) 1928–1929.

15. (a) Zhang J., *et al.*, *Langmuir* **24** (10) (2008) 5277–5283; (b) Chen B. *et al.*, *J Am Chem Soc* **127** (12) (2005) 4166–4167; (c) Zhu Y., *et al.*, *Chem Eur J* **10** (10) (2004) 2421–2427.
16. Liu G., *et al.*, *J Am Chem Soc* **128** (31) (2006) 10103–10110.
17. Müller A., *et al.*, *Anorg Allg Chem* **625** (7) (1999) 1187–1192.
18. Müller A., *et al.*, *Angew Chem Int Ed* **37** (24) (1998) 3359–3363.
19. Botar B.; Kogerler P. and Hill C. L., *Chem Commun* (25) (2005) 3138–3140.
20. Müller A., *et al.*, *Angew Chem Int Ed* **38** (21) (1999) 3238–3241.
21. Todea A. M., *et al.*, *Angew Chem Int Ed* **46** (32) (2007) 6106–6110.
22. Mal S. S. and Kortz U., *Angew Chem Int Ed* **44** (24) (2005) 3777–3780.
23. Müller A., *et al.*, *Angew Chem Int Ed* **41** (7) (2002) 1162–1167.
24. Howell R. C., *et al.*, *Angew Chem* **113** (21) (2001) 4155–4158.
25. Kistler M. L., *et al.*, *J Am Chem Soc* **129** (2007) 6453–6460.
26. Liu G. Cai Y. and Liu T., *J Am Chem Soc* **126** (2004) 16690–16691.
27. Oleinikova A., *et al.*, *Chem Phys Chem* **8** (5) (2007) 646–649.
28. Schmitz K. S., *J Phys Chem B* **113** (2009) 2624–2638.
29. (a) Qiu X., *et al.*, *Phys Rev Lett* **99** (3) (2007) 038104; (b) Das R., *et al.*, *Phys Rev Lett* **90** (18) (2003) 188103; (c) Qiu X., *et al.*, *Phys Rev Lett* **101** (22) (2008) 228101.
30. (a) Oosawa F., *Biopolymers* **6** (11) (1968) 1633–1647; (b) Manning G. S., *J Chem Phys* **51** (1969) 924.
31. Liu G., *et al.*, *J Cluster Sci* **17** (2) (2006) 427–443.
32. Pigga J. M., *et al.*, *Angew Chem Int Ed* **48** (2009) 6538–6542.
33. Kistler M. L. Patel K. G. and Liu T., *Langmuir* **25** (13) (2009) 7328–7334.
34. Liu T., *et al.*, *J Am Chem Soc* **128** (49) (2006) 15914–15920.
35. Pigga J. M., *et al.*, *Langmuir* **26** (12) (2010) 9449–9456.
36. Verhoeff A. A., *et al.*, *Phys Rev Lett* **99** (2007) 066104.
37. Kistler M. L., *et al.*, *Dalton Trans* (26) (2009) 5094–5100.
38. Schäffer C., *et al.*, *Angew Chem Int Ed* **48** (1) (2009) 149–153.
39. Liu T., *Langmuir* **26** (12) (2009) 9202–9213.
40. (a) Liu T., *J Am Chem Soc* **125** (2) (2003) 312–313; (b) Liu G. and Liu T., *Langmuir* **21** (7) (2005) 2713–2720.
41. Zhang J., *et al.*, *J Am Chem Soc* **131** (42) (2009) 15152–15159.
42. Li D., *et al.*, *J Am Chem Soc* **130** (2008) 4226–4227.
43. (a) Neira J. L., *FEBS Journal* **276** (21) (2009) 6110–6117; (b) Stray S. J., *et al.*, *Nat Biotechnol* **24** (3) (2006) 358–362.

44. (a) Ceres P. Zlotnick A., *Biochemistry* **41** (2002) 11525–11531; (b) Zlotnick A., *Virology* **315** (2) (2003) 269–274; (c) Douglas T. and Young M., *Nature* **393** (1998) 152–155; (d) Zlotnick A., *et al.*, *Biochemistry* **35** (1996) 7412–7421; (e) Zhou S. and Standring D. N., *Proc Natl Acad Sci USA* **89** (1992) 10046–10050; (f) Zhao X., *et al.*, *Virology* **207** (2) (1995) 486–494.
45. Cardone G., *et al.*, *Nature* **457** (2009) 694–699.
46. Mishra P. P. Pigga J. and Liu T., *J Am Chem Soc* **130** (5) (2008) 1548–1549.
47. Dolbecq A., *et al.*, *Chem Rev* **110** (10) (2010) 6009–6048.
48. Xu L., *et al.*, *Angew Chem Int Ed* **41** (21) (2002) 4129–4132.
49. Mayer C. R., *et al.*, *Angew Chem Int Ed* **38** (24) (1999) 3672–3675.
50. Song Y.-F., *et al.*, *J Am Chem Soc* **131** (4) (2009) 1340–1341.
51. Weiner H. Hayashi Y. and Finke R. G., *Inorg Chem* **38** (11) (1999) 2579–2591.
52. Guan W., *et al.*, *Inorg Chem* **45** (19) (2006) 7864–7868.
53. (a) Graff A., *et al.*, *Proc Natl Acad Sci USA* **99** (2002) 5064; (b) Stanish I., *et al.*, *Adv Mater* **17** (9) (2005) 1194–1198; (c) Kabanov A. V. and Kabanov V. A., *Adv Drug Delivery Rev* **30** (1–3) (1998) 49–60; (d) Ritchie C., *et al.*, *Nat Chem* **1** (1) (2009) 47–52; (e) Zhou S., *et al.*, *Science* **291** (5510) (2001) 1944–1947.
54. Carlisle R C., *et al.*, *Chem Commun* (19) (2003) 2456–2457.
55. Landsmann S. Lizandara-Pueyo C. and Polarz S., *J Am Chem Soc* **132** (14) (2010) 5315–5321.
56. Song Y.-F., *et al.*, *Chem Eur J* **14** (8) (2008) 2349–2354.
57. Zhang J., *et al.*, *J Am Chem Soc* **130** (44) (2008) 14408–14409.
58. Zhang J., *et al.*, *Chem Eur J* **16** (37) (2010) 11320–11324.
59. Zhang J., *et al.*, *Eur J Inorg Chem* **2010** (20) (2010) 3195–3200.
60. Mal S. S., *et al.*, *Dalton Trans* (25) (2007) 2627–2630.
61. Pradeep C. P., *et al.*, *Angew Chem Int Ed* **48** (44) (2009) 8309–8313.

Chapter 3

DIRECTED ASSEMBLY OF POLYOXOMETALATES ACROSS LENGTH SCALES: FROM MACRO-MOLECULES TO MICROSYSTEMS AND ICHELLS

ANTOINE G. BOULAY, GEOFFREY J. T. COOPER
and LEROY CRONIN*
School of Chemistry, WestCHEM
The University of Glasgow, University Avenue
Glasgow, G12 8QQ, UK.
**www.croninlab.com*

1. Building Blocks to Cation Control

Polyoxometalates (POMs) are discrete, anionic metal-oxide clusters of early transition metals. Composed principally of Mo, W, V and Nb, they represent a large and important class of inorganic molecules and materials,[1] whose relevance is being highlighted in a number of review articles, with an extraordinary reach in terms of applications, physical and chemical properties.[2,3] One of the most interesting aspects of POM chemistry is the underlying deceptively simple synthetic procedures.[4] Indeed, the overwhelming majority of the POMs isolated and reported to date in the literature are formed using straightforward "one pot" and one step conditions. Frequently, in this "one-pot" process, acidification of the precursor metal salts in aqueous solutions in the presence of chosen ingredients such as reducing agents, hetero-elements, counterions etc. leads to the formation of a variety of combinatorially linkable polyoxometalate building blocks in solution.[5] These small building blocks, formed by the edge- and corner-sharing of {MOx} units (where M = Mo, W; $x = 4 - 7$), then self-assemble through condensation processes to form various cluster anions depending on the reaction conditions. In fact, these "one-pot" reactions can hide very complex solution speciation processes arising from the presence of a highly dynamical situation, that depends on a multitude of factors such as pH, ionic strength, reaction time, temperature, counterions, concentration of starting materials, presence of electron donors and their redox potential. However, recently, in our work we have shown how cation exchange reactions can be used not only to control assembly of the building blocks for POMs, but also direct their assembly into clusters, clusters-of-clusters, colloids, micro-structures and cellular compartments (see Fig. 1).

What makes understanding, not to mention control of the synthesis of POM clusters so difficult, is the sensitive dependence of the cluster architectures assembled with the reaction *and* crystallization conditions (see Fig. 2). This means that a barely detectable change in any given parameter

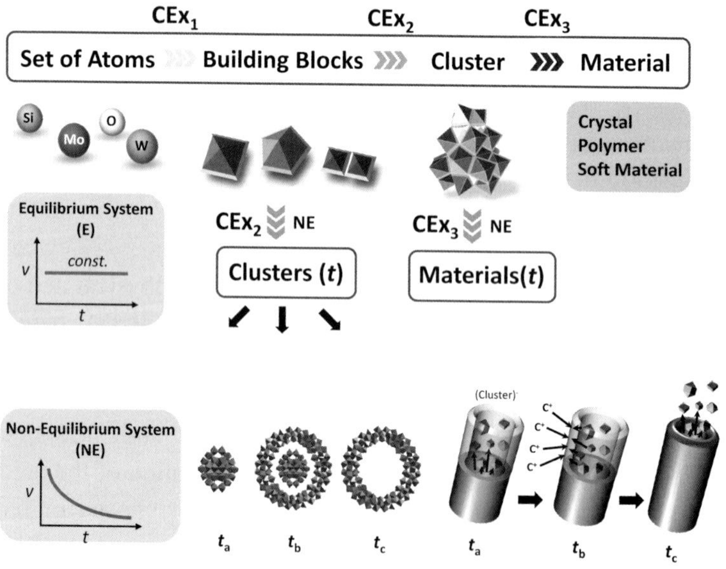

Fig. 1. Schematic representation of the top-down approach of POM synthesis under equilibrium and non-equilibrium conditions. The non-equilibrium approach allows the system to form time-dependant structures otherwise impossible to synthesize and/or characterize (t_a, t_b, t_c).

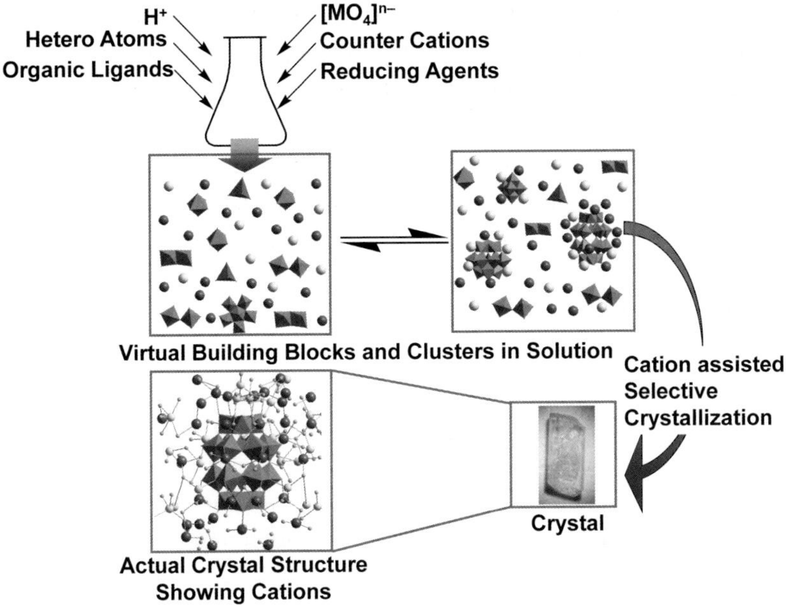

Fig. 2. Schematic representation of the traditional "one-pot" synthesis of POM clusters leading to the formation of various structural archetypes in solution highlighting the role of cations in selective crystallization of a particular product. Yellow and green spheres represent different cations. The top left shows a range of "virtual" geometrically possible units and the top right shows the cations complexing the clusters and pulling certain structures out of solution. **See Color Plate 6**.

can alter the solution speciation affecting the outcome of the reaction.[1] This fact is exemplified in the assembly of reduced Mo clusters; ring shaped "Mo-Blues" and spherical shaped "Mo-browns", in which the building blocks *viz.* $\{Mo_1\}$, $\{Mo_2\}$, $\{(Mo)Mo_5\}$, $\{Mo_8\}$, that are formed in the reaction medium self-assemble to form very large molecular clusters of varying structural diversity ranging from balls, wheels, capped wheels and large "lemon" shaped clusters depending upon the specific reaction conditions.[6] However, the common aspect in these clusters is the "closed" nature of the structures being of spherical or toroidal in topology, and this aspect has been the subject of much discussion.[5]

From a crystal engineering point of view, one important factor that affects the formation of one particular POM species out of vast library of candidates relates to the crystallization process itself. This point is brought into sharp focus when one realizes that, although many researchers focus on the POM framework exclusively, these POMs are poly-anions and cannot exist without the charge balancing cations, which often define the network into which the anion is "complexed" and electro-neutrality is achieved. In this way the cations themselves appear to be able to re-route or control the complex, since almost an infinite number of species and equilibria are available in solution that are removed from solution to the solid state (see Fig. 2). Since the properties of the cations such as size, charge, symmetry, solubility *etc.* are found to modulate the reactivity of POM building blocks, these cations can affect the nature of the product obtained from a POM synthesis.[7] A proper match between the cations and anionic POM species is needed for effective crystallization; hence most often the product isolated in crystalline form may not necessarily be the one with highest abundance in solution.[1b] Figure 3 depicts the distribution of cations, anionic POM clusters and their combined lattice arrangement in a true POM crystal structure, which highlights the vast number of counter cations that are present in the structure.

Modern crystal engineering is a hybrid of supramolecular chemistry and materials chemistry, dealing with the design and synthesis of

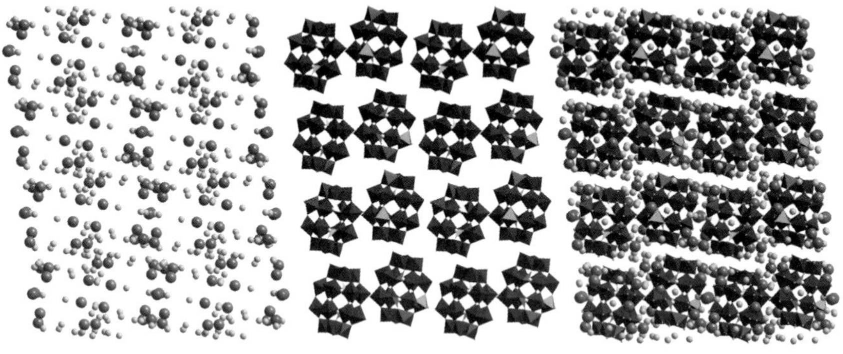

Fig. 3. Packing diagrams depicting the separate and combined arrangements of cations and POM cluster anions in a crystal lattice. *Left*: only cations (blue and green spheres) and solvent water molecules (rose spheres); *Middle*: only POM clusters (violet and yellow polyhedra) and *Right*: POM cluster anions, counter cations and solvent water molecules together in a crystal lattice. **See Color Plate 6**.

crystalline materials having the desired solid state properties. As POM chemistry is now very well established after several decades of research including intense underpinning structural studies, the initial structural and topological curiosity has been replaced by the quest for new tuneable and functional clusters capable of exhibiting novel, and controllable physical properties.[4] This means that the crystal engineering of POM-based materials should be directed towards the incorporation of well-conceived functionalities into such systems in order to develop the materials aspects of POM chemistry. For example POM synthesis, employing solution processing methods, has great potential to allow the assembly of new solid state materials using a molecular design approach.[9–12]

This chapter briefly outlines our recent efforts to control the self assembly of POM-based nanoscale molecules, and their assembly into nanostructures and solid state materials. Given that POMs are gigantic polyanions, the manipulation of the self-assembly process using cations is vital, and perhaps could provide the basis to design, using a crystal engineering type approach, pre-defined POM-based materials.[9] We reasoned that two things are most important here — first, the discovery of novel POM building blocks and second, the direction of their self-assembly in a controlled way to form useful materials. To achieve these targets, we are using a simple strategy, which is based on the use of bulky organic amine cations as counter-ions in POM synthesis.[9–12] The use of such bulky cations helps to isolate self-assembled POM species in one-pot reaction systems thus preventing their rapid aggregation into clusters having stable uniform spherical topology. Also such cations together with other linker units are found to be capable of directing the self-assembly of POM building blocks into extended structures. We termed this approach as "Shrink wrapping"[9–12] and involves mainly the use of organic amines such as hexamethylene tetramine (HMTA), triethanol amine (TEA), *N*,*N*-bis-(2-hydroxyethyl)-piperazine (BHEP), morpholine etc. which are capable of acting as well defined (encapsulating) cations, as well as ligands (although at low pH it is unlikely these amines act as ligands), buffers, and even as redox reagents in some cases (see Scheme 1). Further, we were able to produce a number of discrete iso- and heteropolyoxo clusters as well as many extended architectures using this simple but efficient concept, described briefly in the following sections.

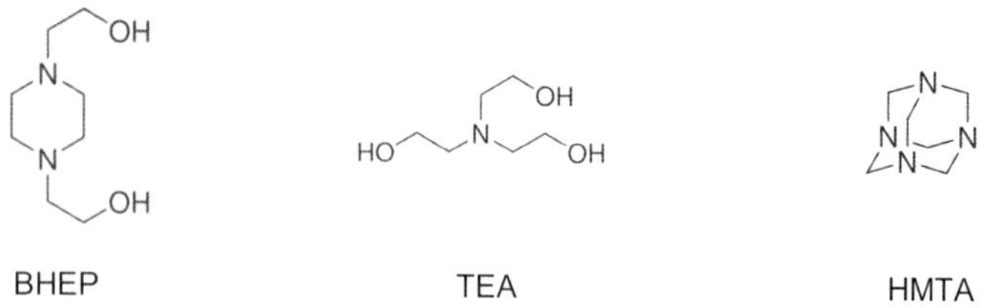

Scheme 1. The various shrink-wrapping cationic precursors used in our studies.[9–12]

1.1. *Background*

The effective control over the self-assembly of $\{MO_x\}$ units in the traditional "one-pot" synthesis has always been the main aim for researchers looking for new POM clusters. In the case of reduced Mo-based clusters often called "Mo-blues", the work of Müller *et al.* showed that it was possible to generate and control the self-assembly of some building units in solution by varying the synthetic conditions.[5] One of such building block is the pentagonal $\{Mo(Mo)_5\}$ unit which is pivotal in the effective tiling of many spherical Archimedean solid structures. These $\{Mo(Mo)_5\}$ building blocks, in the presence of suitable linkers, such as doubly bridging $\{MoV_2O_4(ligand)\}$ units, where "ligand" can be acetate, sulfate etc., leads to giant mixed-valence clusters with diverse topologies such as the spherical icosahedral $\{Mo_{132}\}$, big wheel $\{Mo_{154}\}/\{Mo_{176}\}$, capped cyclic $\{Mo_{248}\}$ and basket shaped $\{Mo_{116}\}$ architectures[5,6] (Fig. 4). One other important achievement with Mo-based clusters, in the context of cluster engineering, is the ability to manipulate the structure, pore size and interior properties of the spherical Keplerate capsules by changing the bidentate ligands of the $\{MoV_2O_4(ligand)\}$ linker.[13] In parallel with these studies in "Mo-blue" chemistry, Pope and co-workers have achieved interesting W-based clusters including the largest tungstate cluster reported so far, by using hetero-transition metal ions as linkers.[14] Apart from Mo and W, which are the two main metals explored in POM chemistry, the engineering of novel cluster types based on other metals like vanadium,[15] tantalum,[16] niobium,[17] uranium[18] and palladium[19] is also being explored with great vigor since these represent possible new routes in the search for new metal-oxide based materials. Further, the pentagonal

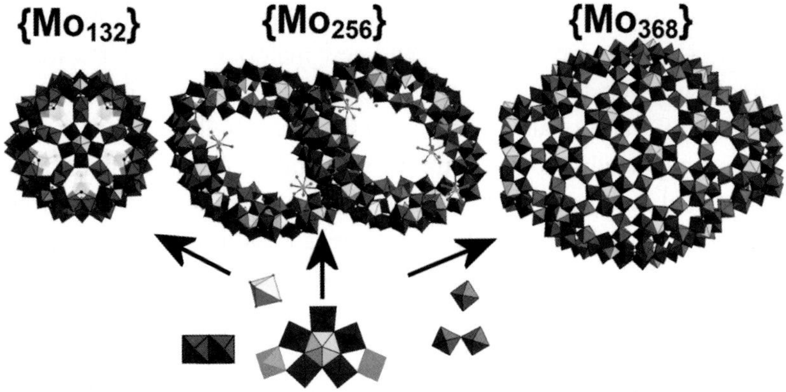

Fig. 4. $\{Mo_1\}$, $\{Mo_2\}$, $\{(Mo)Mo_5\}$ etc. building blocks self-assemble according to the reaction conditions leading to various structurally diverse Mo-blue clusters.[5,6] **See Color Plate 7**.

$\{M(M_5)\}$ building block necessary for the tiling of large Archimedean structures, was recently discovered in the case of W and Nb clusters (rather than just Mo),[20] pointing to the prospect of the isolation of large clusters similar to "Mo-blues" in the case of these metals as well.

The often serendipitous nature of the "one-pot" reactions has prompted researchers to utilize a better defined "building block approach", which utilizes preformed POM clusters as building blocks for the generation of networks or aggregates.[1,4] In this way, POM building blocks are assembled either through M–O–M oxo bridges or by the use of metal–organic complexes as linkers. In fact an almost bewildering number of molybdenum and tungsten oxide-based organic–inorganic hybrid materials are reported by various groups using metal–organic complexes/coordination polymers as linkers between POM clusters, and the number of examples is expanding seemingly at an exponential rate.[21] The metal organic units used as linkers can also act as charge balancing counterions, templates and covalently linked network constituents in these compounds. However, the multiplicity of coordination sites on POM building blocks often makes the product prediction difficult using a "building-block approach" as well similar to the "one-pot" synthetic approach. Further, most of these studies exploit well known clusters and have not led to profoundly new or highly functional materials with new properties. However, many of the reaction systems in this class have used hydrothermal reaction conditions, which has led to some interesting results and new materials that appear to cause a shift in their reactivity from the kinetic to the thermodynamic domain compared to traditional aqueous reactions, such that the equilibrium phases are replaced by more complex metastable phases.[22] Therefore, the hydrothermal method is now commonly employed for the engineering of organic–inorganic hybrid POM materials as a modification of "one-pot" synthesis. Many interesting architectures and frameworks are engineered through hydrothermal synthesis by the careful selection of the metal organic and POM building blocks.[23]

Functionalization of POM clusters with organic units, *via* ion exchange or by direct covalent attachment, is another important strategy towards the development of functional materials as it will enhance the synergistic combination of properties of the polyoxometalates and organic units. By tailoring the organic and/or inorganic part properly, one could design interesting hybrid systems with unique functional properties not normally available in the component parts.[24]

1.2. *Isopolyoxometalates and their derivatives*

Isopolyoxometalates (iso-POMs) are a sub-class of polyoxometalates, which are composed of a metal-oxide framework, but without an internal heteroatom/heteroanion template. As the POM clusters are synthesized by the condensation of $\{MO_x\}$ units (where M = Mo,W; $x = 4 - 7$) at low pH, a number of possible edge and corner-shared $\{MO_x\}$ intermediates are possible, which can then serve as structural components of large polyoxometalates assembled by conservative self-organizing processes. However, it is observed that the self-organization of these building blocks to large structures often requires a template to assemble the cage around, such as a heteroanion, e.g. PO_4^{3-}, thus leading to the formation of heteropolyanions. Because of this, the number of POM structures formed without the templating effect of an heteroanion/heteroatoms is surprisingly limited. For example, compared to the vast class of known heteropolytungstates, only a limited number of isopolytungstates have been reported thus far.[4] It is noted that, solution control by the manipulation of pH as well as cations are very important in the synthesis of iso POMs, and perhaps define an even more interesting target than the heteropolyanions for the use of crystal engineering techniques to control the cluster and crystal architectures.

The use of bulky organic cations to control the self-assembly of building blocks in the synthesis of iso-POMs was successfully leading to the development of a new family of Mo-based iso-POMs. By utilizing protonated hexamethylenetetramine (HMTAH, $C_6H_{13}N^{4+}$) as counterion, the cluster compound $(HMTAH)_{10}[H_2Mo_{16}O_{52}]\cdot 34H_2O$, $\{Mo_{16}\}$, was isolated as crystalline precipitate from an acidified molybdate solution (pH 4) in the presence of a reducing agent in 20–30% yield.[9] This cluster has a "flat" geometry and four of the sixteen Mo centers are one-electron reduced. The main body of this "bat-shaped" cluster consists of a central unit with 12 molybdenum atoms and two "wings" each with two molybdenum atoms (giving the formulation $[\{Mo_{12}\} + 2\{Mo_2\}]$). The role of the "shrink-wrapping" cation in the encapsulation of this rare non-spherical POM can be understood from the very close interaction between surface oxoligands of the cluster anion and the HMTAH cations. There are 18 short hydrogen-bonded cluster surface oxygen-to-cation interactions for the anion. This means that the clusters are extensively linked to the HTMAH cations *via* hydrogen bonds. Also, the anion layers in this compound are effectively separated from each other by HMTAH cation layers. Figure 5 clearly illustrates that the cluster anions are effectively covered, or

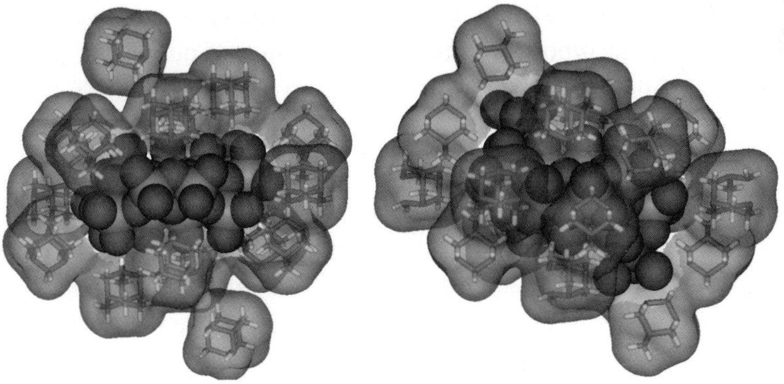

Fig. 5. Model for an actual crystal structure of a POM. The $\{Mo_{16}\}$ isopolyoxometalate cluster anion is closely surrounded by HMTAH cations, effectively "shrink-wrapping" around it.

"shrink-wrapped" by the HTMAH cations in the solid state. Although cation–anion association will be present in solution, encouraged by both electrostatic and hydrogen-bonding effects, the extent of the interactions seen in the present compound in solid state appears to play a crucial role in solution as well. This is because experiments attempting to synthesize the $\{Mo_{16}\}$ cluster under similar reaction conditions (by altering pH value, concentrations, ionic strength, and temperature), but with different large cations (for example $[HN(CH_2CH_2OH)_3]^+$), resulted in the formation of the well-known Dawson-type cluster. Also attempts to dissolve this compound in the presence of other cations appear to result in the decomposition of the cluster anions. These observations strongly support the "Shrink-wrapping" effect of the bulky organic cation HMTAH in the formation of this iso-POM.

This $\{Mo_{16}\}$ cluster anion possesses high nucleophilicity and hence can bind two divalent transition-metal ions (such as Fe^{II}, Mn^{II}, Co^{II}, Ni^{II}, or Zn^{II}) to its framework. This led to a family of isostructural complexes exhibiting the basic topology found for the parent $\{Mo_{16}\}$ cluster with the addition of two $[M^{II}(H_2O)_4]$ moieties yielding a complex of the composition: $[M^{II}_2(H_2O)_8H_2Mo^V_4Mo^{VI}_{12}O_{52}]^{6-}$ (M = Mn, Fe, Co). The encapsulating effect of the "shrink-wrapping" cations is found to prevent any significant inter-cluster magnetic exchange interactions in these compounds.[25] The extension of the "Shrink-wrapping" organic cation strategy to tungsten-based systems has resulted in the trapping and isolation of an interesting isopolytungstate $(TEAH)_9Na_{2-}\{(H_2O)_4K[H_{12}W_{36}O_{12}0]\}\cdot 17H_2O$, which is the largest isopolytungstate reported so far in the literature.[26] This cluster compound was isolated from a reaction mixture comprising Na_2WO_4, $Na_2S_2O_4$ and protonated triethanolamine (TEAH) as organic cation, in water at pH 2. Subsequent experiments have shown that the addition of $Na_2S_2O_4$ is not necessary for the synthesis of this compound. The 3-fold symmetric cluster anion $[H_{12}W_{36}O_{120}]^{12-}$,$\{W_{36}\}$, is comprised of three $[H_4W_{11}O^{38}]^{6-}$, $\{W_{11}\}$, cluster subunits linked together by three $\{WO_6\}$ bridges. The center of this "Celtic ring" cluster has a $\{W_6O_6\}$ ring, that maps extremely well onto the structure of the crown ether 18-crown-6 and holds a K^+ ion at its center.

Focusing on the crown ether like central cavity of $\{W_{36}\}$ cluster anion for host–guest chemistry with alkali and alkali-earth metal ions, complexes of the type $\{MW_{36}\}$ (M = K^+, Rb^+, Cs^+, NH_4^+, Sr^{2+} and Ba^{2+}) have been successfully synthesized.[27] These complexes demonstrate that, $\{W_{36}\}$ cluster anion can act like a type of inorganic "crown ether" with similar preferences to 18-crown- 6, but with much greater rigidity, and therefore, having the potential to distinguish between different cations. In a further extension of this work, a range of protonated aliphatic and aromatic guest cations such as 2-phenethylamine, 4-phenylbutylamine, *p*-xylylene diamine and 1,6-diaminohexane were successfully locked into the crown-ether like cavity of this iso-POM, and their structure directing effects on the resulting framework arrangements are probed. It was found that parts of the organic guest cation protrude from the central binding cavity of the cluster and that the nature of this protruding organic "tail" has a directing ability and is in fact able to control the solid-state structure of the resulting compounds (see Fig. 6).[28]

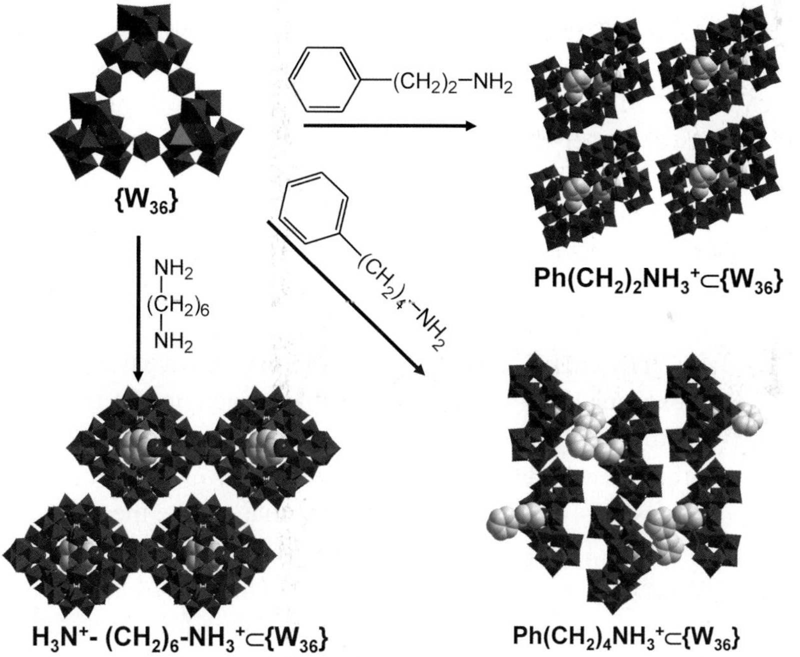

Fig. 6. Illustration of the framework arrangement of some host guest complexes prepared from inorganic crown ether analogue $\{W_{36}\}$ cluster and various amines. The effect of protruding amines in directing the self-assembly leading to various framework arrangements can be visualized from these projections. All the frameworks are viewed along crystallographic a axis.

The synthesis of $\{W_{36}\}$ cluster showed that the presence of inorganic species such as $Na_2S_2O_4$ in the reaction medium could be crucial in the isolation of similar isopolytungstates. But unlike the isopolymolybdates, the influence of such inorganic species on the assembly of isopolytungstates has not been well-documented in the literature. In an attempt to discover radically new cluster architectures in the isopolytungstate family, we have found that, acidification of a solution of sodium tungstate to pH 3.4 in presence of $Na_2S_2O_4$ could lead to the isolation of a "S"-shaped $[H_4W_{22}O_{74}]^{12-}$ cluster anion, $\{W_{22}\}$, and on lowering the pH further (down to pH 2.4) could also lead to a related "§"-shaped $[H_{10}W_{34}O_{116}]^{18-}$ cluster anion, $\{W_{34}\}$.[29] It is noteworthy that we have also observed that these new isopolytungstates could only be isolated in good yield in the presence of the inorganic species Na_2SO_3, even though it is not included in the crystal structure.

All of the above isopolytungstates $\{W_{36}\}$, $\{W_{34}\}$ and $\{W_{22}\}$ contain a basic building block: $\{W_{11}\}$ ($[H_4W_{11}O_{38}]^{6-}$), having a fish-net structure, in which two fused half cubanes share one edge with another half cubane, and has two corner-shared dimeric tungsten-species bridging them together. In $\{W_{36}\}$ cluster, these $\{W_{11}\}$ subunits are linked together by three $\{WO_6\}$ units to give the celtic ring shaped cluster having a crown-ether type cavity at the center. In the $\{W_{22}\}$ cluster, two $\{W_{11}\}$ building units are directly linked together to give the "S" shaped cluster, while in the $\{W_{34}\}$ cluster, two identical $\{W_{11}\}$ subunits are connected together in a *trans* fashion by a $[H_2W_{12}O_{42}]^{10-}$, $\{W_{12}\}$ unit through two μ_{2-}oxo bridges to give an expanded "§"-like architecture. The evolution of the clusters from $\{W_{11}\}$ to $\{W_{22}\}$, $\{W_{34}\}$ and $\{W_{36}\}$ is very important (Fig. 7), as it leads towards a new building block principle for developing new clusters of isopolytungstates. It should be noted here that, unlike the $\{W_{36}\}$ cluster, the $\{W_{22}\}$ and $\{W_{34}\}$ clusters are formed without the presence of any shrink wrapping cations. But it is intriguing to observe that the approximately C_{3v} symmetric $\{W_{36}\}$ cluster anion is formed in the presence of shrink-wrapping cation TEAH having approximately C_{3v} symmetry, the absence of which led to the formation of linear structures such as $\{W_{22}\}$ and $\{W_{34}\}$. This observation could be pointing to the structure directing role of the "shrink-wrapping" cations in the synthesis of such iso-POM species.

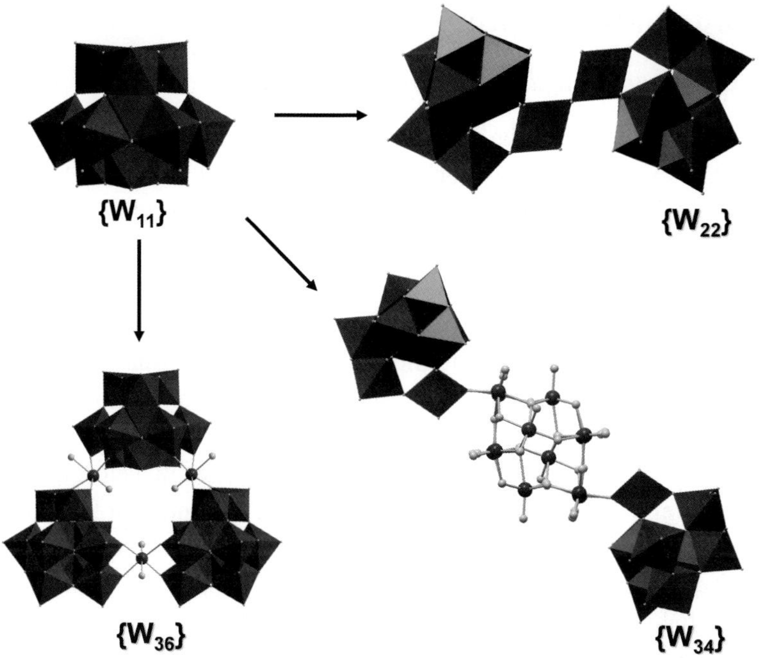

Fig. 7. Formation of three isopolytungstate formed by the different linking modes of same basic building block $\{W_{11}\}$. $\{WO_6\}$ units (violet polyhedra), W, (violet spheres), O (rose spheres). **See Color Plate 7**.

"Shrink-wrapping" cations could also lead to isopolytungstates that are structural analogues of heteropoly acids. For example, a reaction system comprising tungstates and TEAH cations at lower pH values (around 0.8) yielded a fundamentally new type of isopolytungstate isolated as $(TEAH)_6[H_4W_{19}O_{62}]$, $\{W_{19}\}$.[30] This Dawson structural archetype incorporating a $[WO_6]^{6-}$ anion template is a rare and unprecedented cluster type. Furthermore, the stabilization of the $\{WO_6\}$ moiety in a trigonal prismatic coordination environment was new in polytungstate chemistry. In the absence of the bulky organic TEAH cations, under otherwise identical reaction conditions, only the well known $[W_{10}O_{32}]^{4-}$ cluster compound was formed, therefore implying the existence of a crucial cation effect on the formation of this new isopolytungstate.

2. Host-Guest Chemistry to Networks

2.1. *Cage compounds and switchable POMs*

Many POM clusters can be thought of as molecular cages enclosing some particular species at its center. For instance the conventional Dawson archetype incorporates two tetrahedral anions such as PO_4^{3-}, AsO_4^{3-}, or SO_4^{2-} at the center, while the Keggin clusters incorporate only one such anion. Incorporation of a redox-active species inside a POM cage would be highly desirable in this respect, in view of the development of switchable POM molecules. In an attempt to develop a POM based switchable system, we targeted the incorporation of some non-conventional templating units, e.g. sulfite ions, inside a Dawson cage. The hypothesis for this investigation was that the confinement of such species in a closed molecular cage could lead to properties arising from the intra-molecular electronic interactions between the encapsulated anions. Moreover, this could also lead to the manipulation of the physical properties of the $\{Mo_{18}\}$ Dawson type POM cluster. The incorporation of two sulfite ions as templating moieties inside a Dawson-like cluster is achieved from a reaction mixture at pH 4.0 containing metal salts, excess of protonated triethanolamine (TEAH) cations and $Na_2S_2O_4$ as a reducing agent as well as the source of SO_3^{2-}.[10] The product is crystallized as $(TEAH)_6[Mo_{18}O_{54}(SO_3)_2]\cdot 4H_2O$ which incorporates two pyramidal sulfite SO_3^{2-} ions as the central cluster templates.

This new cage compound represents the first Dawson-like $\{M_{18}\}$ cluster incorporating the pyramidal sulfite anion, and the first that includes two such pyramidal anions in the same $\{M_{18}\}$ cage. The two sulfite anions show a short S–S interaction inside the cluster cage and hence unusual electronic properties. This new "non-classical" Dawson-like cluster also shows interesting thermochromic behavior between 77 K (pale yellow) and 500 K (deep red). In an extension of this approach, the first two examples of polytungstate clusters incorporating two sulfite anions, $(TBA)_4[W^{VI}{}_{18}O_{54}(SO_3)_2]$ and $K_7Na[W^{VI}{}_{18}O_{56}(SO_3)_2(H_2O)_2]\cdot 20H_2O$ have been successfully synthesized, where TBA = tetrabutylammonium cation.[31] The cluster $K_7Na[W_{18}O_{56}(SO_3)_2(H_2O)_2]\cdot 20H_2O$ has a curious structure incorporating two terminal water ligands which was unknown for a closed-shell hetero-polyoxometalate cluster (HPOM clusters) of any type. Moreover, this compound demonstrated new electronic properties. The cluster anion, $[W_{18}O_{56}(SO_3)_2(H_2O)_2]^{8-}$, undergoes a unique electron-transfer reaction when heated, in which a structural re-arrangement allows the two embedded pyramidal sulfite (SO_3^{2-}) anions to release up to four electrons (analogous to the "soldiers" hidden inside the "Trojan Horse") to the surface of the cluster generating the sulfate-based, deep blue, mixed valence cluster $[W_{18}O_{54}(SO_4)_2]^{8-}$. Although electron-transfer reactions and structural rearrangements are well-known for hetero-POMs, this is the first example of a coupled structural rearrangement and electron-transfer process, whereby the electrons are "released" from the core of the cluster. It is also the first example of a fully characterized unimolecular reaction involving an HPOM.

The successful trapping of one $\{WO_6\}$ unit inside a Dawson-like cage leading to the formation of the iso-POM $(TEAH)_6[H_4W_{19}O_{62}]$ suggested the possibility of replacing the central $\{WO_6\}$ moiety of this system with other similar $\{XO_6\}$ species, for example X = Pt^{IV}, Sb^{V}, Te^{VI}, or I^{VII}. Encapsulation of such templates in very high or very low oxidation states should significantly affect the acidic, catalytic, and redox properties of the resulting cluster systems. Following a synthetic strategy similar to that of the $\{W_{19}\}$ iso-POM, we were able to embed iodine(VII) into a Dawson-like $\{W_{18}O_{54}\}$ matrix as the periodate species IO_6^{5-} to yield a range of cluster salts, $K_6[H_3W_{18}O_{56}(IO_6)]\cdot 9H_2O$, $(TPA)_6[H_3W_{18}O_{56}(IO_6)]$, and $(TBA)_6[H_3W_{18}O_{56}(IO_6)]$, (TPA = tetrapropylammonium cation).[32] The formation of $[H_3W_{18}O_{56}(IO_6)]^{6-}$ appears, at least empirically, to be induced by the IO_6^{5-} anion (which contains a three-fold symmetric axis), although the cluster is produced in a modest yield. Prompted by this observation, we undertook the synthesis of $[H_3W_{18}O_{56}(IO_6)]^{6-}$ in the presence of three-fold symmetrical TEAH cation, and obtained $(TEAH)_6[H_3W_{18}O_{56}(IO_6)]$ in higher yields. This result is in tune with our previous observations that the formation of 3-fold symmetrical $\{W_{19}\}$ cluster, $[H_4W_{19}O_{62}]^{6-}$, requires the presence of three-fold symmetrical TEAH cations (the well-known $[W_{10}O_{32}]^{4-}$ cluster is formed in the absence of TEAH)[30] and hence highlights the cation effect in the formation of these POM cage clusters.

Looking for other non-conventional switching systems, we also targeted the encapsulation of TeO_6^{6-} units within $\{W_{18}O_{54}\}$ cluster shells to give a tellurate-based Dawson-like anion.[33] Unlike our previous experiences, this did not appear to be straightforward, and necessitated techniques such as cryospray mass spectroscopy (CSI-MS) to screen the reaction systems for suitable cations. Of the cations screened, dimethylammonium (DMAH) and tetrabutylammonium (TBA) were found to be successful, yielding the Te^{VI}-encapsulated Dawson-like cluster anion $[H_3W_{18}O_{56}(Te^{VI}O_6)]^{7-}$ as $Na(DMAH)_6[H_3W_{18}O_{56}(Te^{VI}O_6)]\cdot 14H_2O$ and $(TBA)_7[H_3W_{18}O_{56}(Te^{VI}O_6)]\cdot 4CH_3CN$ respectively. We also investigated the redox properties of $[H_3W_{18}O_{56}(Te^{VI}O_6)]^{7-}$ to see if it was possible to access a Te^{IV}-based species by reducing the inner hetero atom. On addition of solid $Na_2S_2O_4$ to the acidified solution of $[H_3W_{18}O_{56}(Te^{VI}O_6)]^{7-}$, the solution initially turned blue, owing to the reduction of the $\{W_{18}\}$ based cluster shell, and then became pale yellow within seconds. We postulated that the central Te was reduced in this process from Te^{VI} to Te^{IV}. This reduction was confirmed by the isolation of $[H_3W_{18}O_{57}(Te^{IV}O_3)]^{5-}$, by precipitation with TBA, and this material on crystallization from acetonitrile followed by single-crystal X-ray diffraction analysis revealed that the Te^{IV} center shifts by 1.10 Å towards one end of the cluster to form a pyramidal tellurite ion, see Fig. 8. Hence, upon formal two-electron reduction, the

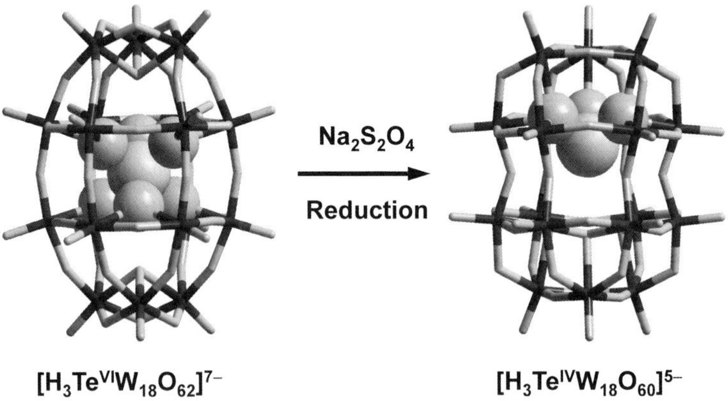

Fig. 8. Structural transformation as a result of the reduction of Te^{VI} to Te^{IV} inside a tungsten based molecular cage. **See Color Plate 8.**

central octahedral $\{Te^{VI}O_6^{6-}\}$ unit is transformed into a pyramidal $\{Te^{IV}O_3\}^{2-}$ unit with the breaking of three Te–O bonds. Two interior "capping" oxo ligands were also lost during this reduction process. This redox reaction can be monitored *in situ* by the use of CSI-MS analyses. Further, it was observed that the inclusion of the tellurium-based heteroanion activates the surface of the $\{W_{18}O_{54}\}$ cluster cage, facilitating the assembly of the clusters to nanoscale structures without the introduction of other transition metal electrophiles. The self-assembled nano structure thus isolated in the solid state, $Na_8(DMAH)_{18}[H_{10}Te^{VI}{}_2W_{58}O_{198}]\cdot 35H_2O$, contains two $\{W_{18}Te^{VI}\}$ anions linked together by two $\{W_{11}\}$ units.

The role of cations is found to be important in the isolation of the above Te-based clusters as well. $Na(DMAH)_6[H_3W_{18}O_{56}(Te^{VI}O_6)]\cdot 14H_2O$ is achieved from a reaction system containing Na_2WO_4 and $Te(OH)_6$ at pH 0.8 in the presence of DMAH. The presence of DMAH is crucial since its absence leads exclusively to the formation of the Anderson-type cluster $[TeW_6O_{24}]^{6-}$, which reiterates the pivotal role of organic cations for the assembly of new POMs.[3] Inclusion of other transition metal based hetero anions as templates inside a Dawson-like cage has also been attempted. This has been achieved in the case of mixed metal (V/Mo and V/W) Dawson-like capsules having a unique composition $\{M_{17}V_3\}$.[34] A family of clusters which can be formulated as $V_2A\{M_{17}V_1\}$ having the α-Dawson cluster framework have been synthesized and the synthetic approach involves the aqueous reaction of $Na_2MO_4\cdot 2H_2O$ (M = Mo or W) and NH_4VO_3 in presence of protonated triethanolamine (TEAH) as "shrink-wrapping" cation. A variety of experimental techniques have been tested to synthesize these compounds including hydrothermal synthesis. It is interesting to note that in all these cases, each of the $\{M_{17}V_3\}$ cluster anions obtained contain two $\{V^VO_4\}$ vanadate templates with one $\{V^{IV}O\}^{2+}$ vanadyl group integrated into the $\{M_{18}\}$ cluster framework. Figure 9 shows a series of Dawson-like cluster cage compounds developed in our group so far incorporating various guest species.

2.2. *POM based architectures and framework materials*

The last couple of decades have witnessed an enormous growth of open framework inorganic materials. The majority of these include metal–organic frameworks containing discrete inorganic building blocks; the

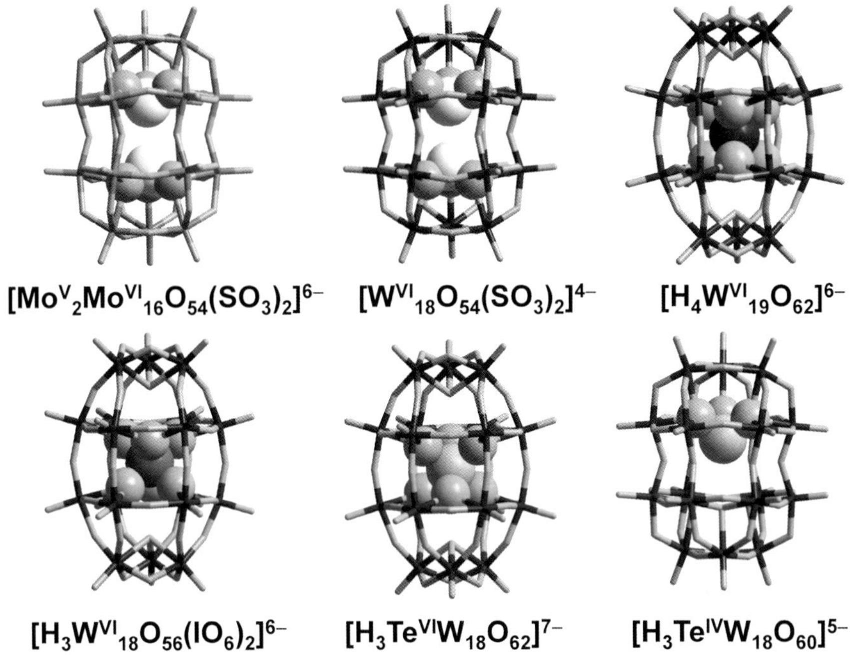

Fig. 9. Some of the Dawson-type cage compounds developed in the Cronin group incorporating various guest systems.

geometry and cavities of such frameworks mainly depending on the nature of the building units.[35] Often, POM clusters are used as large inorganic building blocks in an attempt to expand the lattice of such framework materials.[36] In this respect, we have successfully combined the "shrink-wrapping" strategy and "building block approach" with the help of suitable linker units resulting in diverse polymeric architectures and robust framework materials based on some well-known POM building blocks. For instance, we have utilized silver(I) to act as a linker in POM chemistry due to its versatile range of geometries and bonding modes.[37] In this way, using β-octamolybdate as building unit and bulky TBA as cations, two 1D chain structures, $(TBA)_{2n}[Ag_2Mo_8O_{26}]_n$ and $(TBA)_{2n}[Ag_2Mo_8O_{26}(CH_3CN)_2]_n$ have been formed by the condensation with Ag^I salts.[11] In these 1D chains, the flexible TBA cations nearly completely wrap around the linear chain of linked $[Ag^IMo^{VI}_8O_{26}Ag^I]^{2-}$ units effectively isolating them from one another. The mode in which the $\{Ag(Mo_8)Ag\}$-type "units" assemble to create the overall solid state architecture was also found to be critically dependent on the reaction conditions. On changing the cations from TBA to small cations, the steric shielding and/or "shrink-wrapping" ability of organic cations is reduced leading to a partial disruption of the "insulation" and as a result, the electrophilic polymolybdate chains could be interlinked. This leads to a grid-like structure $(TBA)_{2n}[Ag_2Mo_8O_{26}(DMSO)_2]_n$ and a 2D array, $(HDMF)_n[Ag_3(Mo_8O_{26})(DMF)_4]_n$, where DMSO = dimethylsulfoxide and HDMF = protonated dimethylformamide. In the latter, the relatively bulky TBA ions are exchanged for HDMF cations thereby allowing the chains to condense to a 2D array. Interestingly, the use of Ph_4P^+ ions instead of TBA cations leads to a discrete cluster $(Ph_4P)_2[Ag_2Mo_8O_{26}(DMSO)_4]$. These compounds clearly demonstrate that suitable combination of linkers and bulky organic cations can effectively control the self-assembly of POM building blocks thereby resulting in interesting architectures (see Fig. 10).

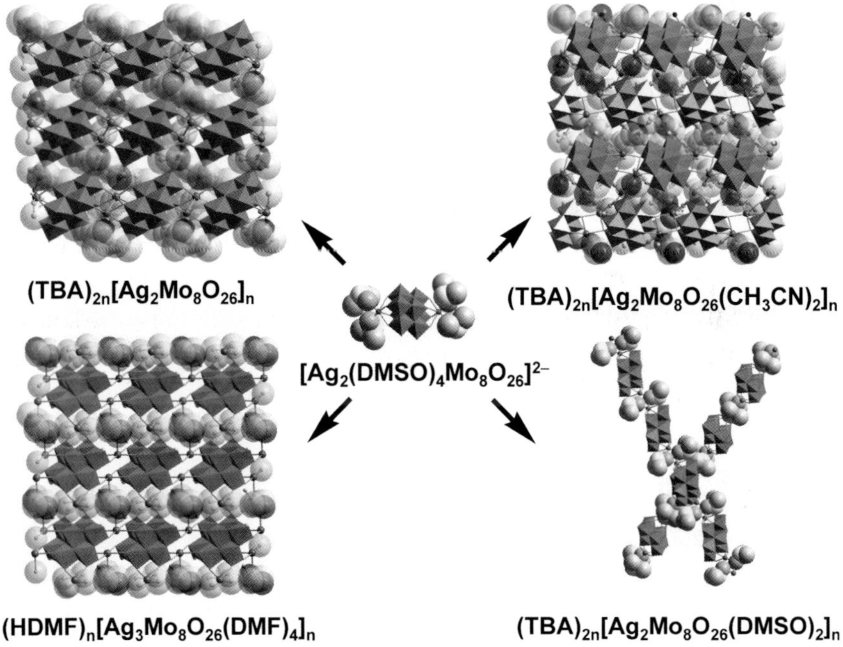

Fig. 10. Summary of the structures isolated from the silver-octamolybdate reaction system. The solvents and counterions are represented in space fill models to highlight their influence in determining the overall topologies achieved which include monomeric structures, 1D chains, grid and 2D network as shown. Color scheme: $\{Mo_8\}$ polyhedra orange, O (rose), C (grey), N (blue), S (yellow). **See Color Plate 8.**

Further studies on silver-octamolybdate systems by varying solvents and counter ions led to a new set of interesting POM architectures.[38] In these structures, formation of $\{Ag_2\}$ linking units is realized and the fine-tuning of Ag–Ag distances in these linkers is achieved with the help of different coordinating solvents. Five compounds based on silver octamolybdate building blocks have been isolated, including an uncommon 0D $(Ph_4P)_2[Ag_2(CH_3CN)_2(Mo_8O_{26})]$, three 1D polymeric chains $([Ag(C_7H_{12}O_2N)(CH_3CN)]_{2n}$ $[Ag_2(CH_3CN)_2(Mo_8O_{26})]_n \cdot 2CH_3CN$, $(Ph_4P)_{2n}[Ag_2(DMF)_2(Mo_8O_{26})]_n \cdot 2DMF$, and $(DMAH)_{2n}[Ag_{2-}$ $(DMF)_2(Mo_8O_{26})]_n \cdot 2DMF$, and a 2D cross-linked network $[(Ag(DMF))_2(Ag(DMF)_2)_2Mo_8O_{26}]_n$. Figure 11 depicts some of these architectures highlighting the crucial role of counter ions and solvent molecules in the isolation of such architectures. Therefore, these studies on silver-octamolybdate systems comprehensively describe the

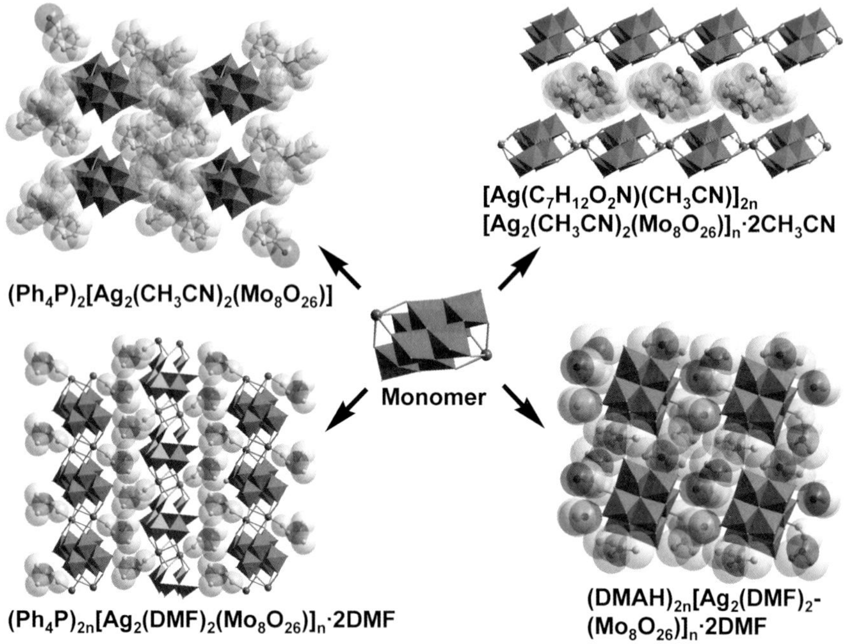

Fig. 11. More examples of silver-octamolybdate based architectures isolated as a result of solvent and counter ions influencing self-assembly. Color scheme: $\{Mo_8\}$ polyhedra (orange), O (rose), C (grey), N (blue), P (pink). **See Color Plate 9**.

synthetic strategies that allow the formation of 0D, 1D and 2D structural assemblies starting from a single POM building block $\{Mo_8\}$, and the flexible silver(I) cation linker.

The above studies on silver-octamolybdate systems have also shown that, careful control over solvent and ligand systems can result in the formation of structurally supporting $\{Ag_2\}^{2+}$ building blocks. Exploiting the potential of such silver(I) dimers to act as linkers between larger isopolyoxometalates, we were able to develop a purely inorganic 3D porous framework.[39] The resulting "POM based open framework" or "POMOF", $[Ag(CH_3CN)_4]A\{[Ag(CH_3CN)_2]_4[H_3W_{12}O_{40}]\}$, is made up of two principal SBUs, protonated α-metatungstate clusters $[H_3W_{12}O_{40}]^{5-}$ ($\{W_{12}\}^{5-}$) and dimeric $\{[Ag(CH_3CN)_2]_2\}^{2+}$ ($\{Ag_2\}^{2+}$) bridging units. These building units are assembled such as to enclose two sets of collinear channels in the crystal lattice. *Td*-symmetric $[Ag(CH_3CN)_4]^+$ units are located in these channels and appear to act as templates in the self-assembly of this framework material. One important key to this framework formation is the ability of the Ag^+ to self-organize into $\{Ag_2\}^{2+}$ dimers with short silver–silver contacts which are stabilized by argentophilic metal–metal interactions. Each dimeric $\{Ag_2\}^{2+}$ motif cross-links four POM building blocks. It was observed here that, in order to encourage the formation of the Ag…Ag interactions, coordinating solvents must be avoided, because the use of such solvents resulted in the capping of Ag centers yielding low dimensional structures. This material shows reversible sorption capabilities allowing the sorption and desorption of small organic molecules. The assembly of such purely inorganic POM-based porous frameworks is highly desirable in materials science, because they are expected to combine the thermodynamic stability of zeolites and mesoporous silicas with the sophistication and versatility of metal–organic frameworks (MOFs).

The effect of the structural features of the organic cations on the supramolecular structure of the POM framework is examined in the case of polyoxomolybdenum(V) phosphate clusters $[CH_6N_3]_{12}[Mo_{12}O_{62}H_{11}NaP_8]\cdot 11H_2O$, $[C_4H_6N_3]_{14}[Mo_{12}O_{62}H_9NaP]\cdot 13H_2O$ and $[C_2H_{10}N_2]_6[Mo_{12}O_{62}H_9Na_3P_8]\cdot 18\ H_2O$ containing sodium-linked dimeric $\{Mo_6\}_2$ moieties as inorganic building blocks and guanidinium, 2-aminopyrimidinium or protonated ethylenediamine respectively as structure-directing amine cations.[40] It was found that in these compounds, the rigid amines guanidinium and

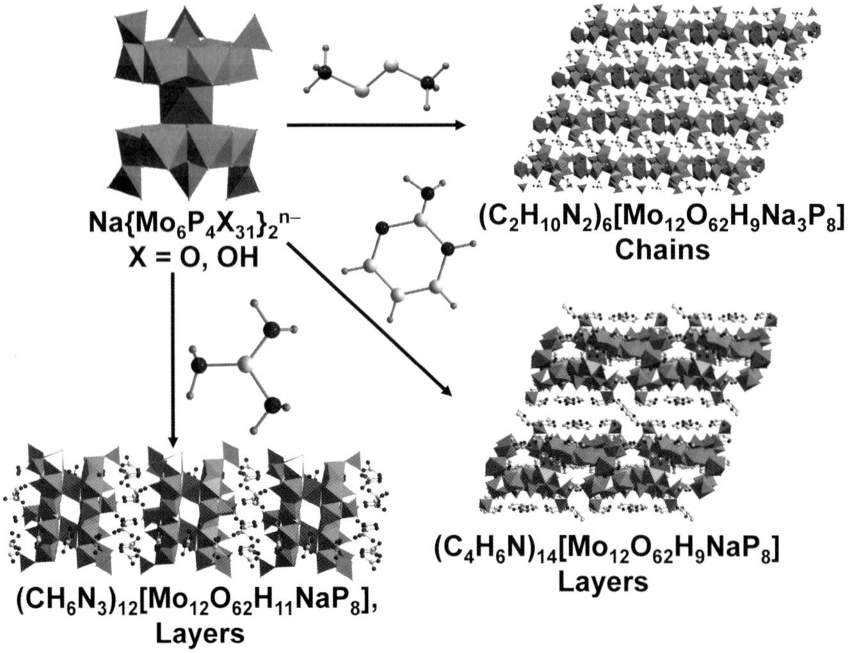

Fig. 12. Summary of the formation of layered or chain-like structures from $\{Mo_6\}_2$ moieties depending on the amine used in the synthesis.

2-aminopyrimidinium induced the formation of layered arrangements, whereas structurally more flexible ethylenediamine resulted in the formation of polymeric chains, Fig. 12. This study is therefore one of the best examples which shows that the geometry, rigidity and hydrogen bonding ability of organic amine cations can influence the overall structure of the POM supramolecular framework.

The transfer of properties from small building blocks to the large POM cluster assemblies is an interesting aspect in POM chemistry that is ripe for expansion/concrete development. For example, Craig Hill and co-workers have achieved chirality transfer from smaller enantiopure organic molecules such as tartrate onto POM clusters leading to chiral, nonracemizing and enantiomerically pure polytungstate clusters.[41] Extending this concept into large POM based supramolecular frameworks, we have successfully developed a chiral framework material having sorption capabilities. We used diprotonated (2)-sparteine as the chiral cation and the dimeric polyoxomolybdenum (V) phosphate anion $[H_{15}Mo_{12}NaO_{62}P_8]^{8-}$ as the POM building block, resulting in the porous chiral hybrid framework $(C_{15}H_{28}N_2)_4(C_9H_6O_6)_1[H_{15}Mo_{12}NaO_{62}P_8]\cdot 10H_2O$.[42] In this chiral framework, each dimeric $\{Mo_6\}$ POM cluster is shielded by four (2)-sparteine cations which form N–H···O hydrogen bonds to the cluster anion (minimum distance *d*N–H···O = 2.67 Å). The structure is further reinforced by a complex network of hydrogen-bonded water molecules and incorporates chiral channels along the crystallographic *a* axis, capable of accommodating organic molecules like benzene tricarboxylic acid and water molecules. The porous nature of this compound is confirmed by reversible water sorption properties.

In our attempt to develop POM based chiral frameworks, we have also used large chiral metal–organic macrocation as one of the starting material. Fe-acrylate based chiral macrocation $[Fe_3(\mu_3\text{-}O)(CH_2\ CHCOO)_6(H_2O)_3]^+$ is reacted with trilacunary Dawson cluster $[a\text{-}P_2W_{15}O_{56}]^{12-}$, $\{P_2W_{15}\}$, under normal reaction conditions. This macrocation is found to decompose slowly in the presence of $\{P_2W_{15}\}$ and in doing so, helps to buffer the reaction medium leading to the formation of a nanosized tetrahedral polytungstate molecule $K_{21}Na_8[KFe_{12}(OH)_{18}(\alpha\text{-}1,2,3\text{-}P_2W_{15}O_{56})_4]\cdot 70H_2O$.[43] This 21 kDa POM cluster is found to be water-stable as was directly observed in solution/gas phase using electrospray mass spectrometry. In the crystal lattice, the tetrahedral anionic cluster units are inter-connected *via* external potassium ions to form an extended 3D network in the solid state with an accessible solvent volume of 6000 $Å^3$ per unit cell. This study was partly inspired by the works of Mizuno *et al.*, who have extensively used similar metal–organic macrocations for example: $[Cr_3O(OOCH)_6(H_2O)_3]^+$ with POM clusters resulting in interesting framework materials.[44]

Looking for new organic cations as shrink-wrapping agents, we used *N,N′*-bis(2-hydroxyethyl)piperazine (BHEP), which is found to act as a tri-functional species being a cation, ligand and buffer in POM synthesis. The ethanol arms attached to the nitrogen donors of the piperazine ring add flexibility to the ligand and form flexible coordination environment for the ligation of transition metals. BHEP can effectively buffer the pH range 4–8 allowing precise pH control. These properties make BHEP an interesting structure directing cation in POM synthesis. Two new POM architectures, the dimeric cluster unit $[\{PMnW_{11}O_{39}\}_2(PO_4)]^{13-}$ and a 2D network-based on $[P_2Mn_4W_{18}O_{68}]^{10-}$ anions are isolated at pH 6.05 and 6.80 respectively using BHEP as buffer, encapsulating cation and ligand.[12] The dimeric cluster unit, formulated as $(H_2BHEP)_5(HBHEP)Na_2[(PMnW_{11}O_{39})_2\{PO_4\}]\cdot 18H_2O$ is composed of two α-$[PMnW_{11}O_{39}]^{5-}$ clusters connected by a bridging phosphate unit that links the Mn^{2+} centers attached to the mono-vacant Keggin entity. This moiety is stabilized, in the solid state, by the surrounding four H_2BHEP ions. The 2D network, $(H_2BHEP)_3Na_4[P_2Mn_4W_{18}O_{68}]\cdot 15H_2O$, is synthesized using a slight increase of pH and temperature in comparison to the dimeric compound. However, the structure of this new compound was entirely different; consisting of two $\{B\text{–}PW_9O_{34}\}$ units sandwiching four manganese centers between them in a belt like fashion. The synthesis of these compounds reiterates the multifunctionality of the BHEP ligand and the importance of solution pH in determining the nature of the resulting POM architectures under otherwise identical reaction conditions.

The important pH buffering capacity of BHEP is revealed in the synthesis of three Co(II)-containing silicotungstates *viz.* $[Co_3(H_2O)(B\text{-}\beta\text{-}SiW_9O_{34})(B\text{-}\beta\text{-}SiW_8O_{29}(OH)_2)]^{12-}$, $[Co_3(H_2O)(B\text{-}\alpha\text{-}SiW_9O_{34})(B\text{-}\beta\text{-}SiW_8O_{31})]^{14-}$, and $[Co_9Cl_2(OH)_3(H_2O)_9(B\text{-}\beta\text{-}SiW_8O_{31})_3]^{17-}$ which have been observed before only as dimeric or higher nuclearity cluster complexes. The synthetic strategy employed involves the reaction of Co(II) ions, the dilacunary $[\gamma\text{-}SiW_{10}O_{36}]^{8-}$ polyanion and the bulky organic cation BHEP in basic media at pH 8.0–9.0.[45] Although the BHEP ligand is not incorporated in these structures, the fact that these clusters do not form in the absence of BHEP in the reaction medium, led to the conclusion that this amine ligand plays a vital role in the isolation of these intermediate clusters, by acting as a buffer at these high pH values.

A prototype of a 3D polyoxometalate framework based on pure metal oxides and constructed without the use of external linkers to connect the polyanionic framework nodes has been achieved by the reaction of the divacant lacunary polyoxometalate $[\gamma\text{-}SiW_{10}O_{36}]^{8-}$ with Mn(II) in the presence of morpholinium cations and potassium permanganate, under strict pH control. This extended modular framework made up of POM building blocks incorporates "active sites" capable of responding to guest inclusion.[46] The resulting pure metal oxide framework, $[(C_4H_{10}NO)_{40}(W_{72}M_{12}O_{268}X_7)_n]$ ($M = Mn^{III}$, $X = Si$), based upon Mn-substituted Keggin-type POM building

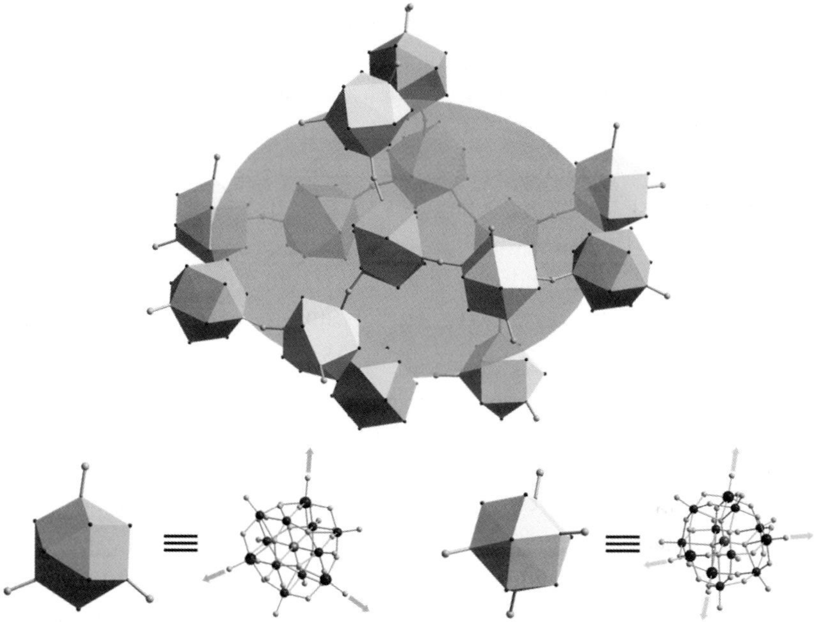

Fig. 13. Illustration of the nanosized pockets in $[(C_4H_{10}NO)_{40-}(W_{72}Mn_{12}O_{268}Si_7)n]$ highlighted by the purple ellipsoid (dimensions: ~ 2.7\2.4\1.3 nm). The trigonal (rose) and tetrahedral (yellow) building block connectors are shown at the bottom; rose and yellow arrows highlight the connecting modes. **See Color Plate 10**.

blocks encloses elliptical pockets of 26.85\23.62\12.93 Å size and house solvent molecules and morpholinium cations in these pockets, Fig. 13. This material can undergo a reversible redox process that involves the simultaneous inclusion of the redox reagent with a concerted and spatially-ordered redox change of the framework. Importantly, this compound can be repeatedly disassembled and reassembled without any structural changes by dissolution in hot water followed by recrystallization. These unique properties define a new class of POM-based materials that bridge the gap between coordination compounds, metal–organic frameworks and solid-state oxides. Furthermore, it has been shown that all the Mn(III) centers in the oxidized form of this compound can be "switched" to Mn(II) using a suitable reducing agent to give the fully reduced framework. Moreover, the redox processes can be precisely followed by the single-crystal to single-crystal transformation between the oxidized and reduced states of the framework.

Extending the building block approach further by using very large building blocks such as a superlacunary $[H_7P_8W_{48}O_{184}]^{33-}$, $\{P_8W_{48}\}$, polyoxoanion and transition metal ions such as Co(II) as linkers, two cyclic cobalt-substituted heteropolyoxometalate frameworks $K_{15}Li_5[Co_{10}(H_2O)_{34}(P_8W_{48}O_{184})]\cdot 54H_2O$ and $K_8Li_{12}[Co_{10}(H_2O)_{44}(P_8W_{48}O_{184})]\cdot 60H_2O$ have been synthesized under strict control of pH and buffer. In both these frameworks, the central crown cavity of $\{P_8W_{48}\}$ is functionalized by a total of six Co(II) ions and additional Co(II) ions are grafted onto then outer rings extending the Co(II)-$\{P_8W_{48}\}$ anions into 1D chains in the former and to a 3D network in the latter.[47] As we have seen from the above examples, the combination of "building block approach" of POM synthesis with "shrink wrapping" cation technique has yielded several novel architectures and framework materials. It is to be noted here that, most of these materials are derived starting from simple POM building blocks such as lacunary Keggin or Dawson cluster derivatives. The most exciting aspect of this approach lies in the possibility of developing some true porous POM based materials starting from large, inherently porous POM clusters such as $\{P_8W_{48}\}$, $\{Mo_{154}\}$ etc. having open cavities at the center.[47]

2.3. *Organic-inorganic hybrids POMs*

An important step in POM chemistry towards functional materials is the development of organic–inorganic hybrid clusters by covalent functionalization which allows the synergistic combination of the properties of metal–oxo clusters and organic units, thereby allowing to fine tune the cluster properties such as solubility, redox properties, biological activities and so on.[24] Out of the various methods used for the covalent functionalization of POM clusters, grafting of tris(hydroxymethyl)aminomethane and its derivatives (denoted hereafter as TRIS) on to clusters is an important approach, as this group of molecules is geometrically well suited for grafting onto Anderson, Wells-Dawson and Lindqvist-type clusters.[48] Many POM hybrids containing TRIS as a linker between organic moieties and POM units are being developed for their potential applications as polymer support, POM based gels, and as light harvesting complexes.[49]

An interesting framework material, assembled purely through hydrogen bonding interactions, is achieved using an Anderson-based organic–inorganic hybrid cluster as the building block.[50] In the development of this framework material, a TRIS tethered pyrene moiety was covalently grafted onto a Mn-Anderson cluster, leading to the isolation of the hybrid cluster $TBA_3MnMo_6O_{18}\{(OCH_2)_3CNH–CH_2– C_{16}H_9\}_2·2DMF·3H_2O$ with properties different from that of pure building blocks. Furthermore, the self-assembly of this $[Mn\text{- Anderson(TRIS-pyrene)}_2]^{3-}$ building block together with TBA cations produces a nanoporous framework with nanoscale solvent accessible 1D channels in the solid state. This material appears to be stable until 240 °C despite being constructed from very weak C–H···O Mo hydrogen bonding interactions. The guest-uptake measurements revealed that the framework is nanoporous and is able to absorb up to 12% by weight of chlorobenzene.

Using TRIS-grafted Mn-Anderson clusters as the building block and Ag cations as linkers, the hybrid compounds $\{Ag_3[MnMo_6O_{18}\{(OCH_2)_3C NH_2\}_2(DMSO)_5]·3(DMSO)\}_n$; and $\{Ag_3[MnMo_6O_{18}\{(OCH_2)_3CNH_2\}_2 (DMSO)_6(CH_3CN)_2]·DMSO\}_n$ have been synthesized. These compounds having 1D chain structure form fibers on the Si surface >10 mm in length with a diameter of *ca.* 0.5 mm indicating that such hybrid clusters have potential for use in the self-assembly of functional materials.[51] Using long alkyl chain cations, the hydrophobicity of certain POM clusters could be increased, thus imparting new materials properties to such clusters. In an attempt to develop amphiphilic hybrids, we have successfully attached long (C-16 and C-18) alkyl chains onto TRIS grafted Mn-Anderson cluster by manipulating the reactivity of the free $–NH_2$ group of the TRIS moiety. Some of these long alkyl chain grafted hybrids indeed exhibited amphiphilicity forming supramolecular vesicles in solution.[52] Further, cations are exchanged from TBA to dimethyldioctadecyl ammonium (DMDOA), inducing new properties to these hybrids such as increased hydrophobicity. It is found that, some of these new materials form interesting microstructures on surfaces and exhibit interesting physical properties including phase transition and nanoparticle formation.[53]

Moving on to larger clusters, we have developed a series of TRIS grafted Dawson-like clusters to study their self-assembly in solid state and in solution. TRIS derivatives containing free $–NH_2$, $–NO_2$ and $–CH_3$ functionalities $H_2N–C(CH_2OH)_3$, $O_2N–C(CH_2OH)_3$ and $H_3C–C(CH_2OH)_3$ were successfully grafted onto the V_3 capped Dawson-like cluster $[P_2V_3W_{15}O_{62}]^{9-}$, resulting in a series of hybrid POM cluster anions having the general formula $[X\text{-}C(CH_2O)_3P_2V_3W_{15}O_{59}]^{6-}$, where X = $–NH_2$, $–NO_2$

or $-CH_3$. It is found that these hybrid clusters are capable of acting as supramolecular building blocks and are capable of self-assembling into larger supramolecular structures through hydrogen bonding interactions between the grafted TRIS moiety and the cluster oxygen atoms in solid state and in solution. Most importantly, it is found that the formation of these supramolecular architectures is controlled by the nature of the grafted H-bonding organic cap and/or the solvent system employed. Crucially, the NH_2-capped cluster anion $[H_2NC(CH_2O)_3P_2V_3W_{15}O_{59}]^{6-}$, on crystallization from acetonitrile, gave a fully H-bonded supramolecular assembly having a distorted tetrahedral structure, $(TBA)_{16}H_8[H_2NC(CH_2O)_3P_2V_3W_{15}O_{59}]_4\cdot 8(CH_3CN)$; while crystallization from DMF led to $(TBA)_4H_2[H_2NC\text{-}(CH_2O)_3P_2V_3W_{15}O_{59}]\cdot 3DMF$ in which H-bonding between cluster and DMF molecules led to the formation of an infinite hydrogen-bonded chain structure in the solid state (see Fig. 14).

The role of the H-bonding ability of the organic caps of these cluster hybrids $[X\text{–}C(CH_2O)_3P_2V_3W_{15}O_{59}]^{6-}$, where X = $-NH_2$, $-NO_2$ or $-CH_3$ in forming multiple aggregates in solution and gas phase was analyzed by

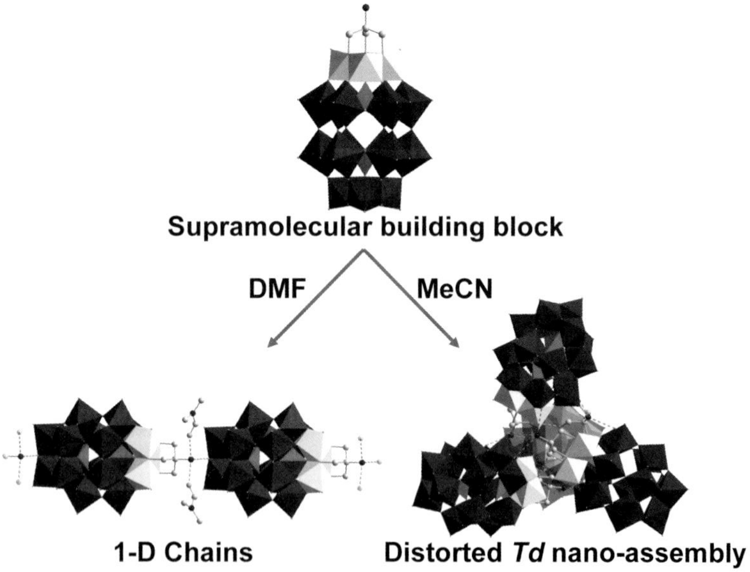

Fig. 14. Solvent dependent formation of different architectures from tris-grafted $\{P_2V_3W_{15}\}$ clusters.

comparing their cryospray mass spectra. The CSI-MS studies confirmed the formation of hydrogen-bonded nano-structures in the solution as well as revealed valuable information about the mechanism of this self-assembly process. For example, the formation of hydrogen bonded monomer, dimer, trimer and tetramers are detected in the solution and gas phase by CSI-MS studies in the case of $-NH_2$ and $-NO_2$ capped hybrid clusters; even higher aggregations such as pentamers and hexamers were detected in solution and gas phase in the case of $-NO_2$ capped cluster (see Fig. 15). The detection of these higher aggregates in solution and gas phase actually points to the formation of a number of virtual clusters in the solution before crystallization of a particular cluster out of the solution phase takes place. Therefore, this study revealed that the covalent grafting of H-bonding organic caps onto POM clusters can precisely control the supramolecular self-assembly of cluster species in the solid state and in solution and also shows the existence of cluster aggregates in solution other than the one that crystallized out ultimately.[54]

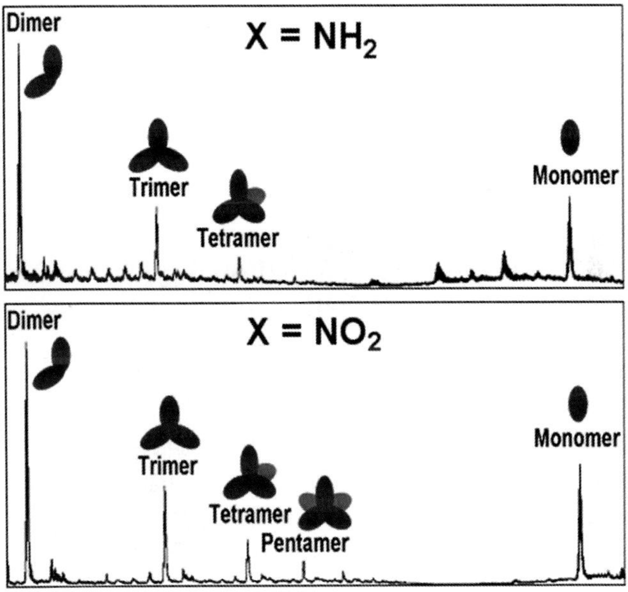

Fig. 15. CSI-MS spectra and supramolecular assemblies of $[X{-}C(CH_2O)_3P_2V_3W_{15}O_{59}]^{6-}$ clusters as a function of the different substituents (X) on the organic cap. Top $X = -NH_2$, bottom $X = -NO_2$.

In further extension of TRIS-Dawson hybrid cluster chemistry, we were successful in developing a class of POM-based "inorganic–organic–inorganic" hybrid clusters derived by grafting linear bis-(TRIS) ligands onto the V_3-capped Dawson-like cluster $[P_2V_3W_{15}O_{62}]^{9-}$. These inorganic–organic–inorganic "dumbbell" hybrids are approximately 3.4 nm long, representing the first examples of such dumbbells based on functional Dawson-like clusters.[55] Structural studies confirmed that these compounds contain two $\{P_2V_3W_{15}O_{62}\}$ cluster anions as dumbbell heads, which are linked together by linear bis(tris) ligands. The TBA cations are found to play a crucial role in controlling the self assembly of these clusters in the solid state. The TBA cations undergo extensive C–H···O hydrogen bonding interactions with the cluster heads of the dumbbells, effectively wrapping around them, hence making the dumbbell heads bulkier compared to their relatively thin middle part. This together with the peculiar zig-zag packing mode of these hybrids, lead to the formation of 1D channels along the crystallographic *b* axis in the crystal lattice. It is found that the diameter of these channels, ~1 nm, also corresponds to the length of the organic linker unit, thus suggesting a possibility of controlling the void space in similar hybrid dumbbell systems by changing the length of the linker unit (see Fig. 16). Furthermore, these hybrids are shown to self assemble in solutions forming vesicles as proved by light scattering and transmission electron microscopy (TEM) techniques.[55]

3. Extended Architectures Based on Transition Metal Salts/Oxides: Tubes, iChells and Blackberries

Given the comparatively large size of the poly-anionic POM clusters and high charge, ion exchange reactions can result in a variety of assembly processes whereby the POM clusters can form super-colloids, micro-tubular architectures, and cellular compartments (see Fig. 17).

3.1. *Tubular systems: Mechanism of formation*

The transformation of POM crystals into the tubular architectures occurs when a solution of a bulky cation on a sparsely soluble crystal allows it to

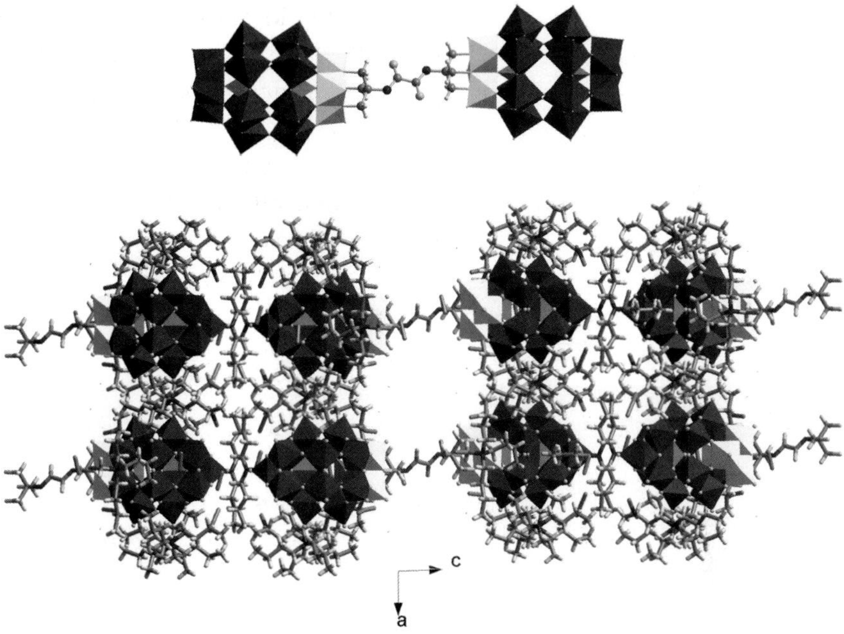

Fig. 16. Combined ball and stick and polyhedral representation of the "inorganic–organic–inorganic" hybrid cluster anion derived by grafting linear bis-(tris) ligands onto the V_3-capped Wells-Dawson-type cluster $TBA_5H_4[P_2V_3W_{15}O_{62}]$ (above) and the packing diagram (below) of the same compound showing the selective gathering of TBA cations around dumb-bell heads leading to voids in the crystal lattice. C (grey), H (white), N (blue), O (rose), W (violet polyhedra), V (yellow polyhedra), P (pink polyhedra). **See Color Plate 11**.

dissolve locally, and produce a hollow membrane mediated by the electrostatically driven aggregation of the POM-crystal and the bulky cation added to the solution which produces an insoluble matrix.[56] Once formed, the permeability of this membrane allows water to pass through and the osmotic pressure inside increases. When the system reaches a critical pressure, the membrane ruptures and the dissolved POM material that is contained within the membrane is ejected through the aperture, whereupon it comes into contact with the cation solution, leading to further aggregation (see Fig. 18). The aggregate is composed of the cations and POM anions bound electrostatically in a charge-balancing ratio.[57] Since water ingress at the crystal maintains a constant outflow of material through the opening, closure of the structure is not possible, and the

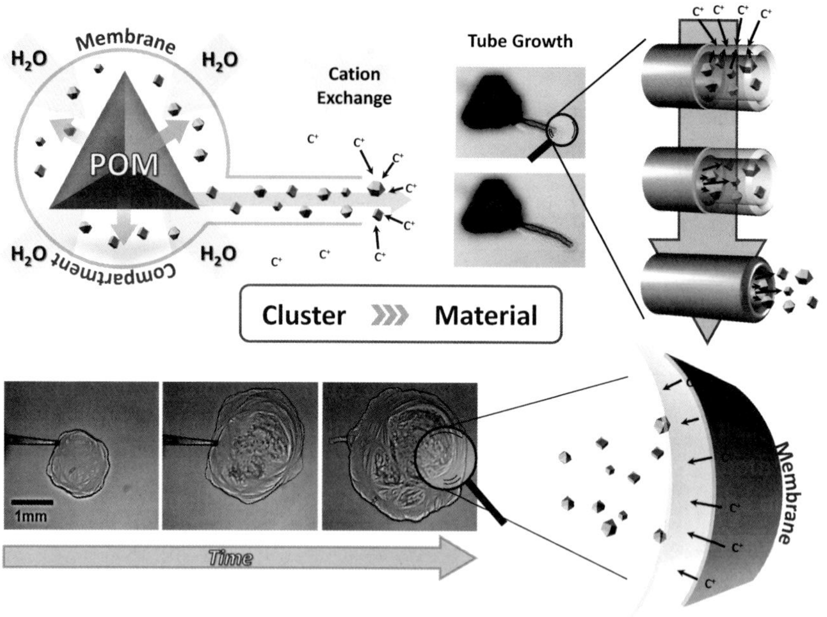

Fig. 17. Representation of how POM clusters can aggregate and form nano/micro-sized materials such as tubes and membranes.

aggregation produces extended tubular architectures. This process is able to transform a large number of crystals simultaneously (Fig. 19), and works for a wide range of POM including the so-called "Keggin Network" species (KN thereafter).[46] These polyoxometalate tubular architectures are always hollow and can be fabricated using a wide variety of clusters including catalytically active components and mixed tubes (fabricated from two or more POMs can also be made).

The flow of the POM-based solution is sustained while crystalline material is still present in the parent membrane, and continues to be dissolved. Consequently, the osmotic engine still pumps water and the system is kept in flux, resulting in continued tube growth until the material housed within the crystal is completely expended. If, for any reason, the flux is weakened or stopped, the tube growth stops immediately. The process of growth could theoretically allow a single tube to propagate until the crystal is entirely consumed, since the flow depletes POM

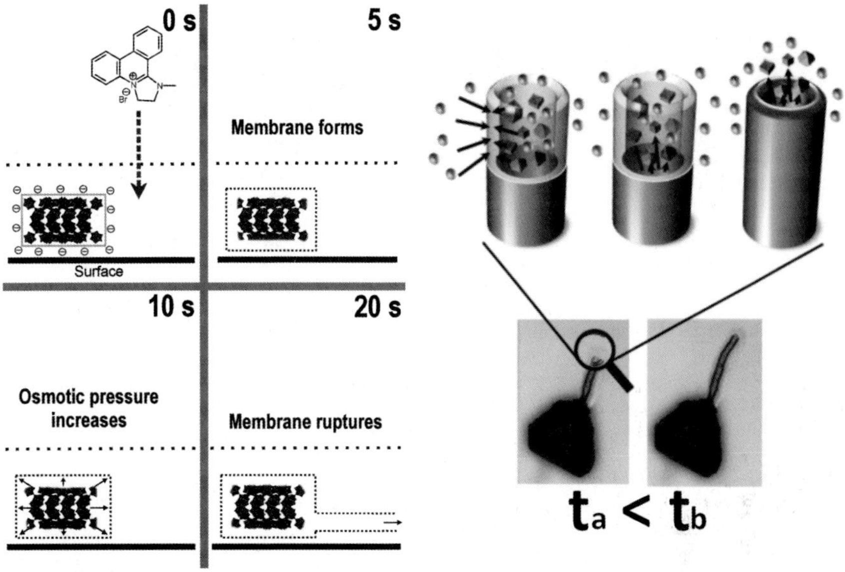

Fig. 18. Schematic representations of the mechanism involved in micro-sized tubular formations based on POM and organic cations. The resulting architecture is amorphous and can extend as far as 2×10^{-5} times the crystal size.

concentration in its neighborhood. This is, however, not the case and the growth often stops within 2–3 minutes (although growth can persist for as long as 30 minutes). This may be due to the pressure in the tube causing a new rupture point, either at the source or somewhere along the tube, in which case, tube growth continues from that new rupture point and the original tube ceases to grow. The likelihood of this occurrence is increased when the growing end of the tube encounters an obstruction on the surface. The diameter of a growing tube can be controlled by changing the concentration of cations, and has been demonstrated in the range of low concentrations for the cation methyl dihydroimidazophenanthridinium, DIP-Me.[59] Increasing the cation concentration leads to a decrease of the tube diameter, whereas addition of pure water (to decrease the overall cation concentration) leads to an increase of the tube diameter.[60] In the limiting case, when the cation concentration is raised too far, tube growth will not propagate/initiate because any rupture in the membrane will be immediately sealed.

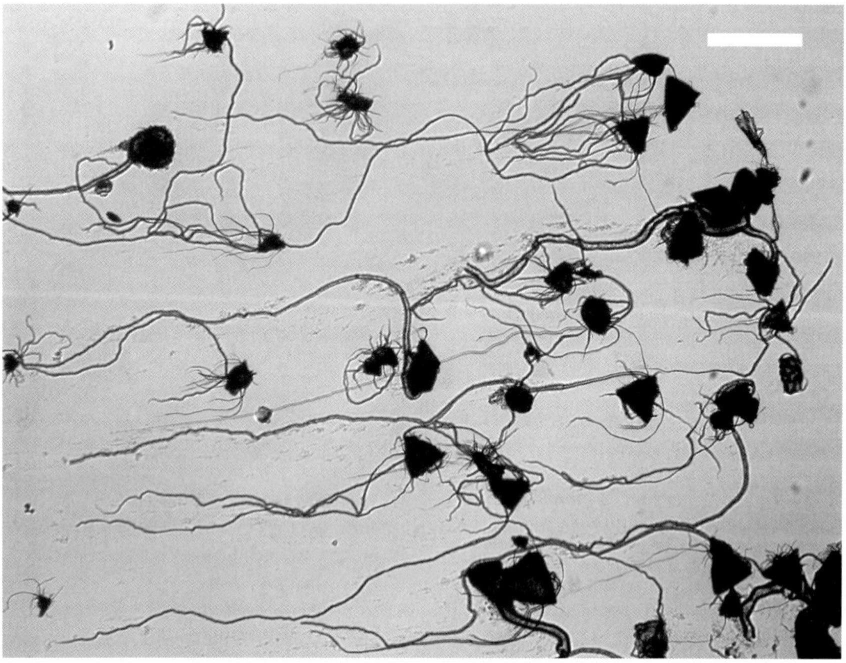

Fig. 19. Picture of a large number of crystals that have all been transformed from crystalline form to crystals that have sprouted very long tubes on addition of large cations to the aqueous solution in which the tubes are immersed. The photograph was taken 15 mins after cation addition. Tubes have grown in a uniform direction due to a bulk flow of solvent in this system (from right to left). The white scale bar is 500 μm long.

These changes can be made in "real-time", while the tube is still growing, and repeated many times to predictably produce a tube with variable diameter. Once the initial membrane has formed around the crystal and tube growth has started, the limiting rate at which water can cross membrane, and therefore the rate of POM dissolution, becomes fixed and is related directly to the surface area of the membrane. Since the membrane forms on the surface of the crystal, the initial crystal size can be used to estimate the membrane surface area and correct for this effect. The growth rate then varies proportionally with cation concentration over the range where tube growth can occur (see Fig. 20). The control of diameter by changing cation concentration can be explained in terms of the availability

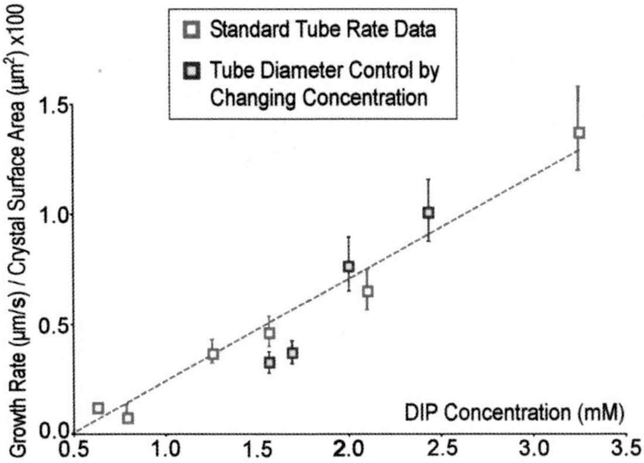

Cation conc. (mM)	Speed/Surface μm/s/μm^2 x100
0.4	0.002
0.6	0.012
0.8	0.007
1.3	0.037
1.6	0.046
2.1	0.065
3.3	0.138

Fig. 20. Plot of growth rate/crystal surface area against the overall concentration of DIP, showing how the diameter is related to the cation concentration. Red squares show the linear decrease in growth rate as the concentration is decreased, and blue squares show rate data for a tube in which the concentration of DIP-Me is varied while it is growing. Note: The crystal surface area is considered as it relates directly to the limiting rate at which water can enter the membrane surrounding the crystal, and therefore the rate of POM dissolution and tube growth. This allows direct comparison of tubes grown from crystals of varying sizes and shapes. **See Color Plate 12.**

of cations to reach a critical level of aggregation required for precipitation of the tube forming material. The increase of the cation concentration in the neighborhood of a growing tube supplies a greater density of cations and so fragments in the plume of ejected POM material will travel less distance before they encounter sufficient cations to aggregate. Once the solid material is aggregated and joined as part of the tube wall, it is relatively inflexible, and so the diameter of the tube decreases. In the reverse scenario, when the cation concentration is reduced, the ejected plume of POM material expands further before the critical aggregation point is reached and so the tube diameter increases. If the concentration of cations is reduced too far, aggregation may occur too far away for fragments to join the walls and the tube may cease to grow.

The mechanism of tube growth is osmotically driven and this is clearly seen by the effect of changing the solution ionic strength.[57] When 2–3 drops of a saturated sodium chloride solution in water is added to a system

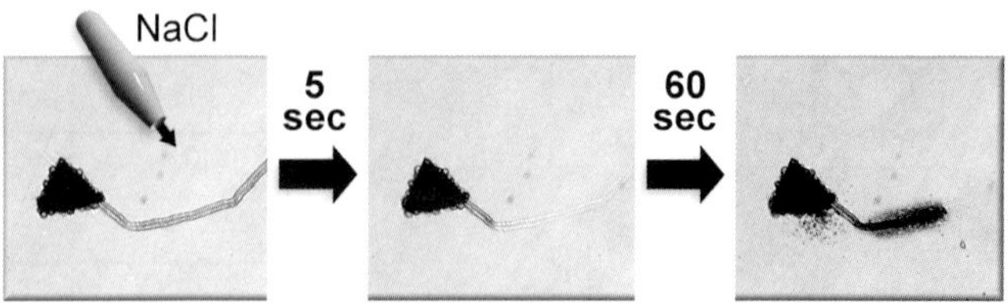

Fig. 21. Addition of aqueous sodium chloride solution to a growing tube (Keggin Network crystals—KN—/ DIP-Me) causes tube growth to temporarily stop as a competitive osmotic flow is initiated. After equilibration of the sodium chloride concentration across the membrane (30–60 s), tube growth continues, not necessarily on the same path.

where tube growth has already initiated, the growth process is temporarily halted (Fig. 21). On addition of the brine outside, ionic strength of the solution is increased to around 13 × the original number and so the osmotic potential across the membrane is reversed, since the region of lowest ionic strength is now the inside of the membrane surrounding the POM crystal. The salt injection causes a second, competitive, osmotic pump to initiate, drawing solution into the membrane surrounding the crystal, through the tubes. This disrupts the osmotic growth mechanism, as no POM material can be ejected against this new flow direction. Competition between the newly added ionic species and the aggregated material in the tube walls leads to local formation of $POM{\cdot}Na^{+}_{x}$ and $DIP{\cdot}Cl^{-}_{x}$ species, which causes some dissolution of the tube walls.

Once the concentration of NaCl inside the membrane and tube has been equilibrated with the surrounding solution the original osmotic pump, whereby dissolved POM material is ejected through the tube, becomes predominant once again. The continued presence of NaCl in the solution no longer influences the osmotic potential directly. However, it does change the solubility properties of the components and the aggregate, leading to significantly thicker/less consolidated tube walls. Without further perturbation, the system will once again continue the morphogenesis until all the crystalline material is expended. The reaction to salt addition, and the recovery of tube growth, demonstrates that the growth process can be considered follow a "siphon" like mechanism, in that the concentration of dissolved POM material is maintained along the whole length of the tube. While the POM material remains available in crystalline form, the local concentration of POM material in solution will tend towards saturation, because of the low solubility of the POM. Once a membrane rupture has occurred, and tube growth is under way, the ingress of water into the membrane is controlled by the removal of saturated solution from the membrane "chamber" through the tube. This means that the POM material is at saturation along the length of the tube, and as a consequence of this, the rate of tube growth remains constant, even as the tube extends several hundred microns from the parent crystal.

The overall solubility of the crystals is critical in tube growth initiation. When the pH of the overall solution is raised after tube growth initiation, thus modulating the solubility of the POM, the rate of tube growth increases as expected. However, if the pH is raised before tube growth has started, only precipitation is observed, as the crystal dissolves too rapidly for the initial membrane-shell to become consolidated.[60] The key properties of the POM material that influence the growth of tubes are the size and solubility of the crystal. These govern both the amount of material that can be dissolved from the crystal surface and the rate at which it can aggregate. The size and shape also directly influence the surface area of the membrane which later controls the rate at which water can enter and therefore the rate of tube propagation. Temperature and pH can be considered as "secondary factors" as they also affect a growth process but do so by modulating the crystal solubility.

It is important to consider the initial stages of tube growth, as this is the critical time where the membrane shell is formed. If the membrane is weak or the dissolution of the crystal is too rapid, there will not be sufficient containment to produce the required osmotic potential for tube growth. Conversely, tube growth will also not initiate if the aggregated membrane is too strong to be broken by the osmotic pressure within, or if the cation concentration is so high that ruptures are instantly "self-healed". The properties of the cation must also be considered. In experiments with larger volumes (> 1 ml) the solution concentration can be considered as a constant throughout the tube growth. Indeed, even if the tube formation depletes the cations in the medium immediately surrounding the growing end of the tube, the point of growth continuously moves forward into areas where the cations are still present. In experiments conducted in droplets, the concentration change over time becomes more significant and indeed tube diameter is seen to increase slightly. When a droplet experiment is continuously supplied with fresh cation solution, so as to truly maintain a constant concentration, no change in tube diameter is observed. However, it is important that the charge and size of both the cations and anions are considered, as the aggregation process is electrostatically controlled. This is because we have established that the aggregation of POM and cation occurs in a ratio that is expected to result in full charge balancing,[56] and it can be tentatively proposed that this modulates both the porosity of the initial membrane and the thickness of tube walls. It is also observed that with the same POM, the use of more highly charged cations gives rise to faster growing, narrower tubes since a lower local cation concentration is required in order to cause aggregation.

3.2. *Particular tubes/minimal systems*

In this section, we focus on the control parameters that can be made to a growing tube. The seeds and more importantly tube from needles will be discussed. Initiation of tube growth requires a sufficient local concentration of cations to be present in order to form the membrane around the POM crystal and it is therefore not possible to grow tubes under normal conditions in pure water or solutions with very low cation concentration. However, since the process only requires a local availability of cations, it is possible to construct a minimal system, where the cations are coated onto the crystal of the POM and are in sufficient concentration to produce tubes. This localization is achieved by depositing a shell of cation material all around the POM crystal using a cation solution in a solvent in which the POM is totally insoluble. The introduction of such a micro sized core-shell particle into pure water gives rise to a situation where initial membrane formation is instantaneous, and the remaining cations diffuse around the neighborhood of the crystal. This 'seed' then proceeds to grow tubes by the normal mechanism (Fig. 22), although due to the limiting availability of cations, the membrane around the parent crystal is weak and so several rupture points occur, leading to multiple tubes. The tubes grow rapidly and, as expected, their diameters increase due to the decreasing concentration of cations as the growing end travels further from the source of cations (precisely the same effect as lowering the cation concentration to control the diameter in

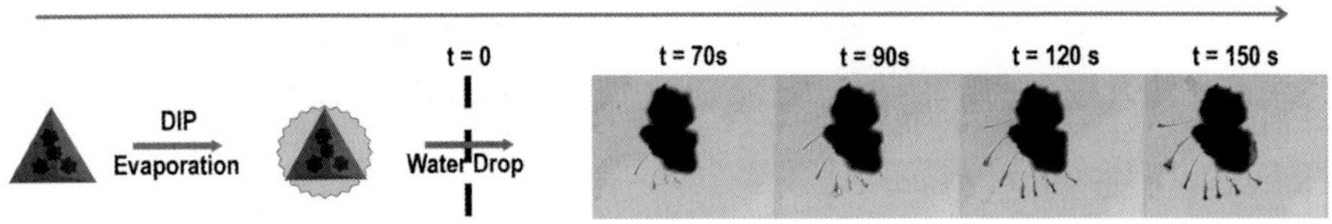

Fig. 22. Description of the "seed" experiment. A crystal of Keggin Net is coated by organic molecules (DIP in this case) through multiples evaporation of a methanolic solution containing the cation. Crystals of this particular POM has proven to be nearly insoluble in such media and end up coated with several layers of DIP. Upon pure water immersion, such a solid releases slowly its organic molecules before starting to dissolve, with the outcome of a micro-sized tube formation. However, in this experiment the amount of organic material is clearly limited and therefore the tubes become bigger and bigger until they stop to grow due to the rarefaction of the DIP molecules.

previous experiments). Growth continues until the tubes reach a region where there is insufficient cation concentration, after which they remain open and dissolved POM material is continuously expelled until the crystal is expended. Because the tubes remain open and material is pumped out through them, the osmotic pressure is continuously relieved and further ruptures to the membrane around the parent POM crystal do not occur.[57]

To explore different avenues for the fabrication of the tubular architectures it was postulated that growth of such tubes could be possible from a narrow aperture involving an extrusion process. To investigate this, crystals of KN compound were dissolved in water and the filtered solution was injected into a cation solution (DIP-Me) *via* a 5 micron aperture needle, using a syringe pump. As predicated, it was possible to fabricate 25 micron diameter tubes up to 500 microns in length. Fabrication was possible both *via* extrusion when the needle was both stationary and moving, demonstrating that such tube 'drawing' could be used to make pre-designed patterns with tubes (Fig. 23). However, the main limitation is once again the solubility, and 'writable' tubes require relatively insoluble POM-based materials; injection of the more soluble POMs did not result in the formation of tubes.[57]

3.3. *Growth control*

The overall solubility of the crystals is critical not only in modulating the rate of tube growth, but also in tube growth initiation. When the pH of the overall solution is raised after tube growth initiation, thus

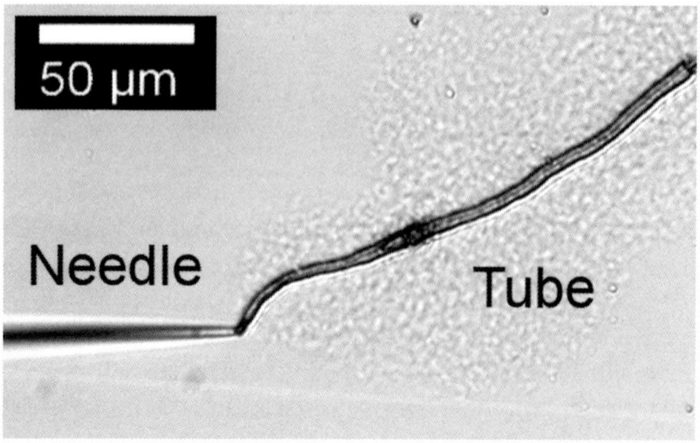

Fig. 23. Description of the "tube from a needle" experiment. Crystals of the Keggin Net specie are dissolved into water to a known concentration before being injected into a solution containing some DIP cations. It is possible to "draw" a particular shape with the tube formed.

modulating the solubility of the POM, the rate of tube growth increases as expected. However, if the pH is raised before tube growth has started, only precipitation is observed, as the crystal dissolves too rapidly for the initial membrane-shell to become consolidated.[59] The key properties of the POM material that influence the growth of tubes are the size and solubility of the crystal. These govern both the amount of material that can be dissolved from the crystal surface and the rate at which it can aggregate. The size and shape also directly influence the surface area of the membrane which later controls the rate at which water can enter and therefore the rate of tube propagation. Temperature and pH can be considered as "secondary factors" as they also affect a growth process but do so by modulating the crystal solubility. It is important to consider the initial stages of tube growth, as this is the critical time where the membrane shell is formed. If the membrane is weak or the dissolution of the crystal is too rapid, there will not be sufficient containment to produce the required osmotic potential for tube growth. Conversely, tube growth will also not initiate if the aggregated membrane is too strong to be broken by the osmotic pressure within, or if the cation concentration is so high that ruptures are instantly "self-healed".

In addition, the overall diameter and growth rate of the tube are not altered by the changes in direction and this is consistent with a growth mechanism where the rate of growth is determined by the surface area of the parent crystal and the concentration of available cations. In order to

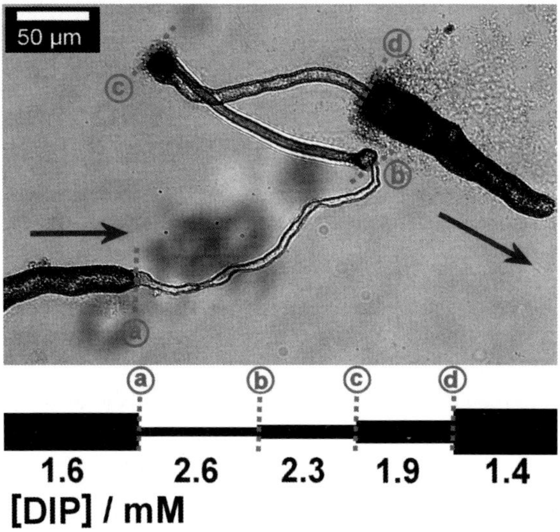

Fig. 24. Size control of a growing tube by altering the available concentration of cations at points (a), (b), (c), and (d). The estimated concentrations at each point are shown in the lower section.

demonstrate this, and to show that the tube diameter can also be controlled, the overall concentration of DIP-Me was altered while the tube was growing. Increasing the concentration of DIP-Me caused the tube to become much narrower and to grow faster while addition of water, thus decreasing the overall concentration of DIP-Me, caused the tube to become much wider and to grow more slowly (Fig. 24).[59]

To achieve direction control, a potential difference of 9 V was applied across opposite pairs of electrodes resulting in localized heating and thus, bulk flow of the solvent through convection. Growing microtubes were observed to always alter their direction of growth so as to grow towards the cathode. The most accurate control was achieved when the tube growth was in the area equidistant between the electrodes and so direction control experiments were exclusively carried out in this area. Application of the potential also caused aggregation of organic material at the anode but ^{1}H-NMR studies of the starting material, and aggregated precipitate, show that the organic framework remains unaltered. However, there is a reduction in the lifetime of tube growth and this due to migration of the charged species towards the electrodes and consequent reduction of concentration in the area in which direction control experiments are carried out.[59] By carefully controlling the direction of the applied electric field and the duration for which it was applied, several different motifs were generated in the growing tube and these fall into two categories. In the first, and most simple example, the growth is continuously switched between two directions which opposite to one another, creating a tight wave-like shape with 180º bends (Fig. 25(a) — top right). In the second, more advanced category, both sets of perpendicular electrodes are used either to create a zig-zag pattern with 90° bends (Fig. 25(b)) by switching between two directions or more complex motifs of 90° bends, for which the tube is steered in all four directions (Figs. 25(c,d)).[59]

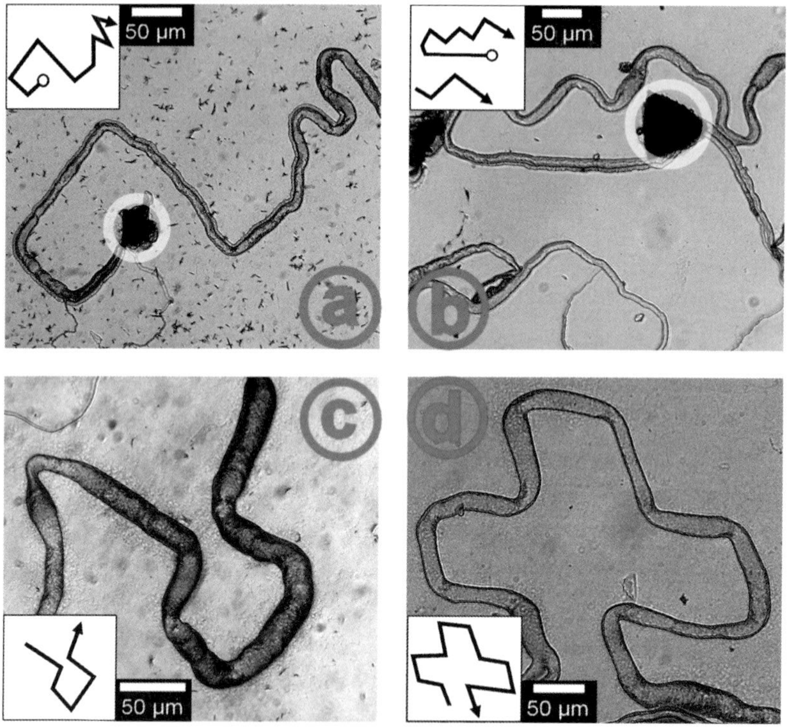

Fig. 25. Motifs "drawn" by controlling the tube growth direction. (a) 90° bends followed by a series of 180° turns. (b) A series of 90° bends to make a zig-zag motif. (c) 90° bends to form a square. (d) A series of 90° turns making a cross shape. Circles indicate the crystals from which tubes are growing. Inset diagrams show the direction of the tube growth.

It will be interesting to see how this discovery is applied since it is now possible, at least in principle to direct the assembly of nano-molecular POM clusters to form micro-tubular architectures with a configurable composition, diameter, and orientation/shape. The ability to use intrinsically catalytically active POMs in the fabrication of the tube walls raises the real possibility of using this discover to develop a new range of micro-devices.

3.4. *Formation of inorganic-chemical cells: iCHELLs*

The formation of membrane material can be achieved *via* an 'extrusion-exchange' mechanism whereby the injection of a solution containing large POM anions along with small cations into a solution containing the organic cations accompanied by small anions.[60] This leads to an ion exchange reaction at the solution interface initiating the fabrication of an insoluble aggregate of the two large species at the boundary (see Fig. 26).

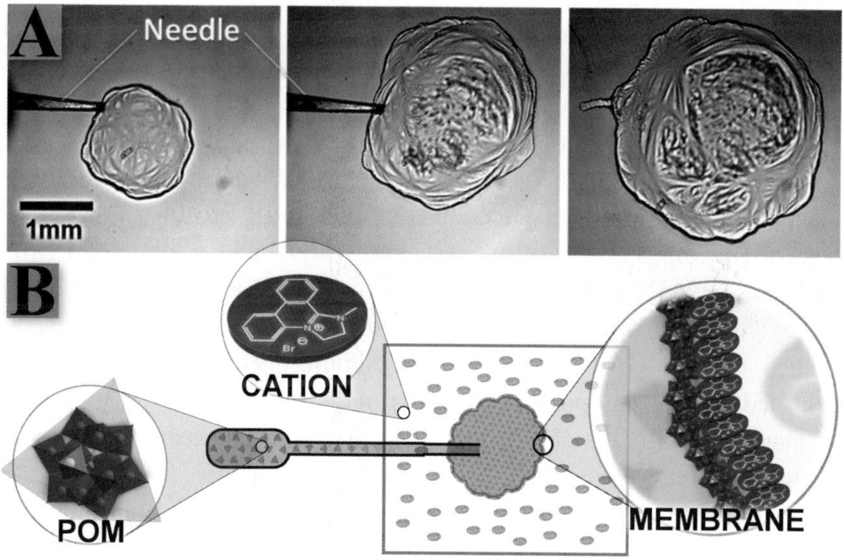

Fig. 26. (a) A sequence of images showing the formation of a 1.2 mm diameter cell as the POM solution (phosphotungstic acid) is injected into the solution of the organic cation (methyl dihydroimidazophenanthridinium, DIP-Me). Needle aperture: ca. 20 μm. (b) Schematic illustration of the "extrusion-exchange" mechanism of membrane formation. One component is injected into a solution of the other, in which cation exchange occurs on the POM, hence leading to aggregation.

This insoluble aggregate differs significantly from simple precipitation due to the morphological persistency of the resulting structure, which shows no permeability to the component species on either side of the membrane, effectively separating the initial solutions. In the general case, while observing the system through an optical microscope, an aqueous solution of the POM, phosphotungstic acid, is injected into a droplet of aqueous cation (in this case a heterocyclic-based phenanthridinum, cation)[58] using a micromanipulator needle resulting in the immediate formation of a membrane which partitions the two solutions. The reverse scenario, in which the cation is injected into the POM solution, also causes membrane formation. The membranes so far produced have been shown to be flexible, encompassing sack-like architectures with controllable diameters that span 50 μm up to several millimeters. These structures can be deflated and re-inflated several times by drawing the contents back into the micromanipulator and then re-injecting. Several such architectures can coexist with one another, allowing bulk manufacture of sack structures, and show no tendency to coalesce upon contact, instead remaining distinct entities. During the fabrication process, or when deliberately damaged, any rupture in the membrane surface is immediately repaired as the two components come into contact with one another.

Membranes can be produced using a range of organic cations (e.g. from heterocylic derivatives of phenanthridiniums (DIPs) to the highly fluorescent $Ru^{II}(bipy)_3(BF_4)_2$ (bipy = 2,2'-bipyridyl)) and a range of POMs such as $\{PW_{12}\}$ and phosphomolybdic acid ($H_3PMo_{12}O_{40}$, $\{PMo_{12}\}$) through to more complex materials including the very large 3.6 nm wheel-shaped $\{Mo_{154}\}$ cluster.[61] This self-assembling system is not, however, confined to the use of polyoxoanions from dissolved crystals as formation of membranous material was also observed upon injection of a simple phenanthridinum-cation, and sodium phosphate (dibasic, Na_2HPO_4), into an aqueous solution of sodium tungstate (Na_2WO_4). Membranes do not form in a similar experiment in the absence of phosphate anions, suggesting that the formation of a structured anion (i.e. a Keggin polyoxometalate POM which can only be templated from phosphate) is required to form the bulk membrane. The phenomenon is also not limited to purely aqueous media, as membranous sack structures can be produced in up to 1:1 water/acetonitrile mixtures and also in aqueous ethanol solutions. This property increases the versatility of the resulting membranes as it allows them to be used to encapsulate organic species which are insoluble in pure water.

The structure of the membranes produced using this modular "extrusion-exchange" process can be studied using scanning electron microscopy of dried samples, and shows that variation of the POM starting material results in large differences in the morphology of the resultant material. Small POM anions, with a low charge of –3, such as the 1.2 nm $\{PW_{12}\}$ produces thinner (1–2 μm), more wrinkled, membrane surfaces, while higher 1.8 nm sized $\{W_{48}\}$ POM clusters, with an overall charge of –40, give a much thicker, more featureless membrane geography. Cracks formed in the membranes, during the drying process in preparation for SEM analysis, show that these membranes can be up to 5 μm in thickness. Closer analysis of the surfaces of both types of membranes reveals similar structures of continuous raised areas in the region of 50–100 nm in width. These raised areas are most pronounced in rough, broken areas of the surface where deep valleys in the material can be observed. We propose that the membrane is composed of discrete nanoparticles which are charge-balanced aggregates of the starting components, where the particle size is controlled by the POM cluster size and charge. Membranes produced from larger, more highly charged clusters are therefore likely to be thicker. Elemental analysis (for C, H and N) of the membranes shows that the composition of the individual components is retained and that the cations and anions aggregate in the correct ratio to exactly balance the overall charge, and this analysis is confirmed by EDX (Energy-dispersive X-ray spectroscopy) analysis. In the specific case of $\{PW_{12}\}$ or $\{PMo_{12}\}$ with the phenanthridinum based-cation DIP-Me, this gives a POM: Cation ratio of 1 : 3, while with the large and highly charged $\{Mo_{154}\}$ cluster, a ratio of 1 : 56 is observed.

The mechanical strength of $\{PW_{12}\}$-based membranes can be investigated using an AFM cantilever to tear the membrane and observing the deflection of the tip as tearing occurs. Initial results indicate that a force of ~110 μN is exerted before the tip tears the membrane when it is derived from simple DIP cations. This is around 100x less than the force required to rupture a biological vesicle wall. The strength of membranes can be enhanced by ~15% by using a larger, more highly charged dimeric DIP cation in place of DIP-Me, probably due to the enhanced cross-linking ability of the cation. This demonstrates a method of controlling mechanical strength of these materials which will be extremely useful when designing and building membrane-based devices.

3.5. *iCHELL chemistry*

Since selective permeability is an essential feature in synthetic hybrid membranes, the materials can be tested by subjecting $\{PW_{12}\}$ containing sacks (in DIP-Me solution) to a flow of ammonium hydroxide solution and solutions of tetraalkylammonium hydroxides of varying chain lengths. Upon contact with any of these ammonium species, $\{PW_{12}\}$ immediately precipitates so it is easy to observe the time taken between addition of the ammonium salt outside of the sack and the precipitation of $\{PW_{12}\}$ inside. Treatment with ammonium hydroxide and tetramethylammonium hydroxide resulted in very rapid precipitation inside the sack. However, when larger tetraethyl- or tetrapropylammonium hydroxide was used, a significant lag before precipitation was observed and when tetrabutylammonium was added, no precipitation was observed, thus demonstrating that the membranes are more permeable to smaller molecules. Injection of either component into the other produces a membranous pouch, the construction of a sack within another sack is possible. These "nested" systems can be used to compartmentalize chemical reactions. As a simple demonstration of this, DIP-Me solution can be injected into a solution of $\{PMo_{12}\}$ and the resulting membrane sack was then further injected with a solution containing $\{PW_{12}\}$ with a small amount of potassium permanganate added, thus producing a smaller sack inside the first. Both remained stable for a period of several hours without merging or leakage, unless physically pushed together. Addition of hydrogen peroxide to the outer $\{PMo_{12}\}$ solution caused discoloration of the potassium permanganate after a short diffusion time, showing that the 'nested' system can be used to compartmentalize chemical reactions since the H_2O_2 is able to pass over both membranes and through the DIP solution. Compartmentalized sack to sack reaction systems would also be possible, where a sacks containing different reagents (able to cross the membrane at different rates) are brought into close proximity with one another, allowing sequences of chemical transformations to occur (see Fig. 27).

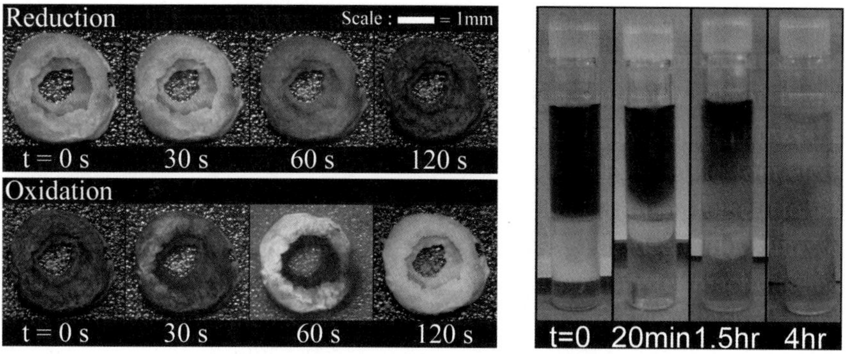

(a) (b)

Fig. 27 (a) A schematic representation of one cell being grown inside another. A salt containing a large organocation (green) is injected into a salt solution containing a large POM anion (blue) forming a membrane. A second POM that contains another POM reagent (red) is injected into the first cell, producing a second membrane. Adding an external reagent (yellow) can then cause a reaction in the inner cell after a diffusion time. (b) Time-lapse images are shown below with the growth of a DIP-Me cell in $\{PMo_{12}\}$, followed by the construction of a $\{PW_{12}\}$ cell inside the encapsulated DIP-Me droplet. The $\{PW_{12}\}$ cell contains potassium permanganate and when hydrogen peroxide solution is added to the outer $\{PMo_{12}\}$ solution, discoloration of the $\{PW_{12}\}$ cell is observed after a few minutes. **See Color Plate 12**.

One of the major advances represented by the development of these membrane systems is to enable incorporation of chemically active components into the material. In the simplest case, the POM component of the membrane material represents a redox active species capable of being reduced to form blue reduced POM species by the action of a suitable reducing agent. After the addition of sodium dithionite ($Na_2S_2O_4$) to a solution containing a $\{PW_{12}\}$-DIP-Me membrane, the membrane quickly switches to the reduced blue form (Fig. 28 (a)). Subsequent washing of this blue material followed by addition of an oxidizing agent (e.g. hydrogen peroxide) led to the rapid loss of the blue color as the membrane was re-oxidized. This process proved to be reversible and could be repeated up to three times before degradation of the membranes. As well as forming closed, sack-like, architectures, membrane formation will occur at any interface between solutions of suitable cations and anions, allowing the formation of layered domains which are effectively phase separated by the intervening POM-DIP-Me membrane. We demonstrate this by forming a membrane across a glass vial, where the pH sensitive indicator,

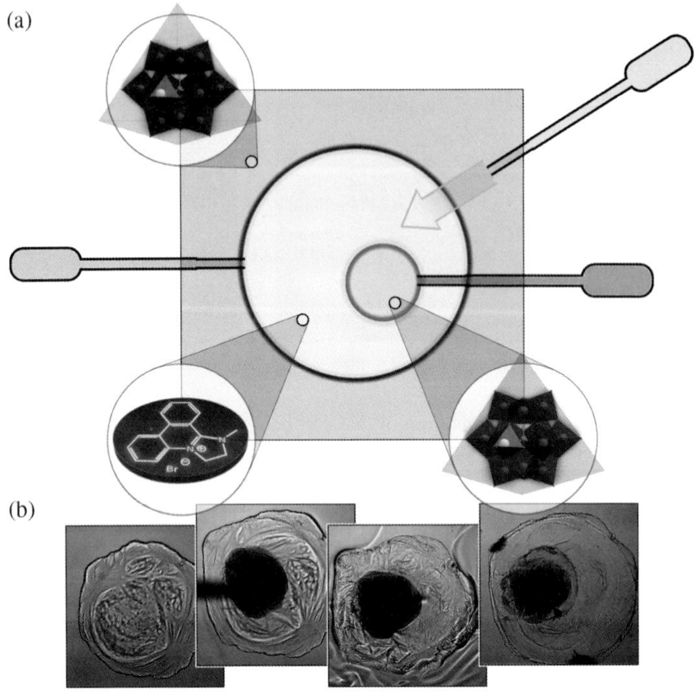

Fig. 28. (a) Addition of dilute sodium dithionite (reducing agent) to a washed membrane in water, results in the reduction of the POM over 120 seconds. After washing with deionized water, addition of hydrogen peroxide solution (30% v/v) to the same system causes the POM to become reoxidized over approximately 120 seconds. This process is fully reversible and can be repeated up to three times without significant membrane degradation. (b) Diffusion of H_3O^+ cations through a $\{PW_{12}\}$-DIP-Me membrane showing the change in color of the bromocresol purple in the upper DIP fraction from purple to yellow indicating the lowering of the pH in this region. **See Color Plate 13**.

bromocresol purple, which appears purple above pH 6.8, but yellow below pH 5.2, is encapsulated in the DIP-Me solution and separated from the inherently acidic POM $\{PW_{12}\}$. A color change from purple to yellow is observed in the cation solution over ~4 h, as the H_3O^+ cations diffuse through the membrane (Fig. 28 (b)).

Once formed, the membrane sacks can be successfully transferred out of their mother liquor, as long as they remain in a sufficiently high ionic strength environment. They are not stable if transferred into pure water, were osmotic pressure causes rapid expansion and rupture of the sack. The supporting of membranes on polymer supports allows a number of further applications to be investigated, as the membrane coated matrix can be easily removed from the initial solution and washed several times without losing their structural integrity. We showed that our membranes can be grown on a supporting hydrogel matrix by sorbing the cationic membrane forming component into an anionic hydrogel material, such as partially deprotonated polyacrylic acid. The cation soaked hydrogel material was then immersed in a solution containing the POM component, forming membranous material on the surface of the hydrogel.

4. Conclusions

There is no doubt that POMs are extremely diverse, but in this chapter we have attempted to explore the unique nature of their high charge and large size. As large anions they are perhaps unique in chemistry, and therefore it is not surprising that POM-based molecules and materials are still throwing up many discoveries of new phenomenon. The other important aspect is the fact that POM structures mostly represent kinetically trapped units, and as such their reactivity can be controlled by successive ion exchange reactions. This is a rather simple yet profound realization since armed with this knowledge it is possible to develop routes to assemble functional materials and devices. The next challenge will be to harness these properties to develop a new generation of functional nano molecular and macromolecular systems.

Acknowledgements

The authors wish to acknowledge the University of Glasgow, WestCHEM and the EPSRC for funding. L.C. thanks the Royal Society/Wolfson Foundation for a merit award and would also like to acknowledge the many diverse and exciting collaborators beyond Glasgow that have contributed to the work discussed in this chapter.

References

1. (a) Pope M. T. and Müller A., *Angew Chem Int Ed Engl* **30** (1991) 34; (b) *Polyoxometalate Chemistry: From Topology* via *Self-Assembly to Applications*,

ed. Pope, M. T. and Müller, A. (Eds.) Kluwer, Dordrecht, 2001; (c) C. L. Hill (Ed.), Special issue on polyoxometalates, *Chem Rev* **98** (1998) 1–387.

2. Long D.-L., Tsunashima R. and Cronin L., *Angew Chem Int Ed* **49** (2010) 1736.
3. (a) Hasenknopf B., *Front Biosci* **10** (2005) 275; (b) Yamase T., *et al.*, *Inorg Chem* **45** (2006) 7698; (c) Kortz U., *et al.*, *Coord Chem Rev* **253** (2009) 2315; (d) Neumann R. and Dahan M., *Nature* **388** (1997) 353; (e) Rhule J. T., *et al.*, *J Am Chem Soc* **123** (2001) 12101; (f) Mishra P. P., Pigga J. and Liu T., *J Am Chem Soc* **130** (2008) 1548; (g) Noro S.-i., *et al.*, *Angew Chem Int Ed* **48** (2009) 8703; (h) Imai H., *et al.*, *J Am Chem Soc* **131** (2009) 13578; (i) Streb C., *et al.*, *Angew Chem Int Ed* **48** (2009) 6490; (j) Fleming C., *et al.*, *Nat Nanotechnol* **3** (2008) 229; (k) Mitchell S. G., *et al.*, *Chem Commun* (2009) 2712.
4. (a) Long D.-L., Burkholder E. and Cronin L., *Chem Soc Rev* **36** (2007) 105; (b) Long D.-L. and Cronin L., *Chem–Eur J* **12** (2006) 3698; (c) Kögerler P. and Cronin L., *Angew Chem Int Ed* **44** (2005) 844.
5. (a) Müller A., Kögerler P. and Kuhlmann C., *Chem Commun* (1999) 1347; (b) Cronin L., Kögerler P. and Müller A., *J Solid State Chem* **152** (2000) 57; (c) Cronin L., *et al.*, *Angew Chem Int Ed* **41** (2002) 2805–2808.
6. Müller A. and Roy S., *Coord Chem Rev* **245** (2003) 153.
7. (a) Contant R., *Inorg Synth* **27** (1990) 109; (b) Contant R. and Ciabrini J.-P., *J Chem Res Synop* (1977) 222; (c) Knoth W. H. and Harlow R. L., *J Am Chem Soc* **103** (1981) 1865; (d) Canny J., Tézé A., Thouvenot R. and Hervé G., *Inorg Chem* **25** (1986) 2114; (e) Kirby J. F. and Baker L. C. W., *Inorg Chem* **37**, (1998) 5537.
8. (a) Desiraju G. R., *Crystal Engineering: The Design of Organic Solids*, Elsevier, Amsterdam, (1989); (b) Braga D., Grepioni F. and Orpen A. G., ed., *Crystal Engineering: from Molecules and Crystals to Materials*, Kluwer Academic Publishers, Dordrecht, (1999); (c) Lehn J.-M., *Angew Chem Int Ed Engl* **29** (1990) 1304; (d) Aakeröy C. B., Champness N. R. and Janiak C. *Cryst Eng Comm* **12** (2010) 22; (e) Braga D., Brammer L. and Champness N. R., *Cryst Eng Comm* **7** (2005) 1.
9. Long D.-L., *et al.*, *Angew Chem Int Ed* **42** (2003) 4180.
10. (a) Long D.-L., Kögerler P. and Cronin L., *Angew Chem Int Ed* **43** (2004) 1817; (b) Goral M., *et al.*, *Dalton Trans* (2009) 6727; (c) Baffert C., *et al.*, *Dalton Trans* (2007) 4599; (d) Baffert C., *et al.*, *Chem–Eur J* **12** (2006) 8472; (e) Yan J., *et al.*, *Inorg. Chem* **49** (2010) 1819; (f) Miras H. N., *et al.*, *Chem Commun* (2008) 4703.

11. Abbas H., *et al.*, *Chem–Eur J* **11** (2005) 1071.
12. (a) Ritchie C., *et al.*, *Chem Commun* (2007) 468; (b) Ritchie C., *et al.*, *Dalton Trans* (2009) 6483; (c) Ritchie C., *et al.*, *Dalton Trans* (2009) 1587; (d) Ritchie, C., *et al.*, *Dalton Trans* (2006) 1712.
13. (a) Müller A., *et al.*, *Angew Chem Int Ed* **42** (2003) 5039; (b) Müller A., *et al.*, *Angew Chem Int Ed* **43** (2004) 4466; (c) Müller A., *et al.*, *Angew Chem Int Ed* **44** (2005) 7757; (d) Müller A., *et al.*, *Angew Chem Int Ed* **45** (2006) 460.
14. (a) Wassermann K., Dickmann M. H. and Pope M. T. *Angew Chem Int Ed Engl* **36** (1997) 1445; (b) Sadakane M., Dickman M. H. and Pope M. T. *Angew Chem Int Ed* **39** (2000) 2914.
15. Müller A., *et al.*, *Angew Chem Int Ed Engl* **30** (1991) 1674.
16. Ohlin C. A., *et al.*, *Angew Chem Int Ed* **47** (2008) 5634.
17. Nyman M., *et al.*, *Science* **297** (2002) 996.
18. Sigmon G. E., *et al.*, *Angew Chem Int Ed* **48** (2009) 2737.
19. Chubarova E. V., *et al.*, *Angew Chem Int Ed* **47** (2008) 9542.
20. (a) Schäffer C., *et al.*, *Angew Chem Int Ed* **48** (2009) 149; (b) Tsunashima R., *et al.*, *Angew Chem Int Ed* **49** (2010) 113.
21. (a) Hagrman P. J., Hagrman D. and Zubieta J., *Angew Chem Int Ed* **38** (1999) 2638; (b) Fang X., Anderson T. M. and Hill C. L., *Angew Chem Int Ed* **44** (2005) 3540; (c) Wang X. -L., *et al.*, *Angew Chem Int Ed* **45** (2006) 7411; (d) Lan Y. Q., *et al.*, *Chem Commun* (2008) 58.
22. Gopalakrishnan J., *Chem Mater* **7** (1995) 1265.
23. (a) Sha J., *et al.*, *Cryst Growth Des* **7** (2007) 2535; (b) Lan Y.-Q., *et al.*, *Dalton Trans* (2008) 3824.
24. (a) Proust A., Thouvenot R. and Gouzerh P., *Chem Commun* (2008) 1837; (b) Landsmann S., Lizandara-Pueyo C. and Polarz S., *J Am Chem Soc* **132** (2010) 5315.
25. Long D.-L., *et al.*, *Dalton Trans* (2005) 1372.
26. Long D.-L., *et al.*, *J Am Chem Soc* **126** (2004) 13880.
27. Long D.-L., *et al.*, *Dalton Trans* (2006) 2852.
28. (a) Streb C. *et al.*, *Chem Eur J* **14** (2008) 8861; (b) McGlone T., *et al.*, *Chem–Asian J* **4** (2009) 1612.
29. Miras H. N., *et al.*, *Angew Chem Int Ed* **47** (2008) 8420.
30. Long D.-L., *et al.*, *Angew Chem Int Ed* **45** (2006) 4798.
31. (a) Long D.-L., *et al.*, *Angew Chem Int Ed* **44** (2005) 3415; (b) Fay N., *et al.*, *Inorg Chem* **46** (2007) 3502.

32. Long D.-L., *et al.*, *Angew Chem Int Ed* **47** (2008) 4384.
33. Yan J., *et al.*, *Angew Chem Int Ed* **48** (2009) 4376.
34. Miras H. N., *et al.*, *Dalton Trans* (2008) 214.
35. (a) Cheetham A. K., Férey G. and Loiseau T., *Angew Chem Int Ed* **38** (1999) 3268; (b) Cheetham A. K., Rao C. N. R. and Feller R. K., *Chem Commun* (2006) 4780.
36. Rodriguez-Albelo L. M., *et al.*, *J Am Chem Soc* **131** (2009) 16078.
37. Villanneau R., *et al.*, *Chem Commun* (1998) 1491.
38. Abbas H., *et al.*, *Cryst Growth Des* **8** (2008) 635.
39. Streb C., *et al.*, *Angew Chem Int Ed* **46** (2007) 7579.
40. Streb C., Long D.-L. and Cronin L., *Cryst Eng Comm* **8** (2006) 629.
41. Fang X., Anderson T. M. and Hill C. L., *Angew Chem Int Ed* **44** (2005) 3540.
42. Streb C., Long D.-L. and Cronin L., *Chem Commun* (2007) 471.
43. (a) Long D.-L., *et al.*, *Chem Asian J* **1** (2006) 352; (b) Pradeep C. P, *et al.*, *Chem Commun* (2007) 4254.
44. (a) Uchida S., *et al.*, *Angew Chem Int Ed* **41** (2002) 2814; (b) Uchida S. and Mizuno N., *J Am Chem Soc* **126** (2004) 1602.
45. Mitchell S. G., *et al.*, *Dalton Trans* (2008) 1415.
46. (a) Ritchie C., *et al.*, *Angew Chem Int Ed* **47** (2008) 6881; (b) Thiel J., *et al.*, *J Am Chem Soc* **131** (2009) 4180.
47. (a) Mitchell S. G., *et al.*, *Cryst Eng Comm* **11** (2009) 36; (b) Mitchell S. G., *et al.*, *Nat Chem* **2** (2010) 308.
48. (a) Khan M. I., *et al.*, *J Am Chem Soc* **114** (1992) 3341; (b) Müller, A., *et al.*, *Chem–Eur J* **4** (1998) 1388; (c) Han J. W. and Hill, C. L., *J Am Chem Soc* **129** (2007) 15094; (d) Hou Y. and Hill C. L., *J Am Chem Soc* **115** (1993) 11823; (e) Zeng H. D., Newkome G. R. and Hill C. L., *Angew Chem Int Ed* **39** (2000) 1771; (f) Song Y.-F., *et al.*, *Inorg Chem* **47** (2008) 9137.
49. (a) Song Y.-F., *et al.*, *J Am Chem Soc* **131** (2009) 1340; (b) Han Y., *et al.*, *Macromolecules* **42** (2009) 6543; (c) Favette S., *et al.*, *Chem Commun* (2003) 2664.
50. (a) Song Y.-F., Long D.-L. and Cronin L., *Angew Chem Int Ed* **46** (2007) 3900; (b) Song Y.-F., Long D.-L. and Cronin L., *Cryst Eng Comm* **12** (2010) 109.
51. Song Y.-F., *et al.*, *J Mater Chem* **17** (2007) 1903.
52. Zhang J., *et al.*, *J Am Chem Soc* **130** (2008) 14408.
53. Song Y.-F., *et al.*, *Chem–Eur J* **14** (2008) 2349.
54. Pradeep C. P., *et al.*, *Angew Chem Int Ed* **47** (2008) 4388.

55. Pradeep C. P., *et al.*, *Angew Chem Int Ed* **48** (2009) 8309.
56. Ritchie C., *et al.*, *Nature Chemistry* **1** (2009) 47.
57. Cooper G. J. T., *et al.*, *J Am Chem Soc* **133** (2011) 5947–5954.
58. Parenty A. D. C., *et al.*, *J Org Chem* **69** (2004) 5934–5946.
59. Cooper G. J. T. and Cronin L., *J Am Chem Soc* **131** (2009) 8368–8369.
60. Cooper G. J. T., *et al.*, *Angew Chem Int Ed* **50** (2011) 10373–10376.
61. Müller A., *et al.*, *Angew Chem Int Ed Engl* **34** (1995) 2122–2124.

Chapter 4

MAGNETIC POLYOXOMETALATES

JUAN M. CLEMENTE-JUAN, EUGENIO CORONADO
and ALEJANDRO GAITA-ARIÑO
Instituto de Ciencia Molecular, Universidad de Valencia,
c/Catedrático José Beltrán, 2, 46980 Paterna, Spain

1. Introduction

Polyoxometalates (POMs) form a class of inorganic compounds that can incorporate a large number of magnetic centers at specific sites of the diamagnetic POM framework. Furthermore, the rigidity of the POM framework guarantees the stability of these systems in solution as well as in the solid state. These magnetic polyoxometalates can be broadly divided into three families.

A first one consists of polyoxometalates encapsulating a number of transition metal ions, which can be connected through oxo bridges forming magnetic clusters of variable nuclearities and high symmetries,[1] often enabling exchange interactions between them. Functionally, they can be referred to as spin clusters, and they provide model examples to study the exchange interactions at the molecular level.

A second family is formed by POMs encapsulating lanthanoid ions. These two families can provide examples of single-molecule magnets. Finally, a third family is constituted by mixed-valence polyoxometalates, which host a number of electrons that are usually hopping throughout the whole POM framework.

In this chapter, we will show that these inorganic molecules provide excellent examples of spin clusters with tunable exchange interactions, single-molecule magnets with tunable quantum spin relaxation times, and switchable spin molecules in which the spin coupling can be tuned through an electric field. Thus, we will show that these systems have a high potential in the different branches of Molecular Magnetism, including the emergent fields of molecular spintronics[2] and quantum computing.[3–5]

2. Spin Clusters

Heteropolyoxometalates are metal-oxide clusters, essentially wolframates and, to a lesser extent, molybdates, which have the special ability of acting as chelating ligands toward 3d-transition metal ions, thereby encapsulating a variety of clusters of these magnetic ions within their own rigid structures. Therefore, the POM ligand imposes coordination geometries, nuclearities and angles for the oxo-bridges between the magnetic metals in the magnetic clusters. Literature offers an extensive number of examples of these magnetic clusters.[6]

The bulky nonmagnetic polyoxometalate framework also guarantees an effective magnetic isolation between clusters, simplifying the theoretical analysis. Furthermore, clusters of higher nuclearity can be constructed from smaller fragments by controlling certain chemical parameters. Thus, highly symmetric clusters with intermediate nuclearities, such as dimers, trimers and tetramers, have been widely studied as model systems for the development of magnetic exchange theories.[7]

A relevant case that illustrates the possibilities offered by polyoxometalate chemistry to conduct detailed studies on magnetic exchange interactions is provided by the cobalt(II) clusters, which exhibit a large spin anisotropy.[8] This d^9 ion, when octahedrically coordinated by the weak ligand field of the polyoxometalates, shows a high-spin $S=3/2$ with an unquenched orbital momentum. The first order spin-orbit coupling then results in an effective spin doublet, $S=1/2$, which is highly anisotropic. Still, the magnetic properties in these compounds have shown to be largely insensitive to the exchange anisotropy and thus, other physical techniques have been needed to extract this information. The key technique in this context has been the inelastic neutron scattering (INS). This spectroscopic technique provides a direct access to the splitting of the low-lying energy levels of a cluster caused by exchange interactions. Therefore, it allows a much deeper and more detailed

Fig. 1. Structure of the $[Co_4(H_2O)(PW_9O_{34})_2]^{10-}$ complex and structure of the Co_4O_{16} cluster and the network of the exchange parameters.

insight into the nature of the magnetic coupling compared with the usual magnetic susceptometry.

In 1992, we started to use this technique to study a tetranuclear rhomb-like Co(II) cluster encapsulated by POMs of formula $[Co_4(H_2O)_2(PW_9O_{34})_2]^{10-}$ (See Fig. 1).[7] Fig. 1. Structure of the $[Co_4(H_2O)(PW_9O_{34})_2]^{10-}$ complex and structure of the Co_4O_{16} cluster and the network of the exchange parameters.

These studies provided for the first time direct evidence about the Co(II)-Co(II) exchange anisotropy.[10] More recently we have shown that INS data are sensitive not only to the magnitude and anisotropy of the exchange parameters but also to the relative orientation of the exchange anisotropy axes. This has been observed in two trinuclear Co(II) clusters derived from the tetranuclear one. The first trimer is an angular Co_3 cluster obtained by replacement of a Co situated in the short diagonal of the rhomb by W (shorthand Co_3W) (Fig. 2, left). It has been isolated in the POM salt $Na_{12}[Co_3W(H_2O)(ZnW_9O_{34})_2]\cdot nH_2O$. The second trimer is a triangular Co_3 cluster in which a Co in the long diagonal of the rhomb has been replaced by Na^+ (shorthand $NaCo_3$) (Fig. 2). It has been isolated in the POM salt $Na_{17}[(NaOH_2)Co_3(H_2O)(P_2W_{15}O_{56})_2]\cdot nH_2O$.

The INS spectra of Co_3W show three cold peaks and three hot ones, while $NaCo_3$ shows up to two cold peaks and two hot peaks (Fig. 3, left).

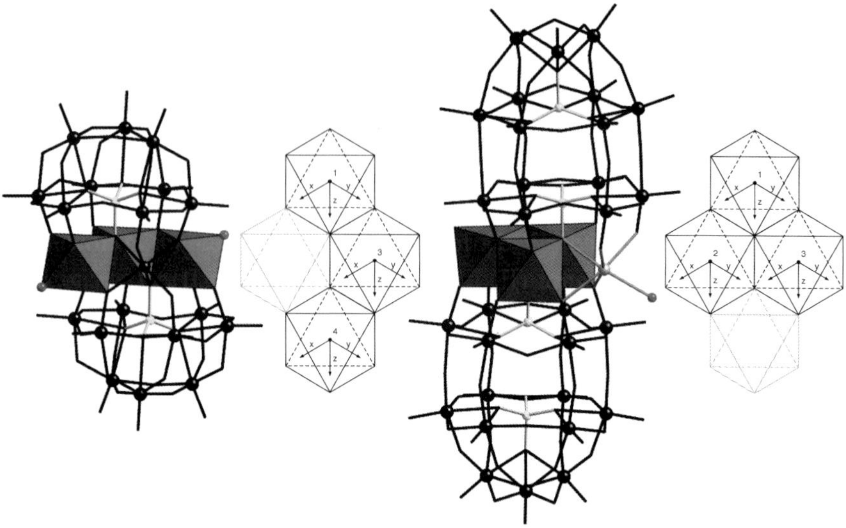

Fig. 2. Polyhedral representation and orientations of the local anisotropy axes of the $[Co_3W(D_2O)(ZnW_9O_{34})_2]^{12-}$ (left) and $[(NaOH_2)Co_3(H_2O)(P_2W_{15}O_{56})_2]^{17-}$ (right) clusters.

From these data, an energy splitting pattern can be obtained (Fig. 3, right). The diagrams show four doublets arising from the three coupled effective spins S = 1/2. These can be closely reproduced by a fully anisotropic exchange model that takes into consideration the molecular symmetry of the system. In this anisotropic model, the orientations of the local anisotropy axes are assumed to be parallel to the W-O bonds and the three components (x, y and z) of each exchange interaction are different (Fig. 2). Making use of the molecular symmetry of the cluster, it is possible to describe the INS spectra of both systems from a set of 5 parameters, which are the following: (i) the three components of the exchange parameters $J^{13}_{(x,y,z)}$ associated with the side of the rhombus (similar for both clusters) and (ii) the two components of the axial exchange $J^{23}_{(x,y,z)}$ through the short diagonal ($J^{23}_x = J^{23}_y$ due to the C_s symmetry). Under these considerations, the best solution for $NaCo_3$ leads to $J^{13}_{(x,y,z)}$ = (1.4±0.5, 8.6±0.5, 10.0±0.5) cm^{-1} and $J^{23}_{(x,y,z)}$ = (6.5±0.5, 6.5±0.5, 3.4±0.5) cm^{-1}.

In conclusion, the possibility of studying these Co clusters through INS has provided the opportunity to obtain a very detailed information on the

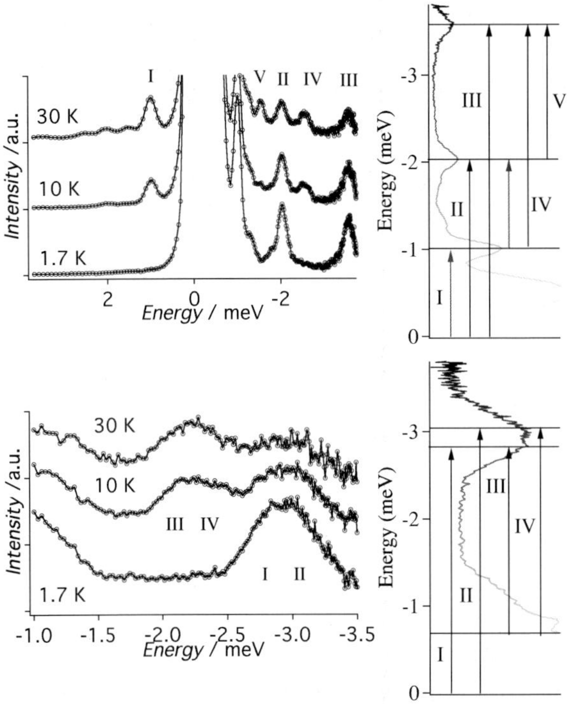

Fig. 3. INS spectra of polycrystaline sodium salts of $[Co_3W(D_2O)(ZnW_9O_{34})_2]^{12-}$ (*top*) and $[(NaOH_2)Co_3(H_2O)(P_2W_{15}O_{56})_2]^{17-}$ (*bottom*) clusters and experimentally determined ground-state splitting with observed transitions represented by arrows.

anisotropic nature of the exchange interactions in Co^{II} systems having different molecular symmetries.[11]

3. Single-molecule Magnets

An important class of magnetic molecular materials is the so-called single-molecule magnets (SMMs). These systems, usually formed by magnetic clusters, exhibit a slow-relaxation of the magnetization at low temperatures (superparamagnetic blocking) giving rise to a magnetic memory effect similar to that found in hard magnets, together with quantum tunneling effects. In the past two decades, these properties have been observed in a variety of molecules, with the magnetic moment mainly being due to d electrons in transition metal ions, mostly high-spin

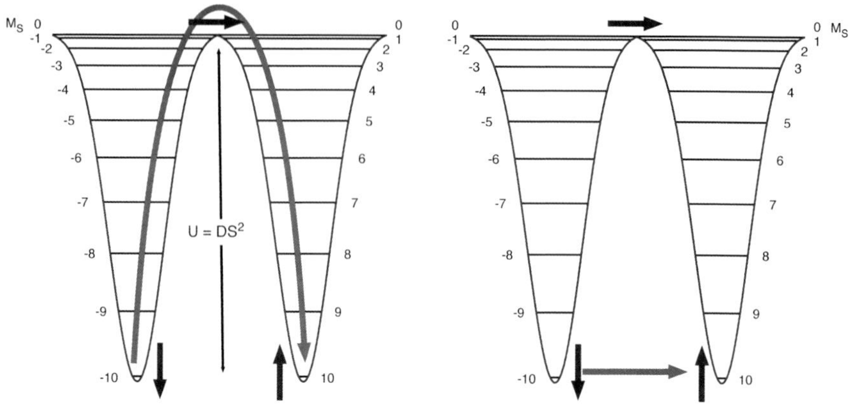

Fig. 4. Energy level scheme for the ground S = 10 multiplet of the Mn_{12} SMM.

anisotropic clusters with a strong exchange that results in a high spin ground state, and a large uniaxial anisotropy.[12] The archetypical molecule of this kind is the so-called Mn_{12} acetate. The magnetic behavior of this SMM can be understood from the spinstructure of the low-lying energy levels (Refs. 13 and 14).

Thus, owing to the uniaxial anisotropy of Mn_{12}, its ground spin state $S = 10$ splits into its M_s components in such a way that the M_s doublet $= \pm 10$ is stabilized and becomes the ground spin state of the cluster, while the rest of spin sublevels becomes more and more excited as $|M_s|$ decreases (Fig. 4). Such a spin structure creates an energy barrier that needs to be overcome to reverse the orientation of the (from $M_s = +10$ to $M_s = -10$). That explains the superparamagnetic blocking observed at low temperatures in these nanomagnets and the occurrence of a magnetic hysteresis. At the same time, the presence of quantum-tunneling effects allows the spin reversal to occur without the need of overcoming the barrier (Figure 4). Among those, polyoxometalates exhibiting SMM behavior have recently been described, both conventional high-spin, high-anisotropy systems[15] and low-spin, isotropical systems where the quantum dynamics is governed by a complex network of antiferromagnetic exchange interactions.[16]

In the last decade, another class of SMMs has been discovered, in which the magnetic moment resides on a single lanthanoid ion octacoordinated by phthalocyaninato anions to produce a pseudoaxial D_{4d} symmetry.[17] In this case, the energy level splitting of the large total angular

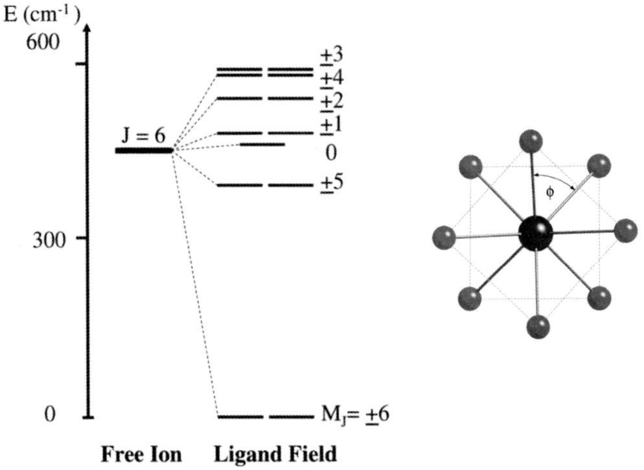

Fig. 5. Energy level scheme for the ground J=6 multiplet of Terbium phthalocyaninato derivative SMM.

momentum J of the Ln^{+3} ion caused by an intense spin-orbit coupling together with the anisotropy of the ligand field around the metal ion, may result in a ground high-M_J doublet and, thus, in the presence of an energy barrier for the J reversal, which gives rise to a slow relaxation of the magnetization. In the case of terbium phthalocyaninato anions, the ground state presents the maximum M_J (±6), with the first excited state around 400 cm^{-1} (Fig. 5). This explains the SMM behavior and the high blocking temperature observed (at ca. 40–50 K). In the present case, Dy and Ho derivatives also behave as SMMs, in agreement with their high M_J ground states.

The main difference between the SMMs based on magnetic clusters and the SMM based on mononuclear lanthanoid complexes is that in the latter the quantum tunneling effects are dominant and difficult to avoid. Thus, even if in the lanthanoid complexes the energy barriers are quite high (above room temperature), the presence of quantum tunneling effects dominates the low temperature magnetic behavior and therefore, it is difficult to keep a memory effect, even at low temperatures. In these quantum systems, a highly symmetric axial crystal field around the lanthanide is crucial in order to keep the purity of the M_J doublets and to ensure a SMM behavior. That is why this kind of behavior has been only observed in the pseudoaxial D_{4d}

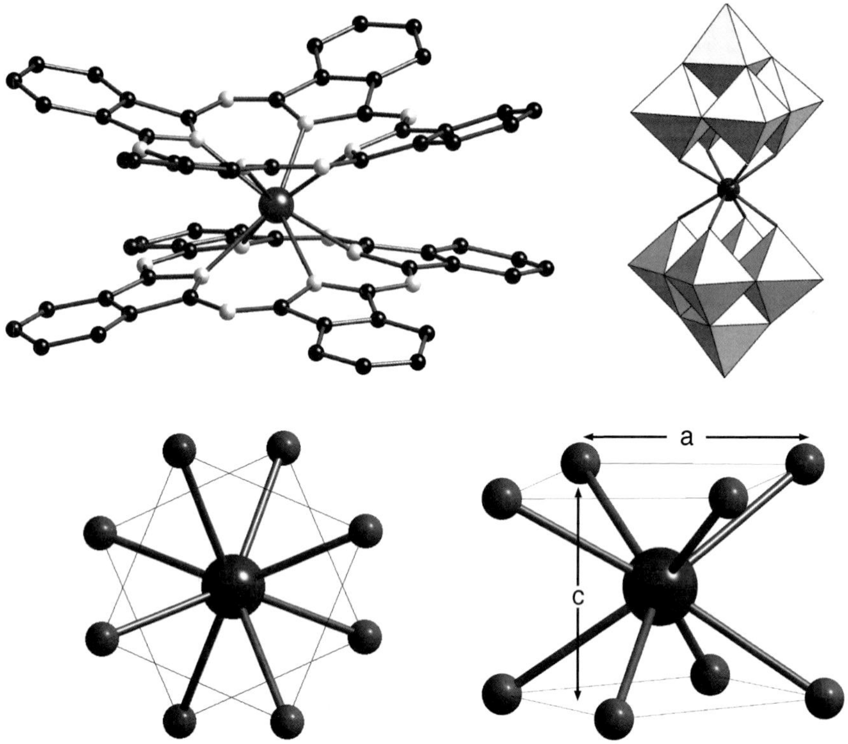

Fig. 6. Above: Structures of phthalocyaninato (left) and POM (right) derivatives. Below: Detail of the coordination structure of approximate D_{4d} symmetry, a square antiprism of side a and height c.

symmetry. It seems obvious that many other lanthanide complexes should satisfy these structural and electronic requisites. Indeed, the lanthanoid coordination sphere of several POM complexes exhibits the same geometry as the phthalocyaninato systems (Fig. 6). As shown in this figure, the structure of these complexes is also approximately D_{4d}. The main difference with the bisphthalocyaninato case concerns the longitudinal distorsion.

Thus, the antiprism formed by 8 nitrogen atoms in the bisphthalocyaninato molecules has a height c=2.76 Å, which is very similar to the square side a=2.80 Å. In contrast, in the polyoxometalate $[Ln(W_5O_{18})_2]^{9-}$ the 8 oxygen atoms form a compressed antiprism, with c=2.47 Å vs a=2.86 Å. This simple picture is slightly complicated if one considers the direction of the lone pairs in the sp2 nitrogen atoms.[18] This small structural variation leads to different LF parameters and, in particular, to a variation of the sign of the axial ZFS parameter of order 2, $A^0_2r^2$. As a consequence, the lowest-lying energy levels for Er(III) and Tb(III) are very different and almost reversed in the two classes of complexes. In fact, while in the case of Er(III) the POM ligands stabilize larger $|M_J|$ values, the phthalocyaninato ligand stabilizes the lowest value, $M_J = \pm 1/2$. In a similar way, in the Tb(III) case, the POM derivatives stabilize the singlet $M_J = 0$, while the doublet with the largest J value, $M_J = \pm 6$, becomes the ground state in the phthalocyaninato derivative. Frequency-dependent out-of-phase signal was detected in ac magnetometry on Lindqvist-type series $[Ln(W_5O_{18})_2]^{9-}$ (Ln(III) = Ho, Er) and Keggin-type series $[Ln(\beta_2\text{-}SiW_{11}O_{39})_2]^{13-}$ (Ln(III) = Dy, Ho, Er, and Yb).[19] Figure 7 shows the ac susceptibility measurements for the case of $[Er(W_5O_{18})_2]^{9-}$. After this first series of POM-based mononuclear SMMs, other examples with more exotic geometries such as a fivefold axis have been described.[20]

On the other hand, one can notice that the magnetic relaxation processes are faster in POMs than in the phthalocyaninato complexes showing SMM behavior. The reason can be related with the smaller separation between the lower-lying energy levels caused by the LF splitting, which in POM derivatives often contain two or more levels separated by less

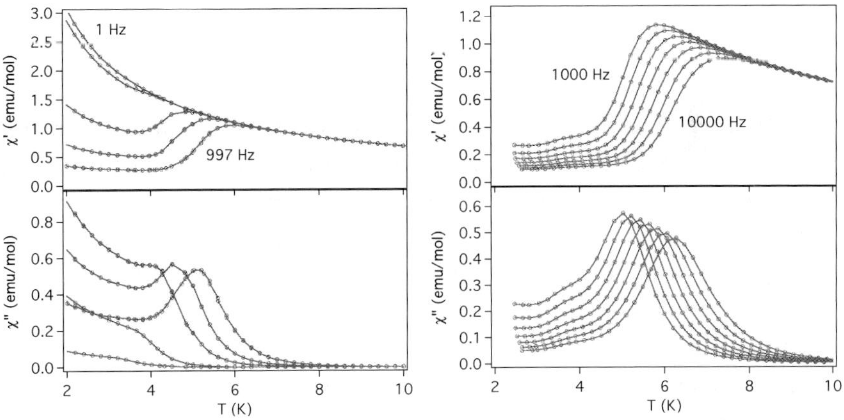

Fig. 7. In-phase and out-of-phase dynamic susceptibility of $Na_9[Er(W_5O_{18})_2]\cdot xH_2O$ (*left*). Low-frequency measurements from top to bottom: 1, 10, 110, 332, and 997 Hz and (*right*) High-frequency measurements from left to right: 1000, 1500, 2200, 3200, 4600, 6800, and 10000 Hz.

than 30 cm^{-1}. Another related question is the observation that, in general, for a given lanthanide, the magnetization of the Keggin-type series relaxes faster than that for the Lindqvist one. This is due to the presence of non-axial distortions in the Keggin-type derivatives (the two coordinating planes are not parallel). In fact, apart from having a non-negligible influence on the sublevel structures, this distortion can also mix the different wave functions, therefore enhancing the quantum tunneling.

A final point that deserves to be discussed concerns the divergence in χ'' observed in this class of mononuclear lanthanide SMMs at temperatures below the blocking temperature. Such a behavior is specific for this class of SMMs and is in sharp contrast to that observed in the cluster-based SMMs (Mn_{12}, for example) in which χ'' tends to vanish below the blocking temperature. In the mononuclear lanthanide SMMs, both processes, the thermally activated relaxation process giving rise to the observation of a maximum in χ'', due to the superparamagnetic blocking of the magnetic moments, and a very fast quantum tunneling process within the $\pm M_J$ ground-state doublet giving rise to a divergence in χ'', seem to coexist. It is remarkable that the spin relaxation in the quantum tunneling regime can be tuned "a la carte" both chemically and physically.

This unique feature has been shown in the $Na_9Er(W_5O_{18})_2$.[21] Thus, it has been found that the quantum tunneling process can be blocked at low temperatures by diluting the magnetic sample in the Yttrium analog, or by applying an external magnetic field. Furthermore, it has also been found that quantum tunneling provides a spin relaxation that is ten orders of magnitude more efficient than what is predicted by accepted theories. Speculative explanations for this fast spin-lattice energy flow have been proposed based on the collective emission of single phonons by a group of spins.

Compared with the phthalocyaninato systems, these POMs present several advantages, starting from their chemical and thermal stability. The POM diamagnetic framework also provides a better magnetic insulation of the lanthanoid, and its rigidity and near-ideal symmetry are helpful for rational design efforts.[22] Furthermore, insofar as the composition of some POM shells only includes W/Mo, O and maybe Si, they can in principle be prepared as nuclear spin-free systems. As the interaction with nuclear spins is a major source of decoherence for electron spins, these molecules are in a nearly unique position to function as model spin-1/2 systems (spin qubits) for sophisticated experiments aiming to address fundamental problems in quantum computing.[23]

Finally, one can mention that there are three-dimensional analogues of these magnetic molecules in which SMM behavior are also expected. These solid-state materials are doped metal oxides or metal halides, like the scheelite-type $Er_xCa_{(1-x)}WO_4$[24] and $LiHo_xY_{(1-x)}F_4$.[25] Those systems, while chemically extended, are expected to be magnetically very similar to mononuclear lanthanoid SMMs, as the magnetic properties and the spin dynamics depend mainly on the coordination sphere of the single lanthanoid ions.

4. Mixed-valence Clusters

POMs provide unique examples in coordination chemistry of large nuclearity mixed valence clusters in which a variable number of electrons can undergo a rapid electron transfer from one center to the other. As in most high-nuclearity mixed-valence systems, the following electronic processes are present in these systems:

- Magnetic superexchange, J, when two spins occupy neighbouring sites, favoring either a ferromagnetic or antiferromagnetic coupling between them
- Electron transfer, t, between adjacent “occupied” and “empty” sites i.e. between two metals with different oxidation state
- Coulombic repulsion, V, between every pair of electrons, that tends to place the electrons far apart and thus distribute them uniformly.

Additionally, POMs have the particularity of being electron sponges, i.e. the same cluster structure can usually host a variable number of electrons. This gives rise to a variety of behaviors, from the strictly diamagnetic cases to the molecules where the number of electrons and thus the effective magnetic exchange can be raised or lowered.

This magnetic control can be achieved either physically by an electrical gating in a single-molecule setup[26,27] or chemically, in bulk, at the synthetic level.[28]

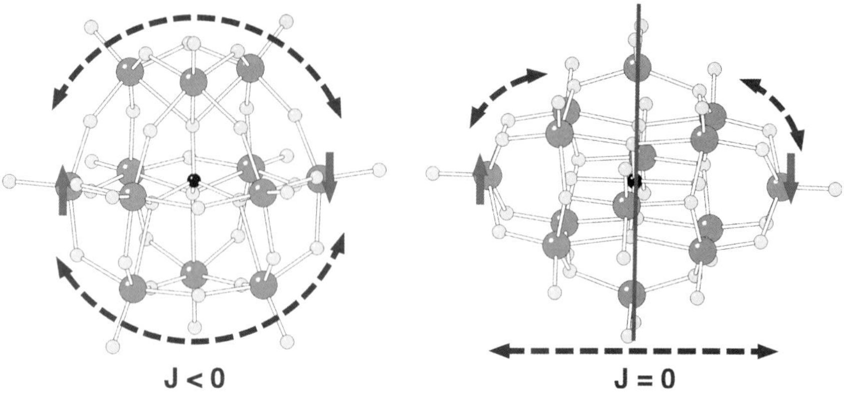

Fig. 8. Structure of the two electron reduced $[PW_{12}O_{40}]^{5-}$ and $[GeV_{14}O_{40}]^{8-}$ anions. Blue arrows show the electron delocalization: complete in the case of W_{12} giving a strong antiferromagnetic coupling between the spins, and partial in the case of V_{14} giving a weak interaction between the spins. **See Color Plate 14.**

A classic example of diamagnetic mixed-valence POM is provided the dirreduced Keggin anion $[PW_{12}O_{40}]^{5-}$ (Fig. 8, left). The diamagnetism of this heteropolyblue was originally suspected to be originated by multiroute superexchange mechanism. Later we demostrated that the large spin-triplet gap could originate from the interplay between Coulombic repulsion and electron transfer, provided the transfer routes had a certain relation in sign and magnitude,[29] and eventually these were estimated by *ab initio* methods and this hypothesis was confirmed.[30] Indeed, while the electrons are far away from each other at all times in the ground state, the possible electron transfer routes are different for singlets and triplets, and the result is a strong stabilization of the singlet. Note that in these high-nuclearity systems, the dynamic Jahn-Teller problem i.e. the coupling between electron transfer and molecular vibrations, is highly complex and its solution requires a symmetry-adapted approach.[31]

Less often, POMs host two electrons that do not interact with each other and thus behave as independent $S=1/2$ spins. That is the case of dirreduced $[GeV_{14}O_{40}]^{8-}$ (Fig. 8, right).[32] This system was studied by *ab initio* calculations on embedded fragments. Surprisingly, both the connectivity of the molecule and the magnitude of most parameters in the Hamiltonian are similar to those of decatungstate, which is diamagnetic.[33] And yet the diagonalization of the t-J–ε Hamiltonian for the whole molecule using the *ab initio* parameters resulted in two independent $S = 1/2$ spins, in full agreement with the experimental result.

The key to explain this apparently anomalous behavior lies in the central square $[GeV_4O_{12}]$ of the cluster, which is sandwiched between the two square pyramids $[V_5O_{14}]$. Because of a combination of poor electron transfer between the square pyramids and a high orbital energy of the central square, whose vanadia have two apical oxo anions instead of one, they act as a barrier that effectively isolates the pyramids.

In view of the electronic versatility of these MV clusters, one can think on the application of an electric field to manipulate the coupling between the spins. This question is an open challenge in molecular nanospintronics that deserves to be investigated both theoretically and experimentally. So far, some theoretical proposals have been developed in this context. A first example is provided by the previous $[GeV_{14}O_{40}]^{8-}$.

Indeed, if a strong enough electric field is applied to this POM, the electronic distribution will change and spins that were previously independent will be forced to approach and to be strongly coupled. A second example is provided by $[PMo_{12}O_{40}(VO)_2]^{q-}$. In this POM two $(VO)^{2+}$ are sitting at opposite tetragonal sites of the Keggin anion and host a $S = 1/2$. On the other hand, the central Keggin structure $PMo_{12}O_{40}$ can act as electron acceptor accommodating a variable number of electrons. As a result, an effective indirect coupling between the $(VO)^{2+}$ spins is expected to appear, which will depend on the number of electrons delocalized over the central Keggin cluster. Such a possibility has been exploited to control the exchange interactions through an electric current applied in a device formed by a single POM connected through a STM setup (Fig. 9). This device has been proposed to serve as a molecular quantum gate between two spin qubits: by a precise control of the magnetic coupling via an electric voltage it would be possible to implement a "square root of NOT" gate, which, together with arbitrary single-qubit rotations (EPR pulses) constitutes a universal set of quantum gates.[34]

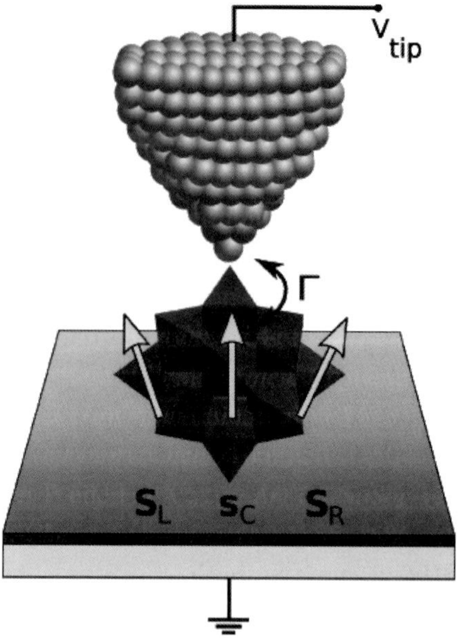

Fig. 9. Schematic drawing of the polyoxometalate $[PMo_{12}O_{40}(VO)_2]^{q-}$ separated by an insulating tunneling barrier from a metallic surface and contacted via a tunnel coupling G to an STM tip at a potential V_{tip}.

5. Conclusions

In this chapter, we have illustrated the richness of polyoxometalate chemistry in providing ideal examples of spin clusters, single-molecule magnets and mixed-valence clusters of increasing nuclearities and topological and electronic complexities. The high symmetry of the polyoxometalate framework facilitates the development of exact quantum-mechanical models from which a clear picture of the relevant parameters involved in the magnetic properties can be extracted. The different cases presented here exemplify this assessment.

As far as the spin clusters are concerned, we have shown that these systems constitute ideal models for studying the exchange interactions at the molecular level, allowing for the study of the exchange anisotropy in orbitally-degenerate metal ions (like Co(II)) using non conventional spectroscopic techniques as the inelastic neutron scattering.

As far as the single-molecule magnets are concerned, we have shown that POM complexes containing a single lanthanide ion can provide unique examples of new SMMs thanks to the high symmetry and ideal magnetic isolation of the lanthanide provided by the structure of the polyanion. A plethora of new SMMs can be foreseen by a suitable choice of POM ligands, providing an excepcional opportunity to investigate and even to tune quantum effects in these molecular nanomagnets.

Finally, as far as mixed-valence clusters are concerned, we have shown that POMs constitute one of the few chemical examples in which it is possible to quantitatively model the magnetic and electronic properties of these complex magnetic systems. Thus, exact solutions have been developed in some cases that provide important hints on the role played by the electron transfer processes on the magnetic properties of these clusters. For example, it has been shown that in these structures, electron delocalization can result in a strong antiferromagnetic coupling between widely separated blue electrons. On the other hand, we have shown that MV POMs can also have an important impact in molecular electronics, as they can provide examples of magnetic molecules in which an electric control of the spin-spin coupling might be possible.

In conclusion, the impact of POMs in molecular magnetism is increasing and evolving towards their applications in nanotechnology. Thus, while magnetic polyoxometalates pose complex theoretical problems and tough experimental challenges in molecular magnetism, these nanomagnets open a window for the manipulation of quantum spins at the molecular level, a nanotechnological feat that would be crucial not only in nanomagnetism but also in molecular spin electronics (spintronics) and quantum computing.

Acknowledgements

We gratefully acknowledge financial support from Spanish MICINN (CSD2007–00010 CONSOLIDER- INGENIO in Molecular Nanoscience, MAT2011-22785, Generalitat Valenciana (PROMETEO program), and the EU (ELFOS project and ERC Advanced Grant SPINMOL) for financial support. A.G.A. thanks the European Union for a FP7-PEOPLE-2007–4-1- IOF contract.

References

1. Kortz, U., *et al.*, *Coord Chem Rev* **253** (2009), 2315–2327
2. Camarero, J. and Coronado, E., *J Mat Chem* **19** (2009) 1678–1684.
3. Lehmann, J, Gaita-Ariño, A, Coronado, E. and Loss, D. *J Mat Chem* **19** (2009) 1672–1677.
4. Clemente-Juan, J. M., Coronado, E. and Gaita-Ariño, A. *Chem Soc Rev* **41** (2012) 7464–7478.
5. Martínez-Pérez, M. J., *et al.*, *Phys Rev Lett* **108** (2012) 247213.
6. Long, D.-L. Burkholder, E. and Cronin, L., *Chem Soc Rev* **36** (2007) 105–121 and references therein.
7. Clemente-Juan, J. M. and Coronado, E., *Coord Chem Rev* **193–195** (1999) 361–394.
8. Palii, A., *et al.*, *Chem Soc Rev* **40** (2011) 3130–3156.
9. Gómez-García, C. J., *et al.*, *Physica B* **181** (1992) 238.
10. Andres, H., *et al.*, *J Am Chem Soc* **121** (1999) 10028–10034.
11. Clemente-Juan, J. M., *et al.*, *Inorg Chem* **44** (2005) 3389–3395.
12. Sessoli, R., *et al.*, *Nature* **365** (2009) 141–143.
13. Friedmam, J. R., *et al.*, *Phys Rev Lett* **76** (1996) 3830–3833.
14. Thomas, L., *et al.*, *Nature* **383** (1996) 145–147.
15. Compain, J. -D., *et al.*, *Angew Chem* **121** (2009) 3123–3127.
16. Chiorescu, I., *et al.*, *Phys Rev Lett* **84** 15 (2000), 3454–3457.
17. Ishikawa, N., *Polyhedron* **26** (2007) 2147–2153.
18. Baldoví, J. J., *et al.*, *Dalton Trans* **41** (2012) 13705–13710.
19. AlDamen, M.A., *et al.*, *Inorg Chem* **48** (2009) 3467–3479.
20. Cardona-Serra, S., *et al.*, *J Am Chem Soc*, **134** (2012) 14982–14990.
21. Luis, F., *et al.*, *Phys Rev B* **82** (2010) 060403.
22. Baldoví, J.J., *et al.*, *Inorg Chem* **51** (2012), 12565–12574.
23. Stamp P. C. E. *et al.*, *J Mat Chem* **19** (2009) 1718–1730.
24. Bertaina, S., *et al.*, *Nat Nanotechnology* **2** (2007) 39–42.
25. Giraud, R., *et al.*, *Phys Rev Lett* **87** (2001) 057203.
26. Bosch-Serrano, C., *et al.*, *Chem Phys Chem* **13** (2012) 2662–2665.
27. Bosch-Serrano, C., *et al.*, *Phys Rev B* **86** (2012) 024432.
28. Botar, B., *et al.*, *Angew Chem Int Ed* **48** (2009) 9080–9083, and references therein.
29. Borrás-Almenar, J. J., *et al.*, *Chem Phys* **195** (1995) 1–15.

30. Suaud, N., *et al.*, *Chem Eur J* **10** (2004) 4041–4053.
31. Tsukerblat. B., *et al.*, *Int. J Quant Chem* **112** (2012) 2957–2964.
32. Kortz, U., *et al.*, *J Cluster Sci* **17** (2006) 143.
33. Clemente-Juan, J.M., *et al.*, *J PhysChem A* **111** (2007) 9969–9977.
34. Lehmann, J., *et al.*, *Nat Nanotechnology* **2** (2007) 312–317.

Chapter 5

MAGNETISM OF KEPLERATES

PAUL KÖGERLER

Institute of Inorganic Chemistry, RWTH Aachen University, 52074 Aachen, Germany and Peter Grünberg Institute, PGI-6, Research Centre Jülich, 52425 Jülich, Germany

1. Introduction

Polyoxometalate chemistry has proven a thoroughly fertile ground for the design of molecular spin clusters.[1–4] Following early breakthroughs in the early 1990s, in particular with reduced polyoxovanadate(IV) clusters such as the seminal $[V^{IV}_{15}As^{III}_{6}O_{42}(H_2O)]^{6-}$ polyanion,[5] an archetypal example for a frustrated quantum spin system[6] with a considerable coherence lifetime,[7] the development of magnetically functionalized polyoxomolybdates and polyoxotungstates was accelerated by the emerging synthetic concepts of virtual building blocks, characterized by specific rigid shapes and connectivity modes and present in dynamic 'virtual libraries' in aqueous solutions.[8] Their seemingly deliberate use and combination served as the conceptual foundation for the generation of a large family of magnetic polyoxometalates, frequently comprising discrete or networked spin arrays of unusual symmetries and connectivities that are not readily accessible via conventional coordination chemistry. It did not take long for these systems to attract the attention of condensed matter physics: Low-dimensional spin arrays of high symmetry promise intriguing physical characteristics ranging from the potential onset of cooperative phenomena to spin frustration effects.[9,10] Herein, we will briefly summarize some key developments of magnetically functionalized Keplerates, a class of highly symmetrical spherical polyoxometalate cluster molecules which continue to reveal fascinating magnetic phenomena of purely molecular origin since their discovery nearly 15 years ago.

2. From Pentagonal Building Blocks to Keplerates

The family of donut-shaped $\{Mo_{57}M_6\}$ clusters was one of the first that showcased the surprising potential of polyoxomolybdates, themselves diamagnetic, to function as versatile scaffold structures into which a wide range of paramagnetic heterometal ions could be integrated.[11,12] The cluster anions comprise three $[Mo_{15}(MoNO)_2O_{58}(H_2O)_2]^{20-}$ $(=\{Mo_{17}(NO)_2\})$ building blocks that are interlinked to a triangular structure via three $[O_3Mo^V(\mu_2\text{-}OH)(\mu_2\text{-}OH_2)_2Mo^VO_3]^{2-}$ $(= \{Mo^V{}_2\})$ groups, situated on the equatorial plane, as well as via six cationic heterometal centers, e.g. VO^{2+}, $Fe^{2+/3+}$, Cu^{2+}, or mixtures thereof (Fig. 1).[13] Note that the shell-shaped, C_{2v}-symmetric $\{Mo_{17}(NO)_2\}$ fragments, which are essential to the curved geometry of the resulting $\{Mo_{57}M_6\}$ clusters, illustrate the "building block" character of such substructures: they are also the main constituents of the (diamagnetic) $[Mo_{36}(NO)_4O_{108}(H_2O)_{16}]^{12-}$ $(=\{Mo_{36}(NO)_4\})$-type anion where they are joined via two additional $[MoO_2]^{2+}$ groups.[14] Not surprisingly, several $\{Mo_{57}M_6\}$-type clusters can be derived from reactions of $\{Mo_{36}(NO)_4\}$-type anions with transition metal cations. Also note that the Mo(4d) electrons of the $\{Mo^V{}_2\}$ groups are fully spin-paired, as the short Mo···Mo distances (ca. 3.3 Å) resulting from the face-sharing arrangement of the two condensed Mo^VO_6 octahedra in these groups translate into metal-metal interactions.

In the $\{Mo_{57}M_6\}$ structures the two M_3 triangles are part of hexagonal $(\text{–}Mo^{VI}\text{–}O\text{–}M\text{–}O\text{–})_3$ fragments that represent the superexchange pathways relevant for the intra-triangle exchange and simultaneously form the two 'pores' to the interior cavity of the cluster anions. The coordination to the relevant oxo positions of the $\{Mo_{17}(NO)_2\}$ groups provides a near-regular square O_4 coordination environment for each metal cation that is completed by one ($M = VO^{2+}$, Cu^{II}) or two ($M = Fe^{II}$, Fe^{III}) terminal water ligands. The resulting trigonal-prismatic M_6 fragment inherits the full D_{3h} symmetry of the donut-shaped $\{Mo_{57}\}$ host framework. The intramolecular magnetic superexchange between the six heterometal spin centers is efficiently mediated by M–O–Mo–O–M (ca. 6.6 Å) or longer M–O–Mo–O–Mo–O–M (ca. 9.6 Å) bridges, which give rise to significant exchange energies even over M···M distances approaching one nanometer.[1] Depending on the nature of the heterometal, the coupling can be antiferromagnetic or ferromagnetic.[1,15]

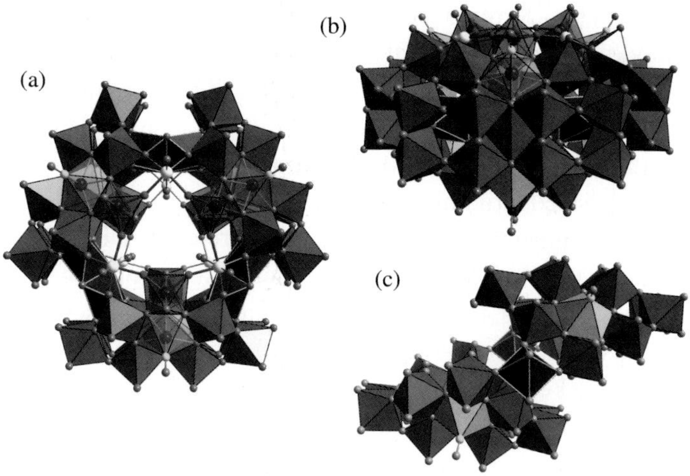

Fig. 1. Structure of the $[H_3Mo_{57}V_6(NO)_6O_{183}(H_2O)_{18}]^{21-}$ (= $\{Mo_{57}V_6\}$)-type cluster anion, determined from single-crystal X-ray diffraction analysis of its ammonium salt, $(NH_4)_{21}\{Mo_{57}V_6\}\cdot 65H_2O$. The Mo atoms and MoO_n polyhedra of the $\{Mo_{17}(NO)_2\}$ groups are shown in blue (the MoO_7 pentagonal bipyramids in light blue); the $\{Mo^V{}_2\}$ groups as dark gray polyhedra. O: light gray, N: green, V: yellow spheres. (a) View along the molecular C_3 axis in central projection highlighting the frontal of the two $(\text{–V–O–Mo–O–})_3$ pores by brown bonds and red O atoms. The coordination polyhedra of the Mo positions belonging to this pore as well as to the adjoined seven-coordinated Mo sites are shown transparent. This pore motif represents the general superexchange pathway of the M_3 groups in both the $\{Mo_{57}M_6\}$ structures as well as the $\{Mo_{72}M_{30}\}$-type Keplerates. (b) Side view approx. perpendicular to the view in (a), showing the symmetrical structure of the arched $\{Mo_{17}(NO)_2\}$ building block which consists two $\{Mo_8(NO)\}$ groups joined by an additional MoO_6 octahedron (purple). (c) Structure of the $\{Mo_{36}(NO)_4\}$ anion consisting of two $\{Mo_{17}(NO)_2\}$ groups fused to two MoO_6 octahedra (dark grey). **See Color Plate 15.**

Although the $\{Mo_{57}M_6\}$ clusters sparked considerable interest when they were first isolated, they harbor a structural motif that was soon thereafter established as the critical ingredient for numerous high-nuclearity polyoxomolybdate structures: The $\{Mo_{17}(NO)_2\}$ groups comprise two $\{Mo_8(NO)\}$ units fused together by an additional Mo atom. The $\{Mo_8(NO)\}$ groups themselves contain a central pentagonal (ON) MoO_6 bipyramid, sharing its equatorial edges with five MoO_6 octahedra. (The situation is analogous for the non-nitrosylated $\{Mo_{17}\}$ groups in the $[Mo_{36}O_{112}(H_2O)_{16}]^{8-}$ = $\{Mo_{36}\}$ isopolyoxomolybdate, where the

$[MoNO]^{3+}$ groups are replaced by $[MoO]^{4+}$ groups.[16]) Due to this connection mode, this nearly planar, regular pentagonal $\{(Mo)Mo_5\}$ building block is compact and stable. In the $\{Mo_8(NO)\}$ building blocks, two additional MoO_6 groups share corners with the nearly planar (usually slightly domed) central $\{(Mo)Mo_5\}$ fragment. Following the discovery of the $\{Mo_{17}\}$-based structures, attempts were made to selectively break this less rigid connection mode in order to release, and subsequently employ, the central $\{(Mo)Mo_5\}$ fragments as pentagonal building blocks. This goal was furthermore motivated by the discovery of the giant mixed-valent wheel-shaped $\{Mo_{154}\}$- and $\{Mo_{176}\}$-type cluster anions[17–19] as the key constituents of soluble 'molybdenum blue', i.e. the intensely colored species that rapidly self-assemble upon reduction of sufficiently acidified aqueous molybdate solutions.[20,21] These (diamagnetic) cluster types underscored the important role of building blocks in polyoxomolybdate chemistry: the wheels can (formally) be constructed from sets of 14 or 16 identical $\{Mo_8\}$ (or $\{Mo_8(NO)\}$), $\{Mo_2\}$ and $\{Mo_1\}$ building blocks. Moreover, lacunary derivatives lack some of the $\{Mo_2\}$ groups that define the inner rims of the wheels,[22] underscoring that their exclusively corner-sharing connection modes with the $\{Mo_8\}$ groups represent the weakest links in these clusters. The concept of transferable building blocks in molybdenum blue chemistry was later on supported furthermore by an even larger lemon-shaped $\{Mo^{V/VI}{}_{368}\}$-type cluster that again is constructed exclusively from $\{Mo_8\}$, $\{Mo_2\}$ and $\{Mo_1\}$-type groups.[23]

In mid-1998, A. Müller and co-workers then succeeded in isolating a polyoxomolybdate structure in which the $\{(Mo)Mo_5\}$ pentagon fragments for the first time assumed the intended role of discrete pentagonal building blocks.[24] Using reducing but slightly less acidic reaction conditions that favor smaller polyoxomolybdate building blocks and providing a high concentration of acetate, which supports the formation of carboxylate-bridged $\{Mo^V{}_2\}$ groups, the group was able to crystallize an ammonium salt of a spherical polyoxomolybdate consisting of twelve pentagonal $[Mo^{VI}{}_6O_{21}(H_2O)_{21}]^{6-}$ groups, the edges of which were interlinked by thirty of the $\{Mo^V{}_2\}$ groups (Fig. 2).

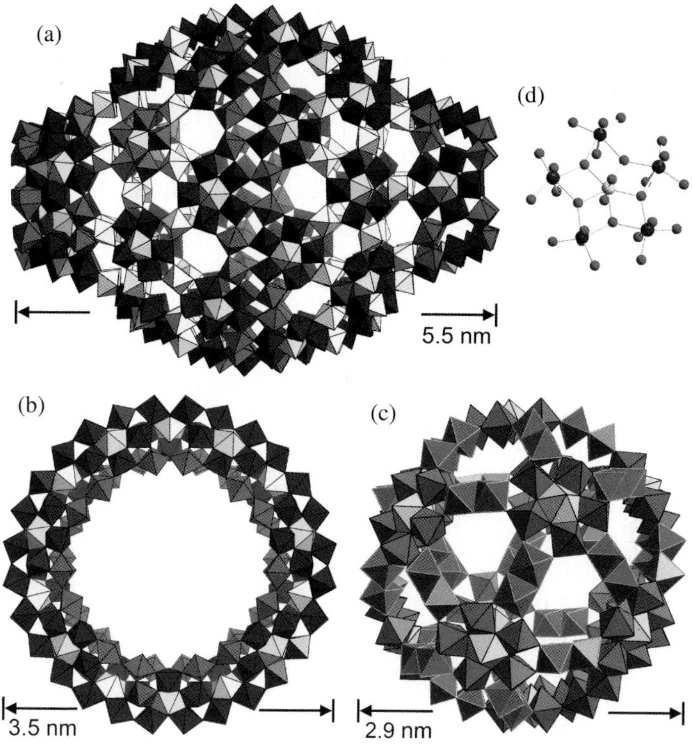

Fig. 2. Comparison of the mixed-valent polyoxomolybdate(V/VI) cluster anions $\{Mo_{368}\}$ (a), $\{Mo_{154}\}$ (b), and $\{Mo_{132}\}$ (c) which are all based on the pentagonal $\{(Mo)Mo_5\}$ building block (light/dark blue) that is displayed in ball-and-stick representation in (d). Red: $\{Mo_2\}$-type groups, yellow: mononuclear Mo linker groups. Additional ligand groups (sulfate in the case of $\{Mo_{368}\}$, carboxylates in the case of $\{Mo_{132}\}$) are omitted for clarity. **See Color Plate 16**.

The resulting, highly (I_h) symmetric $\{Mo_{132}\}$ cluster sphere was immediately heralded as a new "organizational form of matter" and dubbed a Keplerate, highlighting the analogies to the (speculative) model of the solar system devised by Johannes Kepler at the end of the 16th century, in which the spheres of the planets' orbits are inscribed within and circumscribed around by the five platonic solids.[25] In the case of the Keplerate cluster structures, the twelve Mo positions central to the $\{(Mo)Mo_5\}$ groups define the vertices of an icosahedron, the center points of the twenty pores of the structure (see below) define a dodecahedron, and the center points of the thirty linker groups define an icosidodecahedron.

3. The $\{Mo_{72}Fe_{30}\}$ System: From Classical to Quantum Spin Models

Following the first Keplerate, the $\{Mo_{132}\} = \{(Mo^{VI})Mo^{VI}_5\}_{12}\{Mo^{V}_2\}_{30}$ cluster, it soon became clear that the highly anionic $\{(Mo)Mo_5\}$ fragments could be effectively retained under oxidative conditions that caused degradation of the $\{Mo^{V}_2\}$ groups — as long as the acidity of the reaction solution was sufficiently high and as long as other cationic linker groups were present to replace the $\{Mo^{V}_2\}$ groups. In 1999, this effective substitution of linker groups was successfully proven for the direct reaction of $\{Mo_{132}\}$ with Fe^{III} at pH 2.0 and using air oxygen as oxidizing agent, resulting in the first magnetic Keplerate structure, the neutral cluster $\{Mo_{72}Fe_{30}\} = [Mo^{VI}_{72}Fe^{III}_{30}O_{252}(H_2O)_{91}L_{12}L'_2L'']$ (L = acetate, $L' = [Mo^{VI}_2O_7(H_2O)]^{2-}$, $L'' = [H_2Mo^{VI}_2O_8(H_2O)]^{2-}$) which crystallized as the hydrate $\{Mo_{72}Fe_{30}\}$·ca.150H_2O.[26] In this cluster the thirty Fe^{III} centers — a record spin center nuclearity at that time — reside in nearly regular octahedral FeO_6 coordination environments (Fe–O: 2.0–2.1 Å), with two *trans*-positioned terminal H_2O ligands completing the square-planar O_4 environment defined by four oxo positions belonging to two neighboring $\{(Mo)Mo_5\}$ groups, and represent spin-only ($^6A_{1g}$, $S = 5/2$, no orbital momentum) centers. The Fe ions here themselves define the center position of the $[Fe(H_2O)_2]^{3+}$ linker groups and thus span an Fe_{30} icosidodecahedron, an Archimedean solid with 60 nearly identical edges (Fe···Fe: 6.4–6.5 Å) consisting of 20 corner-sharing equilateral Fe_3 triangles grouped around the twelve regular pentagonal building blocks (Fig. 3).

Within the Fe_{30} substructure each Fe atom has four nearest neighbors. As in the $\{Mo_{57}M_6\}$ clusters, the triangular Fe_3 motifs are part of the $(–Mo^{VI}–O–Fe–O–)_3$ rings which and are also nearly isostructural to the ones found in the $\{Mo_{57}M_6\}$ structures. Note the various additional ligand groups (acetate, $\{Mo_2\}$) coordinate to the $\{(Mo)Mo_5\}$ groups from the inside of the cluster shell in a disordered fashion (see Ref. 27 for a detailed description of these $\{Mo_2\}$ groups L′ and L″).

The first temperature-dependent magnetic susceptibility measurements of $\{Mo_{72}Fe_{30}\}$·150H_2O (1.8–290 K, 0.5 Tesla) indicated weak antiferromagnetic coupling between the 30 Fe^{III} centers, and down to approx. 10 K the compound can be well reproduced by a Curie-Weiss-type expression,

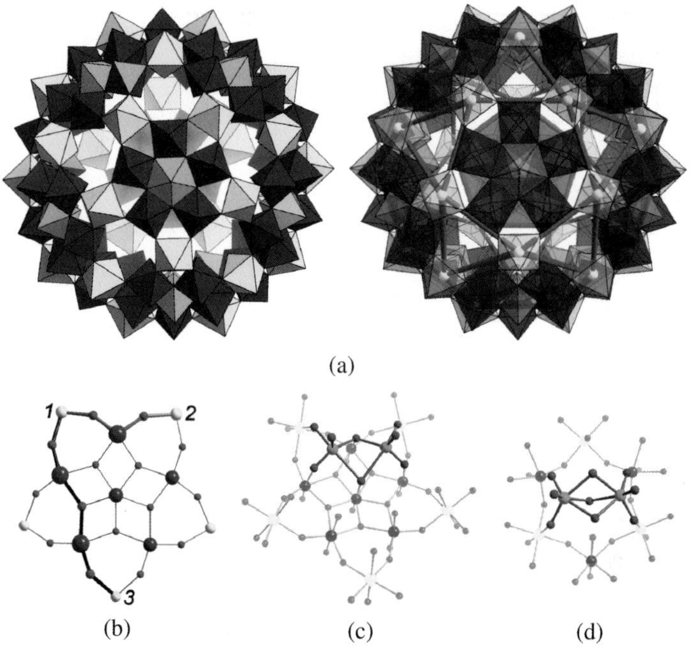

Fig. 3. (a) Polyhedral representation of the (charge-neutral) $\{Mo_{72}Fe_{30}\}$ cluster (left) and with its inscribed Fe_{30} icosidodecahedron (red lines, right). Acetate and $\{Mo_2\}$-type ligands coordinating to Fe and Mo positions from the inside of the cluster sphere are omitted. (b) Ball-and-stick representation of a substructure of the $\{Mo_{72}Fe_{30}\}$ cluster consisting of an $\{(Mo)Mo_5\}$ group with five surrounding Fe centers (Mo: blue, Fe: yellow, O: red). The nearest-neighbor exchange pathway between Fe positions 1 and 2 via a –O–Mo–O– bridge is indicated by thick gray bonds, the next-nearest neighbor exchange pathway to Fe position 3 by thick dark gray bonds. (c, d) Structural details and coordination modes of the $[Mo_2O_7(H_2O)]^{2-}$ and the $[H_2Mo_2O_8(H_2O)]^{2-}$ ligands, (L′ and L″), respectively. **See Color Plate 17**.

$1/\chi = (T - \theta)/C$, with $\theta = -21.6$ K and $C = 127.9$ cm^3 K mol^{-1}, in accordance to 30 spin-5/2 centers with $g_{iso} = 1.974$ (Fig. 4).[26,28] This, plus the field-dependent magnetization measured at 1.8 K (0.1–5.0 Tesla) that indicated the presence of a small mononuclear Fe^{III} impurity (approx. 0.3 centers per formula unit), pointed to a singlet ground state. Yet, antiferromagnetic coupling meant that the Fe_{30} icosidodecahedron represented a highly symmetric *geometrically frustrated* spin polytopes — in its topology, a molecular analogue of a spin-frustrated Kagomé lattice. Was there no magnetic feature that could justify the excitement the prospect of such

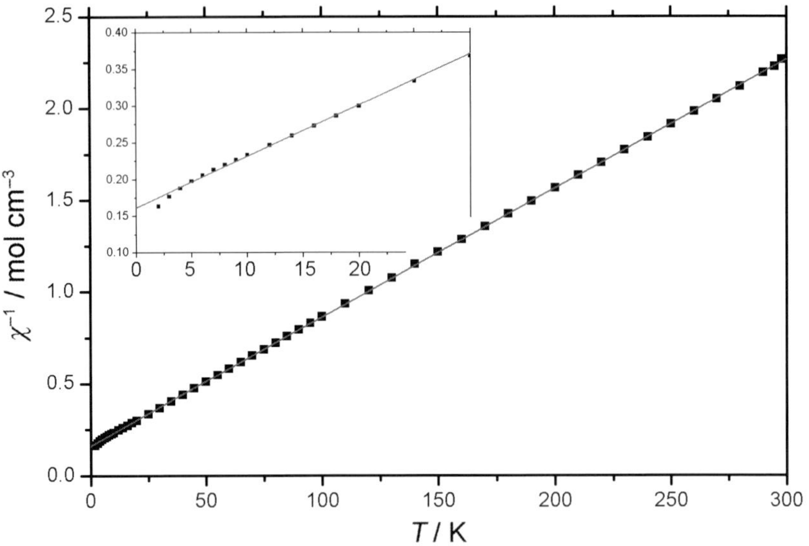

Fig. 4. Temperature dependence of the reciprocal susceptibility χ^{-1} of $\{Mo_{72}Fe_{30}\}$·ca.150 H_2O at 0.5 Tesla. Inset: Low-temperature region showing deviations from a linear behavior below approx. 5 K.

a magnetically frustrated spin icosidodecahedron had caused? In light of the weak exchange coupling, it became clear that susceptibility measurements at temperatures lower than the standard base temperature of a ^{4}He-cooled SQUID magnetometer were needed. In 2000, R. Modler and co-workers succeeded in measuring the low-field susceptibility down to 120 mK using a $^{3/4}$He dilution cryostat.[29] However, even at this temperature the susceptibility showed no sign of decreasing to zero, as would be expected for a diamagnetic ground state but on the contrary appeared to approach a finite value, extrapolated to $\chi_0 = 4.56$ cm^3 mol^{-1} (corrected for diamagnetic and impurity contributions).[28] In order to explain this observation, a microscopic model was needed that explains how 30 spins arrange on the surface of this highly symmetric cluster structure. However, since the Hilbert space dimension for this system $(2S + 1)^N = 6^{30}$ is far too large to numerically solve the quantum mechanical eigenvalue problem for a spin Hamiltonian describing the Heisenberg-type interactions, a simplified approach based on non-quantized description of the spin moments of the Fe^{III} centers was necessary.

First results came from analytical graph theory analysis of a generic regular icosidodecahedron of *classical* spin vectors by M. Luban and M. Axenovich.[30] A first assumption of this model was that the exchange connectivity is restricted to the nearest-neighbor interactions (transmitted by Fe–O–Mo^{VI}–O–Fe bridges), i.e. the 60 edges of the Fe_{30} icosidodecahedron. This implies that the next-nearest neighbor superexchange (transmitted by more extended fragments across the {(Mo)Mo_5} groups) is negligible. The model also assumes all nearest-neighbor exchange interactions to be identical, i.e. characterized by a single exchange energy J. This description exploits the fact that the icosidodecahedron is decomposable in terms of triangles and is three-colorable, i.e. three colors suffice to distinguish each of its 30 vertices from its four differently colored nearest neighbor vertices (Fig. 5). Importantly, the graph-theoretical approach yields a specific geometrically frustrated classical ground state configuration that is characterized by three subsets, each containing ten spin vectors of identical orientation. These three subsets differ in their

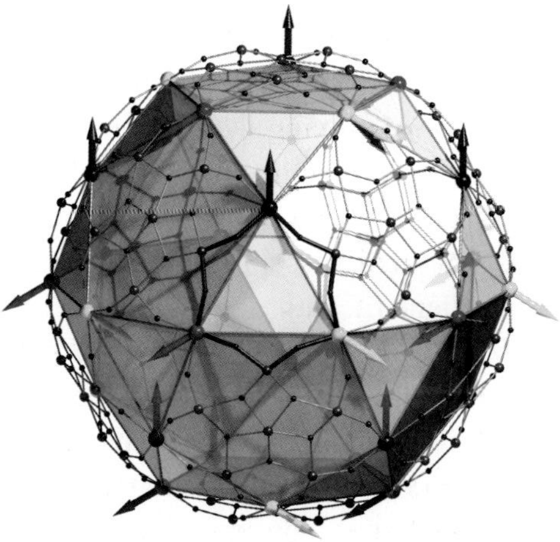

Fig. 5. Superposition of the highly symmetrically frustrated classical ground state configuration onto the {$Mo_{72}Fe_{30}$} cluster structure (terminal oxygen positions and ligands omitted). The ten classical spin vectors of each of the three sublattices (red, green, blue) assume a relative orientation of 120°. Note that the three-colorability of the spin icosidodecahedron results in a configuration where any given spin site is surrounded by four nearest neighbors, all of which belong to different sublattices. One of the 20 triangular $(\text{–Fe–O–Mo–O–})_3$ pores is highlighted by thick brown bonds (cf. Fig. 1). **See Color Plate 18.**

relative vector orientation by 120° (this is complemented by an equivalent 240° configuration, both of these chiral configurations are degenerate). Note that due to the three-colorability of the icosidodecahedron, the assignment of the thirty spins to the three sublattices results in a configuration in which any given spin interacts with two neighboring spins of each of the other two sublattices. Furthermore, the model predicts a low-temperature susceptibility limit of $\chi_0(T \rightarrow 0\,\mathrm{K}) = C/(2|J_c|/k_B) = 4.67$ cm^3 mol^{-2} (k_B here denotes the Boltzmann constant), in very good agreement with the experimental observations. Note that this value corresponds to the classical exchange energy J_c that characterizes the coupling of the classical spin vectors: In this formulation, the quantum mechanical spin operators S_n of the $S = 5/2$ Fe^{III} centers are approximated by scaled classical vectors, $[S(S+1)]^{1/2}\,e_n$, where e_n is a classical unit vector, free to point in any direction of space. Heisenberg-type interaction energies between two such moments n and m then amount to $-J\,S(S+1)\,e_n \cdot e_m = -J_c\,e_n \cdot e_m$. The critical advantage of describing the Fe^{III} centers as classical spin vectors with no quantization constraints of the spin moments is that the Fe_{30} system can now be characterized by an approximate Monte Carlo method that simulates the thermodynamic properties of the molecular ensemble of 30 classical spins as a function of temperature and external magnetic field. This involves solving the equations of motion governing the orientations of the individual unit vectors, coupled to a heat reservoir, which are described by coupled deterministic nonlinear differential equations. Applied to the low-field susceptibility data down to 120 mK this method yields an excellent fit with $J = -1.09$ cm^{-1} for the previously derived isotropic g factor of $g_{\mathrm{iso}} = 1.974$ (note that J here conforms to the quantum mechanical Hamiltonian of the type $H = J\mathrm{S}_i \cdot \mathrm{S}_j$) and approaches the finite χ_0 value that also was derived from the graph theory approach (Fig. 6). Moreover, the Monte Carlo simulation also reproduces the same frustrated classical ground state configuration of three spin sublattices that differ by 120° in their angular direction. As a remark, while the analytical graph theory approach cannot rigorously exclude the possibility of noncoplanar ordering which is degenerate with the planar configuration described, the Monte Carlo simulations showed that all spins are indeed coplanar.

For the evolution of the spin configuration in an increasing external magnetic field, the Monte Carlo results also match the predictions of the

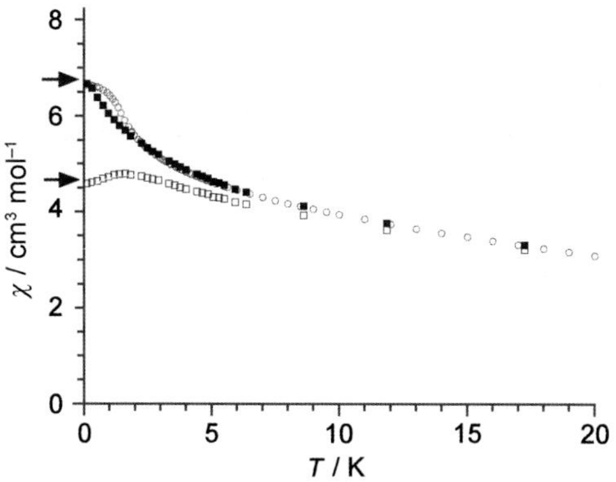

Fig. 6. Temperature dependence of the molar susceptibility χ of $\{Mo_{72}Fe_{30}\}\cdot ca.150H_2O$ at 0.5 Tesla (redrawn from ref. 28), showing that χ approaches a finite value at 0 K. Red circles: experimental data, blue open squares: classical Monte Carlo simulational data, blue solid squares: sum of simulational data plus contribution of paramagnetic impurity. Lower blue arrow: low-temperature limit χ_0 of the classical Heisenberg model. Upper blue arrow: value of χ_0 plus the impurity correction extrapolated to $T = 0$ K. **See Color Plate 18.**

graph theory approach. In the zero-field limit all spin vectors are coplanar and nearest neighbors assume the relative orientation of 120°. With increasing field, the spin vector directions of each of the three sublattices start to cant and gradually align with the external field vector until full alignment, i.e. saturation, is reached at a field of $H_{sat} = (6JS/g_{iso}\mu_B) = 17.7$ Tesla, all while the 120° orientation of the three sublattices is maintained when the spin vectors are projected onto a plane perpendicular to the external field vector — akin to opening or closing an umbrella (Fig. 5).[28] As a consequence, the net magnetization of the Fe_{30} will increase linearly up until H_{sat} and will remain constant ($M/g_{iso} = 75\ \mu_B$) for $H > H_{sat}$. Interestingly, exactly this behavior was observed in a pulsed-field (0–60 Tesla) experiment at 0.46 K. Yet, how could a system display entirely classical properties in what is essentially its corresponding quantum temperature regime (quantum effects in antiferromagnetically coupled magnetic molecules are typically negligible only at temperatures higher than $0.25|J_c|/k_B$, approx. 3.4 K for $\{Mo_{72}Fe_{30}\}$)? Thus, $\{Mo_{72}Fe_{30}\}$ was dubbed a mesoscopic paramagnet in which classical behavior extends down to extraordinarily low temperatures.[28]

To account for the apparent absence of quantum mechanical effects, an approximate quantum model that reproduces the key findings of the two

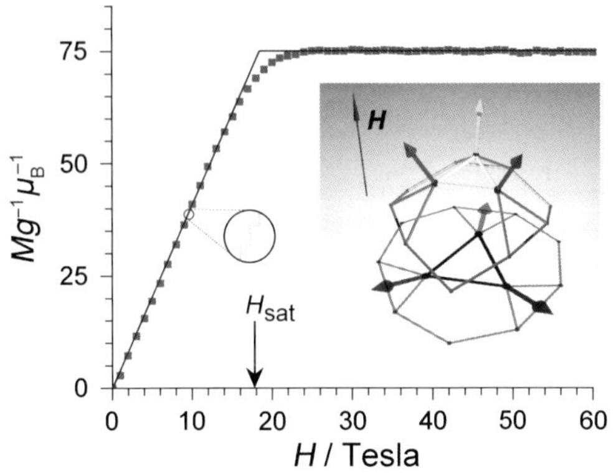

Fig. 7. Field dependence of the molar magnetization M of $\{Mo_{72}Fe_{30}\}$·ca.150H_2O at 0.46 K. Red squares: experimental data, dark blue graph: linear increase expected for the classical spin vector model until the system saturates at $H_{sat} = 17.7$ Tesla. The equidistant quantum spin level crossing steps are indicated by the magnified circular inset. Note that the actual experimental magnetization data density was much higher than shown here and would have allowed identification of the expected level crossing steps. Inset: Simplified representation of the 'umbrella' spin phase (of classical spin vectors) is shown on a section of the Fe_{30} spin array. This spin phase is characterized by a continuous alignment of the spin vectors of the three (red, green, blue) sublattices with an external magnetic field vector, while the relative 120° angular orientation is retained in the projection perpendicular to the H vector. **See Color Plate 19.**

classical spin-based approaches was devised.[31,32] One important assumption incorporated into this model was the finding that for many (non-frustrated) molecular antiferromagnets of N spin centers (with spin quantum numbers S_i) their lowest magnetic excitations can be derived from the rotation of the Neél vector, i.e. a series of states characterized by a total spin quantum number S that runs from 0 to $N \times S_i$. In plots of these magnetic levels as a function of S these lowest S states form rotational (parabolic) bands with eigenvalues proportional to $S(S+1)$.[31–33] While this feature is most evident for non-frustrated systems, the idea of rotational bands can be used to approximate the lowest section of the magnetic excitation spectrum of $\{Mo_{72}Fe_{30}\}$. To then replicate the relative spin orientation of the classical three-sublattice configuration (in the absence of an external field) as well as the rigorous classical ground state energy derived from analytical graph theory, the classical vector operators S_A, S_B and S_C for the three sublattices are simply replaced by quantum-mechanical angular momentum operators, and the resulting effective Hamiltonian now is reduced to interaction between the sublattice *A*, *B*, *C* net spins,

$$H_{\text{eff}} = -J\frac{A}{N}(\mathrm{S}_A\,\mathrm{S}_B + \mathrm{S}_B\,\mathrm{S}_C + \mathrm{S}_C\,\mathrm{S}_A)$$

where the factor A is a function of the spin quantum numbers of the individual spin centers and is specific to the particular spin polygon ($A = 6$ for a spin icosidodecahedron)[29] and $N = 30$ is the total number of spin centers.

The resulting eigenvalues then are a function of the net spin moments of the sublattices,

$$E = \frac{J}{10}\left[S(S+1) - \sum_{a=A,B,C} S_\alpha(S_\alpha+1)\right]$$

The "ground state rotational band" then is assembled from the lowest energy levels for each value of *S*, and is followed by first excited rotational bands that is energetically separated by from the ground state band by 5*J* (Fig. 8).

An external field, accounted for by an additional Zeeman term in the Hamiltonian, lifts the degeneracies of the individual $M_s = -S, -S+1, \ldots,$

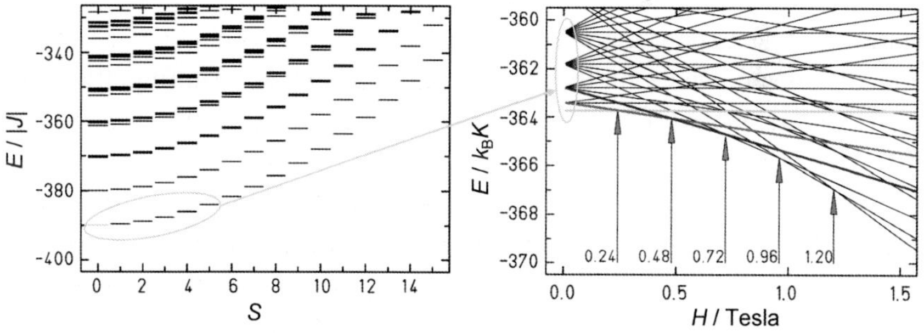

Fig. 8. *Left*: Lowest and some excited parabolic "rotational bands", i.e. sets of spin multiplet eigenvalues, as a function of the total spin quantum number *S* resulting from the approximate quantum model for $\{Mo_{72}Fe_{30}\}$. Note that the ground state is a singlet (green). This model results in a singlet ground state (green level). *Right*: Zeeman splitting caused by an external field separates the M_S substates. The lowest substates intersect at equal field increments of 0.24 Tesla, causing spin level crossing (compare magnified steps in Fig. 7). **See Color Plate 19.**

$+S-1$, $+S$ substates belonging to any S state. Importantly, the parabolic character of the first rotational band causes equidistant spin-level-crossing: the $S = 1$, $M_S = -1$ substate (red graph in right panel of Fig. 8) undercuts the singlet ground state (green graph) at 0.24 Tesla, only to be undercut by an $S = 2$, $M_S = -2$ state (blue graph) at 0.48 Tesla, and so forth until the system eventually saturates in an $S = 75$, $M_S = -75$ state at H_{sat}. In the hypothetical case of zero Kelvin, these crossings cause the magnetization to increase in discrete steps of one Bohr magneton at each crossing field. With increasing temperatures these discrete steps continually erode and a Brillouin curve is approached. However, such quantum steps are not observed even at temperatures as low as 0.46 K, at least in pulsed-field measurements where the magnetic field is ramped to its maximum value within several dozen milliseconds: While these pulsed-field magnetization measurements indeed confirm the expected saturation field H_{sat}, they also display a continuous, i.e. non-quantized, increase in M vs. H, in accordance with the above-mentioned classical spin vector model.

This lack of observable quantum features was first attributed to the very small lowest energy separations such as the singlet-triplet energy difference into which the exchange energy factors only as $J/10$, meaning that transitions between these lowest levels should only be observable at temperatures of a few dozen Millikelvin. It took inelastic neutron scattering measurements on deuterated $\{Mo_{72}Fe_{30}\}$ samples at 65 mK to probe the energy gaps between the first and the first excited rotational band states.[34] These experiments were the first to confirm both the eigenvalues and the degeneracies of the lowest involved levels resulting from the approximate quantum model for $\{Mo_{72}Fe_{30}\}$ (Fig. 9). Yet it became clear that the excitation spectra at 65 mK, 1.8 K, 2.8 K, and 6.5 K could only be successfully reproduced with the rotational band model if a certain broadening is introduced. This reflects a partial lifting of the high degeneracy of several of the levels involved in the allowed transitions, for example the levels of the first excited rotational band are $27(2S+1)$-fold degenerate.

The three-sublattice model was more also supported by recent diffuse neutron scattering experiments utilizing polarization analysis, where the spin correlations derived at 1.5 K agree well with the simulation of the

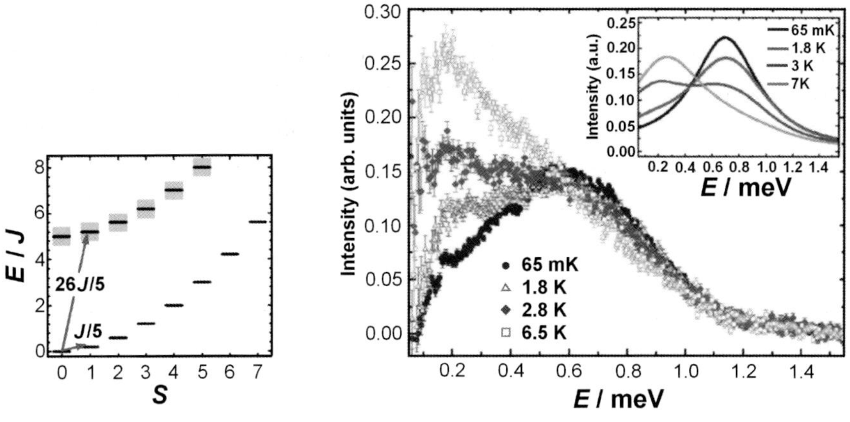

(a)

(b)

Fig. 9. (a) Energy levels calculated for the lowest and the first excited rotational band for {$Mo_{72}Fe_{30}$} (compare Fig. 8). Red arrows indicate allowed transitions from the $S = 0$ ground state to states characterized by $S = 1$. The shading of the energy levels of the first excited rotational band represents potential deviations from the ideal single-J model due to the fact that the high degeneracy of these levels can be partially lifted by a variety of perturbations. (b) Temperature evolution of the magnetic contribution of the inelastic neutron scattering intensity of deuterated {$Mo_{72}Fe_{30}$}, compared to the theoretical scattering patterns for different temperatures (inset), calculated using the quantum rotational band model and assuming identical probabilities for all allowed transitions. **See Color Plate 20.**

Fourier transform of the spin pair-correlation function using the three-sublattice configuration.[35]

An unexpected feature was observed in the field-dependent differential susceptibility at low temperatures: While a linear increase in M vs. H (due to thermally washed-out equidistant level crossings) should translate into a constant differential susceptibility dM/dH below the saturation field, above which M cannot increase anymore and dM/dH should be zero, a broad local minimum (a 'dip') in dM/dH is observed at approx. 5 Tesla. Based on further classical spin vector simulations, this feature was assigned to a thermally induced transition between different 'spin phases', specifically from states belonging to the three-sublattice 'umbrella' configuration of spin vectors with their 120° angular orientation to states that are derived from a distinct three-sublattice configuration.[36] The latter is again characterized by a frustrated arrangement: the collinear spin vectors of two sublattices are aligned parallel ('up') to the external field vector while the spin vectors of the remaining third sublattice is aligned antiparallel ('down'). The classical Monte Carlo simulations yielded the energy of any given spin phase as a function of the external magnetic field: while the 'umbrella' spin phase represents the energetically most stable one up until saturation, where it transforms into a fully field-aligned 'up-up-up' phase, the energy curve of the 'up-up-down' spin phase gets sufficiently close to that of the 'umbrella' phase at $H_{sat}/3$, so that these phases 'compete' energetically. This means that in a macroscopic sample at finite temperatures the 'up-up-down' spin phase is populated according to the Boltzmann distribution, while the 'umbrella' spin phase is concurrently de-populated. Since the 'up-up-down' phase is, per its definition, magnetically stiff (its total magnetic moment remains independent of magnetic field changes) it is characterized by $dM/dH = 0$, and an increasing population of its states therefore decreases the total differential susceptibility in the proximity of $H_{sat}/3$.

Building on the low-temperature inelastic neutron scattering data and, in particular, the field-dependent low-temperature differential susceptibility data for $\{Mo_{72}Fe_{30}\}$, several additional classical spin simulations

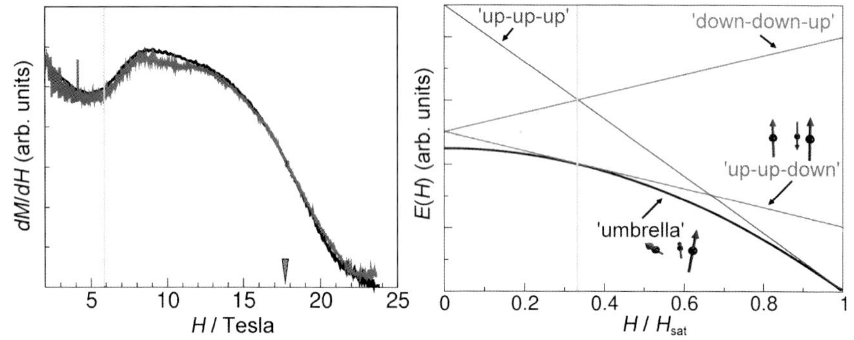

Fig. 10. *Left*: Field dependence of the experimental differential susceptibility dM/dH of $\{Mo_{72}Fe_{30}\}$ at 0.42 K. Black and red graphs indicate increasing and decreasing external field in the pulsed-field measurements, respectively. A local minimum is visible slightly below one-third of the saturation field H_{sat} ($H_{sat}/3$, indicated by a vertical green line; a gray arrow indicates H_{sat}). *Right*: The competing spin phases in $\{Mo_{72}Fe_{30}\}$ (all of them characterized by three sublattices). While the 'umbrella' spin phase is the most stable phase below saturation, the 'up-up-down' spin phase approaches the energy of the 'umbrella' phase in the vicinity of $H_{sat}/3$. Thus the 'up-up-down' spin phase is thermally populated at finite temperatures, which reduces the overall differential susceptibility. **See Color Plate 20.**

were performed in order to establish the origin of the pronounced level broadening effects, i.e. the apparent lifting of the degeneracies of the multiplet states that appears to be the main contribution to the erosion of detectable quantum effects. The underlying measurements, probing d*M*/d*H* at quasi-static external magnetic fields and temperatures down to 60 mK, revealed that below 5 K the differential susceptibility displays an unexpectedly pronounced dependence on both *T* and *H*.[37] In order to successfully reproduce this entire parameter space, the simulations had to take into account not a single common nearest-neighbor exchange energy *J*, but a distribution of values for *J* that reflect small structural variations of the 60 nearest-neighbor exchange pathways (Fig. 11). It should be noted that other attempts to explain limited features of d*M*/d*H*, specifically the broadening of the minimum in d*M*/d*H* vs. *H*, have been considered in the literature. For example the 'washed-out' magnetization saturation around H_{sat} was interpreted as a sign that the effective temperature of the Fe_{30} spin system in the pulsed-field experiments is actually higher than the sample temperature, however this was ruled out since a steady-field measurement reproduced the magnetization curves obtained by the pulsed-field technique.

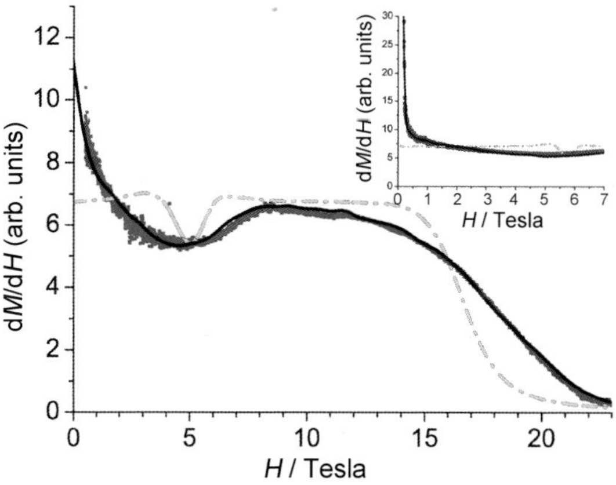

Fig. 11. Experimental (red dots) and simulated (black graph: multi-*J* model, green dashed graph: single-*J* model) field dependence of d*M*/d*H* for $\{Mo_{72}Fe_{30}\}$ at 420 mK and 60 mK (inset). Multi-*J* simulations were performed for an ensemble of 100 Fe_{30} spin clusters, where the average *J* (i.e. averaged over all 60 nearest-neighbor Fe···Fe contacts per molecule) was allowed to vary by ±15% over the entire ensemble, while the spread of individual J values within each $\{Mo_{72}Fe_{30}\}$ molecule was allowed to vary by ±40%. **See Color Plate 21**.

Furthermore, classical Monte Carlo simulations also were able to evaluate the effects of magnetic anisotropies, Dzyaloshinsky-Moriya, and dipole-dipole interactions. The simulational studies of these mechanisms have shown that they only can result in very minor changes to the width of the minimum in d*M*/d*H* for any reasonable choice of model parameters.[38]

It remains unclear if the obtained probability distribution for the nearest-neighbor exchange interaction is caused by a variation in the intramolecular distances between nearest-neighbor magnetic ions.[37] The actual cluster geometry could not account for such a wide variation in *J*, the standard deviation amounts to 0.045 Å (0.70%) and the maximum deviation to 1.4% of the average Fe-Fe distance of 6.449 Å, accumulated for all 60 Fe-Fe nearest-neighbor distances. Also the O–Mo–O bond angle variations (that should influence the total orbital overlap affecting the exchange energy due to the spatially anisotropic character of the (unoccupied) Mo(4d) orbitals) are relatively small, with an angular range of 103.6°–106.6°, which in part is caused by crystallographic disorder of Mo positions. Overall, a refined Heisenberg model accounting for the proposed limited variation the exchange constants would likely need to include additional exchange mechanisms, e.g. next-nearest neighbor couplings.

Given the success of the multi-*J* model in conjunction with the geometrical spin frustration in $\{Mo_{72}Fe_{30}\}$, another open question is whether $\{Mo_{72}Fe_{30}\}$ exhibits any of the characteristics that traditionally define a spin glass. A kink in the ac susceptibility at 5.25 K, consistently observed for several bias fields (albeit for thus far only one ac modulation frequency), constitutes a first sign of spin glass characteristics.[37] Spin freezing (i.e. a significant slowing of the magnetization dynamics) below 1 K has been linked to the exchange disorder and observed in ^1H NMR line widths.[39] An important task in this context is the isolation of derivatives of the classical $\{Mo_{72}Fe_{30}\}$ cluster structure that is either more regular or that is characterized by more pronounced structural deviations from the I_h-symmetric ideal. One example for such a variations is the species $[Mo_{72}Fe_{30}O_{252}(CH_3COO)_{20}(H_2O)_{92}\{Mo_6O_{19}\}]^{2-}$ that does not contain the $\{Mo_2\}$-type ligands (L′ and L″) and contains O_h-symmetric hexamolybdate as a supramolecular template, but low-temperature magnetization studies show a behavior nearly identical to that of the original $\{Mo_{72}Fe_{30}\}$

(the high-temperature susceptibility data also show only insignificant differences).[40] In 2010, the discovery[41] of the formation conditions favoring the polyoxotungstate(VI) analogues of the pentagonal building blocks, i.e. $\{(W)W_5\}$ groups, resulted in the $\{W_{72}Fe_{30}\}$-type cluster anion $[\{(W)W_5O_{21}(SO_4)\}_{12}(Fe(H_2O))_{30}(SO_4)_{13}(H_2O)_{34}]^{32-}$. This cluster species displays weaker Fe···Fe coupling ($J = -0.84$ cm^{-1} in the single-J description, i.e., a 23% decrease), though low-temperature magnetization measurements are pending.

The comparably high redox stability of polyoxomolybdate-based Keplerates in combination with reaction solution monitoring using Raman spectroscopy recently allowed the isolation of a highly reduced $\{Mo_{72}Fe_{30}\}$ derivative, $[\{Mo^{V/VI}{}_6O_{21}(H_2O)_3(SO_4)\}_{12}(Fe^{III}(H_2O)_2)_{30}]^{22-}$. In this species, 16 e$^-$ are fully delocalized over the partially reduced $\{(Mo)Mo_5\}$ building blocks (Robin Day class III), yet the resulting cluster structure remains nearly identical to the $\{Mo^{VI}{}_{72}Fe^{III}{}_{30}\}$ species.[42] Importantly, while the Mo(4d) electrons are spin-paired, they strongly weaken the magnetic exchange: Whereas $\theta = -21.6$ K and $J = -1.09$ cm^{-1} for $\{Mo^{VI}{}_{72}Fe^{III}{}_{30}\}$, for the 16-fold reduced $\{Mo^{V/VI}{}_{72}Fe_{30}\}$ species these values decrease to $\theta = -10.4$ K and $J = -0.53$ cm^{-1}, i.e. a relative change of approx. 50% in the exchange energy J. This decrease is explained with a change in the frontier orbital energies of the molybdate building blocks caused by the partial reduction which, in turn, reduces the overlap with the magnetic Fe(3d)-centered orbitals.

Finally, the family of magnetic Keplerate-type cluster compounds can also be extended by exploiting the Lewis-acidic character of the Fe^{III}-OH_2 groups in $\{Mo_{72}Fe_{30}\}$. Condensation reactions, partially proceeding at room temperature in the crystalline state, connect the spheres to one- or two-dimensional networks comprising Fe-O-Fe or Fe-O-Mo intercluster links with significant consequences for the magnetic properties of these materials.[27,43,44] In case of the formation of the one-dimensional chain compound $Na_2[\{(Mo^{VI})Mo^{VI}{}_5\}_{12}Mo^{V}{}_6Fe^{III}{}_{24}(CH_3COO)_{20}O_{258}(H_2O)_{82}]\cdot$ca.80 H_2O, six Fe spin centers are formally replaced by Mo^V centers, the latter are diamagnetic due to Mo(4d)-Mo(4d) spin pairing and statistically 'dilute' the Fe_{30} spin array, effectively breaking its symmetry.[45] Within the linear cluster chains, neighboring $\{Mo_{72}Mo_6Fe_{24}\}$ clusters are interlinked by two Fe–O–Mo–Fe bridges with a geometry that is very similar to that of the intra-cluster bridges. Importantly, the reduced θ value of -13 K (vs. -21.6 K for $\{Mo_{72}Fe_{30}\}$) is directly proportional to the number of Fe–O–Mo–O–Fe contacts, which depends on the positions within the M_{30} icosidodecahedron that the non-magnetic Mo linker sites are assuming. This is an example for structural information not readily obtainable from e.g. single-crystal X-ray diffraction due to disorder problems, for example if the non-magnetic linker atoms are clustered together or are randomly distributed, that can be derived from magnetic properties.

Theory activities in the area of Fe_{30} spin clusters are ongoing, too, and are in particular driven by the development of more accurate quantum mechanical descriptions — at least for the lowest section of the magnetic excitation spectrum — e.g. by the use of Density Matrix Renormalization Group methods for matrix diagonalization.[46]

4. The Quantum-spin Keplerates $\{Mo_{72}V_{30}\}$ and $\{Mo_{72}Cr_{30}\}$

Despite its conceptual limitations, the quantum mechanical rotational band model, based on the three-sublattice spin structure, can successfully explain various low-temperature phenomena of $\{Mo_{72}Fe_{30}\}$ — i.e. where the energetically lowest set of multiplet states are exerting a dominating influence. Recently it was shown that the minimal energies $E(S)$ of anti-ferromagnetic molecules with cuboctahedral and icosidodecahedral spin structure increase proportional to the total spin S (and not proportional to $S(S+1)$) above a certain total spin. For an icosidodecahedron based on quantum spin centers (e.g. $s=1/2$) this results in large magnetization jumps of *multiple* Bohr magnetons before the system saturates.[47,48] Furthermore, a multitude of singlet states below the first multiplet has been predicted for a quantum-spin icosidodecahedron. Although first postulated for antiferromagnetically coupled icosidodecahedral spin arrays, such features are common to frustrated spin systems such as Kagomé lattices or pyrochlore networks (of corner-sharing tetrahedra) and other frustrated magnetic molecules that feature triangles arranged around vertices together with regular polygons (pentagons in the case of Keplerates, hexagons in the case of Kagomé lattices). The resulting magnetic features were attributed to the fact that the relative ground states of

M_S subspaces are linked to product states of independent localized magnons for large M_S values.[47,48]

In order to probe these predictions for a true quantum spin icosidodecahedron, realized as a Keplerate, various attempts were performed to synthesize derivatives of $\{Mo_{72}Fe_{30}\}$ in which the 30 spin-5/2 Fe^{III} centers are replaced by magnetic centers with significantly smaller spin quantum numbers, in analogy to what has been demonstrated for the $\{Mo_{57}M_6\}$ family of polyoxomolybdates. Interestingly, a projected $\{Mo_{72}Cu^{II}{}_{30}\}$ structure is expected to exhibit ferromagnetic coupling between nearest-neighbor $s = 1/2$ Cu^{II} centers (in principle resulting in an $S = 15$ ground state), at least based on the ferromagnetic coupling within the trigonal $Cu_3Mo_3O_6$ pores in the corresponding $\{Mo_{57}Cu_6\}$ derivative.[15] In case of $s = 1/2$ vanadyl (VO^{2+})-based Keplerates, the first attempts to adapt them to reaction conditions under which Keplerates could potentially be formed used reducing conditions in the presence of excess vanadate that favored the formation of anions such as $\{Mo^{VI}{}_{72}Mo^{V}{}_{8}V^{IV}{}_{22}\}$ or "incomplete" Keplerate derivatives such as $\{Mo_{75}V_{20}\}$. In 2005 a discrete and intact $\{Mo_{72}V_{30}\}$-type cluster anion $[K_{10}(Mo_6O_{21}(H_2O)_3(SO_4))_{12}(VO)_{30}(H_2O)_{20}]^{26-}$ was published,[49,50] which showed a slightly oblate structure with clear symmetry decrease from approx. I_h to approx. D_{5d} (Fig. 12).

This $\{Mo_{72}V_{30}\}$ cluster anion exhibits strong antiferromagnetic coupling: If an approximate single exchange-constant model is assumed, the resulting exchange energy is estimated from quantum Monte Carlo simulations of the high-temperature low-field susceptibility to $J_{eff} \approx -140$ cm^{-1}. However, there are also clear signs of energetically low-lying multiplet states: Starting from an (apparent) singlet ground state, the magnetic moment rises rapidly even at lowest temperatures, and the susceptibility reaches a maximum at approx. 11 K (Fig. 12).

Assuming identical V···V interactions, confined to nearest-neighbor contacts, the high value of J_{eff} implies that the energies of the lowest multiplet states should be separated by energies comparable to the exchange energy. However, lower-energy exchange must be present in the $\{Mo_{72}V_{30}\}$ cluster to explain the immediate onset of χT at low temperatures. To probe the reason for this disparity, attempts were made to isolate a corresponding system with ideal I_h symmetry. A synthesis method employing high K^+ and sulfate concentrations was found to allow

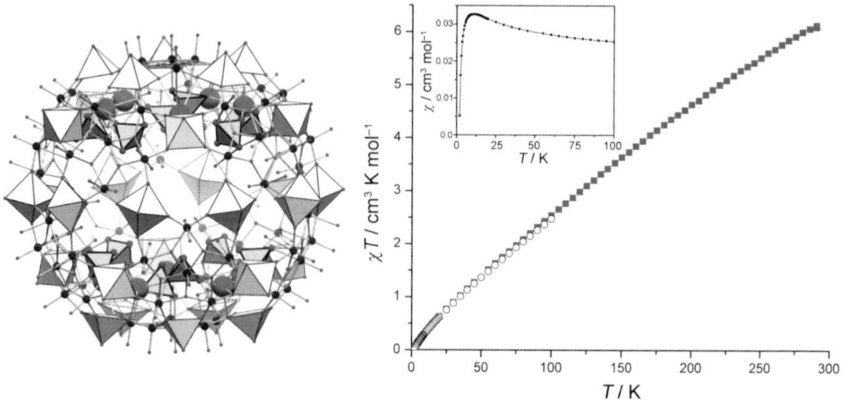

Fig. 12. *Left*: Structure of the $\{Mo_{72}V_{30}\}$ polyanion, emphasizing structural features that cause deviations from an ideal icosahedral geometry of the metal skeleton: The two coplanar $(KSO_4)_5$ rings (K: large purple spheres, sulfate: green tetrahedra) that coordinate to the inner surface of the cluster, specifically to 10 of the 12 $\{(Mo)Mo_5\}$ building blocks. Also note that ten outer V positions reside in a square-pyramidal environment, while the remaining 20 V centers adopt an octahedral coordination geometry. Mo: blue, O: red spheres. *Right*: Experimental susceptibility data for the $\{Mo_{72}V_{30}\}$-based compound (grey squares: 0.1 Tesla, open circles: 1.0 Tesla, corrected for diamagnetic and temperature-independent paramagnetism contributions). The immediate onset of χT even at 2 K and the maximum in χ vs. T at approx. 11 K (inset) indicate the presence of low-lying multiplet states that can be thermally populated even at low temperatures. **See Color Plate 21.**

access to mixed-metal Keplerates of the type $\{Mo^{VI}_{72}Fe^{III}_{x}V^{IV}_{y}Mo^{V}_{z}\}$ (with compositions $x = 7$, $y = 15$, $z = 8$; $x = 11$, $y = 11$, $z = 8$; $x = 20$, $y = 8$, $z = 2$) that exhibit intramolecular ferrimagnetic coupling and high-spin ground states (Fig. 13).[51,52]

Importantly, these clusters incorporate multiple K^+ cations that coordinate from the outside to the oxo sites of the pores of the $\{Mo_{72}M_{30}\}$ framework (the same K^+ attachment mode is seen for the above-mentioned $\{Mo^{VI}_{72}Mo^{V}_{8}V^{IV}_{22}\}$-type Keplerate cluster). The $\{Mo^{VI}_{72}Mo^{V}_{8}Fe^{III}_{x}V^{IV}_{y}\}$ clusters also comprise tridentate sulfate groups that coordinate to pentagonal $\{(Mo)Mo_5\}$ groups from the inside of the cluster sphere, yet, contrary to the $\{Mo_{72}V_{30}\}$ structure, are found to be *evenly* disordered in the resulting solid-state structure and thus do not form $\{KSO_4\}_n$ rings that could impose symmetry-reducing structural effects on the cluster shell. Employing the same synthesis strategy, the molecular assembly

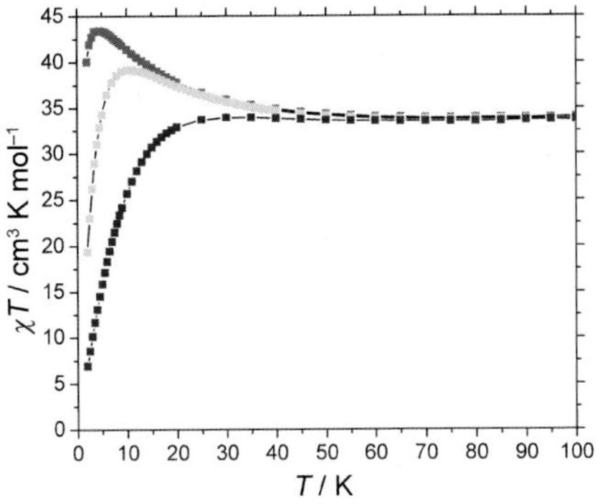

Fig. 13. Field-induced suppression of the ferromagnetic coupling as visible from the field dependence of χT curves for $K_{23}Na_4[\{Mo^{V/VI}_6O_{21}(H_2O)_3(SO_4)\}_{12}(V^{IV}O(H_2O))_{15}$ $(Fe^{III}(H_2O)_2)_7(Mo^{V}O(H_2O))_8]\cdot 150H_2O$ (left) and $K_{21}Na_2[\{Mo^{V/VI}_6O_{21}(H_2O)_3(SO_4)\}_{12}(V^{IV}O(H_2O))_{11}$ $(Fe^{III}(H_2O)_2)_{11}(Mo^{V}O(H_2O))_8]\cdot 150H_2O$ (right). 0.1 Tesla: red, 1.0 Tesla: green, 5.0 Tesla: blue data points. For 0.1 Tesla, the maxima are reached at 4.5 K for the $\{Mo_{72}V_{15}Fe_7Mo_8\}$ and at 5.5 K for the $\{Mo_{72}V_{11}Fe_{11}Mo_8\}$ compound, where the χT values correspond to $S=9$ and $S=11$ total spin states, respectively. **See Color Plate 22.**

$[\{Mo^{VI}_{72}V^{IV}_{30}O_{282}(H_2O)_{66}(SO_4)_{12}\}$-$K_2$-$\{Mo^{VI}_{114}Mo^{V}_{28}O_{432}(OH)_{14}(H_2O)_{58}\}]^{60-}=\{Mo_{72}V_{30}\}K_2\{Mo_{142}\}$ is observed in the solid-state structure of $K_{41}Na_{11}(VO)_4\{Mo_{72}V_{30}\}K_2\{Mo_{142}\}\cdot$ca.$500H_2O$. This assembly features a virtually I_h-symmetric $\{Mo_{72}V_{30}\}$ cluster linked via two K^+ cations to a molybdenum-blue type $\{Mo_{142}\}$-type ring cluster (Fig. 14).[53] Due to substantial temperature-independent paramagnetism, the mixed-valence $\{Mo^{V}_{28}Mo^{VI}_{114}\}$ ring only contributes a small diamagnetic constant to the overall molar susceptibility. In analogy to the discrete mixed-metal $\{Mo^{VI}_{72}Mo^{V}_8Fe^{III}_xV^{IV}_y\}$ species, the sulfate groups appear to again coordinate to the inner Keplerate surface in an evenly disordered fashion in the single-crystal X-ray diffraction-derived structure.

Upon subtraction of the contribution of detached vanadyl cations in the crystal lattice, the resulting susceptibility data nearly matches that of the isolated, oblate $\{Mo_{72}V_{30}\}$ cluster: even at 2 K, χT immediately increases with increasing temperature for both compounds, again indicating the presence of low-lying multiplets, a fact in apparent contradiction to the high absolute J value that would tend towards large energetic spacings between the ground state and the higher multiplet states. Possible splittings due to a decreased symmetry of the V_{30} icosidodecahedron in the

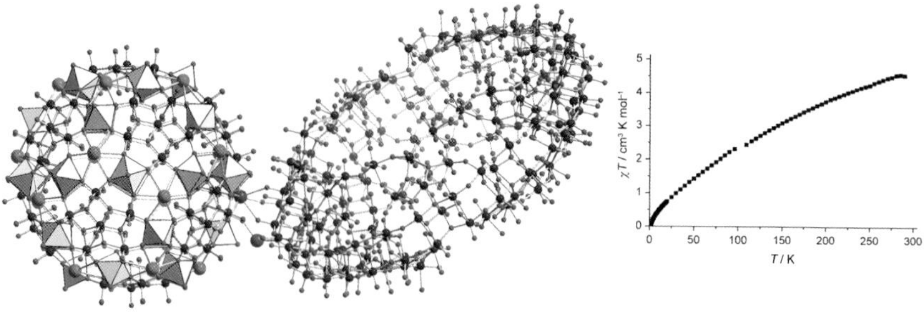

Fig. 14. *Left*: Simplified structure of the "ring-sphere" assembly $\{Mo_{72}V_{30}\}K_2\{Mo_{142}\}$ as it occurs in the solid state. Potassium cations coordinated to the Keplerate moiety as purple spheres, the two K^+ positions interlinking the cluster sphere and the cluster ring are shown as grey spheres. V: yellow octahedra, Mo: blue, O: red spheres. The sulfate groups in the Keplerate sphere are fully disordered and are not shown. *Right*: Temperature dependence of χT of $K_{41}Na_{11}(VO)_4\{Mo_{72}V_{30}\}K_2\{Mo_{142}\}\cdot ca.500H_2O$ at 0.1 Tesla (corrected for the contributions of the four vanadyl cations disordered in the crystal lattice). **See Color Plate 22.**

oblate $\{Mo_{72}V_{30}\}$ structure can thus not be the primary reason for the presence of low-lying multiplets. A more likely source for these multiplets is next-nearest neighbor coupling, which for the strongly coupled vanadyl-based systems potentially becomes significant.

Another system that formally is derived from the spin-1/2 icosidodecahedron is the V_{20} structure occurring in the cluster anion $[Mo^{VI}_{75}V^{IV}_{20}O_{270}(H_2O)_{70}(NaSO_4)_{10}]^{20-}$ (= $\{Mo_{75}V_{20}\}$) that comprises a substructure/fragment of the complete V_{30} icosidodecahedron with two V_5 pentagons on opposite sites at the polar regions replaced by (diamagnetic) molybdate groups (Fig. 15).[54] Ten isosceles V_3 triangles are connected via corners to an equatorial band to form a 'sawtooth' spin ring. Formally missing are two sets of each five polar triangles to join to the top and the bottom parts of the V_{20} spin belt to complete the icosidodecahedron.

Interestingly, the magnetism of the $\{Mo_{75}V_{20}\}$ structure, which remained largely unexplored until recently, has been carefully investigated by 190 GHz EPR measurements in 2012.[55] The study concludes that the frustration in $\{Mo_{75}V_{20}\}$ is different from that in $\{Mo_{72}V_{30}\}$, as $\{Mo_{75}V_{20}\}$ represents a quasi-one-dimensional object, but in addition the magnetic centers are not equivalent. The individual V_3 fragments are best

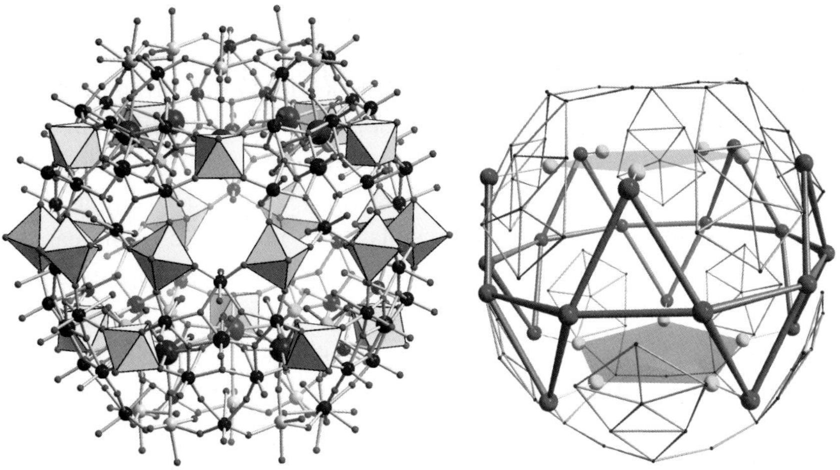

(a) (b)

Fig. 15. (a) Structure of the $\{Mo_{75}V_{20}\}$ cluster anion highlighting the two $\{NaSO_4\}_5$ rings similar to the pentagonal $\{KSO_4\}_5$ rings in $\{Mo_{72}V_{30}\}$ (V: yellow octahedra, Na: purple, S: green, Mo: blue, O: red spheres; Mo positions with occupancy factors of 0.5 are shown as gray spheres). (b) Simplified presentation of the metal skeleton of the $\{Mo_{75}V_{20}\}$ cluster from which the 10 alternating V_3-type isosceles triangles (red spheres, thick gray bonds) can be recognized. **See Color Plate 23**.

modeled as isosceles triangles with a pronounced difference in the two exchange energies ($J_2/J_1 - 2$; J_1 specifies the coupling between equatorial V sites, J_2 between equatorial and outer V sites) that relates it to sawtooth chains with flat bands of one-magnon energies. However, the giant magnetization steps of 50% of the saturation magnetization expected for these chains cannot be observed for $\{Mo_{75}V_{20}\}$ as the strong antiferromagnetic coupling means that they occur at magnetic fields not accessible by today's possibilities.

In 2007, the family of Keplerate-type $\{Mo_{72}M_{30}\}$ cluster structures was extended by a derivative $\{Mo_{72}Cr_{30}\}$, isolated as $[Mo_{72}Cr_{30}O_{252}(CH_3COO)_{19}(H_2O)_{94}\{Na(H_2O)_{12}\}]\cdot ca.120H_2O$, in which 30 $s=3/2$ Cr(III) centers represent the spin constituents.[56] This compounds exhibits antiferromagnetic coupling that is slightly stronger than in the case of $\{Mo_{72}Fe_{30}\}$ and the nearest-neighbor exchange coupling has been determined by Quantum Monte Carlo simulations (limited to $T > 30\,K$ due to the 'negative sign problem' that limits the applicability of the quantum Monte Carlo method in the case of frustrated ground states,[57] and reliable results are available only above a system-specific threshold temperature) to $J = -6.0\ cm^{-1}$ with an isotropic g factor of $g = 1.96$. If the approximate quantum model developed for $\{Mo_{72}Fe_{30}\}$ is employed, a saturation field of $H_{sat} \approx 60$ Tesla results. Very much like in the case of $\{Mo_{72}Fe_{30}\}$, a relatively wide distribution of J values is necessary to reproduce all experimental low-temperature differential susceptibility data, with a spread of ±50% in the 60 nearest-neighbor exchange energies within each $\{Mo_{72}Cr_{30}\}$ cluster.[37]

In summary, the magnetism of polyoxomolybdate or polyoxotungstate-based Keplerate clusters is unique in that these systems thus far represent the only quasi-zerodimensional spin-frustrated systems in which the emergence of phenomena (such as spin glass behavior) commonly associated with frustrated two-dimensional (e.g. Kagomé) or three-dimensional (e.g. pyrochlore networks of corner-sharing tetrahedra) solid-state structures can truly be witnessed. Wide-reaching chemical control allows variation of the spin centers as well as variation of the polyoxometalate fragments that mediate the superexchange. Due to their nearly unmatched complexity in the field of molecular magnetism they will continue to pose (rewarding) challenges to chemists and physicists alike.

References

1. Müller A., *et al.*, *Chem Rev* **98** (1998) 239–271.
2. Clemente-Juan J. M. and Coronado E. *Coord Chem Rev* **193–195** (1999) 361–394.
3. Müller A., *et al.*, *Coord Chem Rev* **222** (2001) 193–218.
4. Kortz U., *et al.*, *Coord Chem Rev* **253** (2009) 2315–2327.
5. Gatteschi D., *et al.*, *Nature* **354** (1991) 463–465.
6. Kögerler P., Tsukerblat B. and Müller A. *Dalton Trans* **39** (2010) 21–36.
7. Bertaina S., *et al.*, *Nature* **453** (2008) 203–206.
8. Müller A., Kögerler P. and Kuhlmann C. *Chem Commun* (1999) 1347–1358.

9. Gatteschi D., Sessoli R. and Villain J. *Molecular Nanomagnets* Oxford University Press Oxford (2006).
10. Schnack J. *Dalton Trans* **39** (2010) 4677–4686.
11. Müller A., *et al.*, *Angew Chem Int Ed Engl* **33** (1994) 849–851.
12. Müller A., *et al.*, *Chem Eur J* **4** (1998) 1000–1006.
13. Müller A., *et al.*, *Inorg Chim Acta* **271** (1998) 9–12.
14. Müller A., *et al.*, *Anorg Allg Chem* **620** (1994) 599–619.
15. Kögerler P. and Müller A. *J Appl Phys* **93** (2003) 7101–7102.
16. Krebs B. and Paulat-Böschen I. *Acta Cryst* **B38** (1982) 1710–1718.
17. Müller A., *et al.*, *Angew Chem Int Ed* **34** (1995) 2122–2124.
18. Müller A., *et al.*, *Angew Chem Int Ed* **35** (1996) 1206–1208.
19. Müller A., *et al.*, *Angew Chem Int Ed* **37** (1998) 1220–1223.
20. Müller A. and Serain C. *Acc Chem Res* **33** (2000) 2–10.
21. Botar B., Ellern A. and Kögerler P. *Dalton Trans* **41** (2012) 8951–8959.
22. Müller A. *et al.*, *Z Anorg Allg Chem* **625** (1999) 1960–1962.
23. Müller A., *et al.*, *Angew Chem Int Ed* **41** (2002) 1162–1167.
24. Müller A., *et al.*, *Angew Chem Int Ed* **37** (1998) 3360–3363.
25. Kemp M., *Nature* **393** (1998) 123.
26. Müller A., *et al.*, *Angew Chem Int Ed* **38** (1999) 3238–3241.
27. Müller A., *et al.*, *Solid State Sci* **2** (2000) 847–854.
28. Müller A., *et al.*, *Chem Phys Chem* **2** (2001) 517–521.
29. Schnack J., Luban M. and Modler R. *Europhys Lett* **56** (2001) 863–869.
30. Axenovich M. and Luban M. *Phys Rev B* **63** (2001) 100407/1–4.
31. Schnack J. and Luban M. *Phys Rev B* **63** (2001) 014418/1–7.
32. Exler, M. and Schnack, J. *Phys Rev B* **67** (2003) 094440/1–4.
33. Anderson P. W., *Basic Notions of Condensed Matter Physics*, Addison-Wesley, Boston, 1997.
34. Garlea V. O., *et al.*, *Phys Rev B* **73** (2006) 024414/1–5.
35. Fu Z., *et al.*, *New J Phys* **12** (2010) 083044/1–15.
36. Schröder C., *et al.*, *Phys Rev Lett* **94** (2005) 017205/1–4.
37. Schröder C., *et al.*, *Phys Rev B* **77** (2008) 224409/1–8.
38. Hasegawa M. and Shiba H. *J Phys Soc Jpn* **73** (2004) 2543–2549.
39. Schröder C., *et al.*, *J Phys Condens Matter* **22** (2010) 216007/1–6.
40. Müller A., *et al.*, *Chem Commun* (2006) 3066–3068.
41. Todea A. M., *et al.*, *Angew Chem Int Ed* **49** (2010) 514–519.
42. Botar B., *et al.*, *Angew Chem Int Ed* **48** (2009) 9080–9083.

43. Cronin L., Kögerler P. and Müller A. *J Solid State Chem* **152** (2000) 57–67.
44. Müller A., *et al.*, *Angew Chem Int Ed* **39** (2000) 1612–1613.
45. Müller A., *et al.*, *Angew Chem Int Ed* **41** (2002) 579–582.
46. Ummethum J., Schnack J. and Läuchli A. M. *J Magn Magn Mater* **327** (2013) 103–109.
47. Schulenburg J., *et al.*, *Phys Rev Lett* **88** (2002) 167207/1–4.
48. Schmidt R., Richter J. and Schnack J. *J Magn Magn Mater* **295** (2005) 164–167.
49. Müller A., *et al.*, *Angew Chem Int Ed* **44** (2005) 3857–3861.
50. Botar B., Kögerler P. and Hill C. L. *Chem Commun* (2005) 3138–3140.
51. Botar B., *et al.*, *Chem Commun* (2005) 5621–5623.
52. Botar B., *et al.*, *Eur J Inorg Chem* (2009) 5071–5074.
53. Botar B., Kögerler P. and Hill C. L. *J Am Chem Soc* **128** (2006) 5336–5337.
54. Müller A., *et al.*, *Chem Commun* (1999) 1885–1886.
55. Oshima Y., *et al.*, *Phys Rev B* **85** (2012) 024413/1–11.
56. Todea A. M., *et al.*, *Angew Chem Int Ed* **46** (2007) 6106–6110.
57. Troyer M. and Wiese U.-J. *Phys Rev Lett* **94** (2005) 170201.

Chapter 6

POLYOXOMETALATES AS LIGANDS FOR FUNCTIONAL LANTHANOID COMPLEXES

CHRIS RITCHIE and COLETTE BOSKOVIC
School of Chemistry,
The University of Melbourne,
Victoria, 3010, Australia

1. Introduction

Considerable current interest in the chemistry of complexes of lanthanoid (Ln) metals owes much to the unique chemical and physical properties of the trivalent lanthanoid ions. These properties have given rise to a number of important applications, including as magnets (in computer hard drives, speakers and in the motors of cordless tools) and as phosphors (in televisions and fluorescent lights). The trivalent lanthanoid ions have relatively large ionic radii and thus often exhibit high coordination numbers. They are hard Lewis acids and in their coordination complexes tend to bind to ligands with hard Lewis donor atoms, particularly oxygen. Because the 4f orbitals that contain the valence electrons do not extend out far enough from the nucleus to interact significantly with ligand orbitals, the lanthanoid-ligand bonds have significant ionic character.

For some time, the strongest permanent magnets available have been alloys of the early lanthanoid metals neodymium and samarium. However it is only recently that the later lanthanoid metals (terbium, dysprosium, holmium and erbium) have been considered more promising than d-block metals for the development of new single-molecule-magnets (SMMs).[1] These are species that display slow magnetic relaxation of molecular origin and quantum tunnelling of the magnetization, giving rise to potential applications as the memory units for high density magnetic data storage or the qubits for quantum computers.[2,3] SMM behavior has now been observed for numerous lanthanoid complexes, first for mononuclear complexes such as the $[Pc_2Ln]^-$ family, where *Pc* = phthalocyanine,[4–6] and subsequently for polynuclear Ln_n and mixed 3d-4f complexes.[7] The energy barrier to magnetization reversal for these molecular species is associated with the ligand field splitting of the ground electronic term of the lanthanoid single ions, i.e. an energy barrier between $\pm M_J$ substates.[8] As a result, lanthanoid-based SMMs are also referred to as "single ion magnets". In nearly all reported examples of lanthanoid-based SMMs, the coordination sites of the lanthanoid ions have been predominantly occupied by oxygen and nitrogen donor atoms from organic ligands.

Trivalent lanthanoid ions also display unique photophysical properties, including sharp and characteristic transitions in the visible or near-infrared regions and long excited-state lifetimes of up to milliseconds. Because of these properties, fluorescent lamps and cathode ray tubes have employed rare earth phosphors for many years. More recently, highly luminescent lanthanoid complexes with organic ligands have shown considerable promise for applications in biosensors, fluorescence immunoassays, and as luminescent probes for time-resolved microscopy and bio-imaging.[9,10] Even more recently, attention has turned towards luminescent hybrid organic-inorganic lanthanoid species to combine the properties of solid state phosphors and molecular coordination complexes to access versatile multifunctional materials.[11,12] Regardless of the host lattice or ligand environment, highly luminescent lanthanoid-based zero-dimensional or extended systems require sensitization to overcome the low molar absorption coefficient of the Laporte forbidden f-f transitions ($\varepsilon < 1$–$10\ \mathrm{L\ mol^{-1}\ cm^{-1}}$). The sensitization process that populates the metal excited states can proceed via a number of mechanisms. In rare earth phosphors the sensitization often involves absorption by the host material, or a second metal ion, with resulting energy transfer through the lattice to the emitting lanthanoid ion. In molecular lanthanoid coordination complexes, the sensitization generally involves population of the triplet excited state of a conjugated organic ligand, which possesses an intrinsically long lifetime, allowing efficient energy transfer to the emitting lanthanoid ion. The lanthanoid(III) ions with energy levels suitable for emitting visible light are: samarium(III), europium(III), terbium(III) and dysprosium(III), with europium(III) (red emission) and terbium(III) (green emission) the most extensively studied because of their longer lifetimes.

Polyoxometalates (POMs) are polynuclear metal-oxygen complexes of the early transition metals that are comprised of metal-centered MO_n polyhedra linked by shared corners, edges or faces, often with additional heteroatoms incorporated within the cluster. They possess considerable structural and compositional versatility, allowing the tuning of their electronic structure and consequently their functionality. The oxygen-rich surface of defect, or lacunary, POMs renders them excellent inorganic ligands for oxophilic d- and f-block metals, giving rise to a wide variety of coordination complexes. A number of lanthanoid complexes with POM ligands (Ln-POMs) have been reported and the synthesis and structures of many of these have been described in recent reviews.[13,14] The incorporation of organic groups into POMs in a variety of ways is also a significant area of investigation directed towards the development of novel hybrid organic-inorganic materials.[15]

As the majority of SMMs possess organic ligands bound to the open shell metal centers that give rise to the magnetic behavior, the novelty of the SMMs $[Ln(W_5O_{18})_2]^{9-}$ (Ln = Ho and Er) and $[Ln(\beta_2\text{-}SiW_{11}O_{39})_2]^{13-}$ (Ln = Dy, Ho, Er and Yb) reported by Aldamen *et al.*[16,17] arises from their employment of diamagnetic POMs as inorganic ligands. Their steric bulk renders POMs particularly promising supporting ligands for SMMs because they can help to ensure magnetic isolation of the individual molecules from one another. In addition, they can be prepared as nuclear spin free systems, with nuclear spin a major source of the quantum decoherence that would be detrimental to the use of SMMs in quantum computing.

Ln-POMs can also be considered as discrete molecular analogs of rare-earth phosphors. Photoexcitation of the O→M (M = Mo or W) ligand to metal charge transfer (LMCT) bands of the POM ligands can lead to intramolecular energy transfer from the O→M excited states to excited energy levels of the lanthanoid(III) ion, thereby sensitizing lanthanoid(III) emission. Comprehensive studies from Yamase and coworkers have provided considerable insight into the molecular mechanisms of these energy transfer processes.[18,19] The majority of these studies have been performed on Eu-POMs, which have provided the most examples of luminescent Ln-POMs. Relatively low emission intensities at room temperature for the other Ln-POMs have been attributed to radiationless deactivation via Ln(IV)-W(V) charge-transfer states for Pr- and Tb-POMs and cross-relaxation based on concentration quenching, with a large critical distance for Sm-, Dy- and Ho-POMs. The emission characteristics of the lanthanoid(III) ions are highly dependent on the nature of the POM ligand, the symmetry and coordination geometry at the lanthanoid center and the number of aqua co-ligands. Indeed, POM ligands bound to lanthanoid(III) ions display a varying number of coordination bonds, but are frequently polydentate, minimizing the co-ligation of aqua and similar ligands, which contain high frequency O–H and N–H oscillators that otherwise tend to quench the luminescence.

Our ongoing research is focused on the use of lacunary POMs with heteroatoms such as arsenic(III) that possess a stereoactive lone pair of electrons. The targeting of these species as building blocks to construct larger molecules arises from their inability to form closed plenary structures. The lone pair in the precursor helps to direct the self-assembly process, with the formation of larger, more complex structures following reaction with electrophiles. Our synthetic strategy also targets the inclusion of carboxylate-containing co-ligands to access "ternary" lanthanoid-organic-POM (Ln-org-POM) complexes. The organic carboxylate co-ligands can stabilize novel structures and serve as antennas to sensitize the lanthanoid luminescence. The aim of this article is to provide a cohesive review of our general synthetic approach, to present the structures of the Ln-org-POMs that we have obtained to date and to discuss the interesting magnetic and luminescent properties displayed by some of these complexes.

2. Synthesis and Structures

Ln-POMs have been explored for many years, with numerous examples and a number of different structural classes known. The structurally simplest class of Ln-POMs contain mononuclear lanthanoid centers "sandwiched" between two monolacunary POM ligands, with examples including $[Ln(W_5O_{18})_2]^{9-}$, $[Ln(\beta_2\text{-}SiW_{11}O_{39})_2]^{13-}$, $[Ln(\alpha_2\text{-}P_2W_{17}O_{61})_2]^{17-}$.[13,14] Another fairly widespread class of Ln-POMs display cryptate-like structures where lanthanoid centers are encapsulated within POMs replacing the alkali metal cations in $[(NaOH_2)\subset P_5W_{30}O_{110}]^{14-}$, $[Na\subset Sb_9W_{21}O_{86}]^{18-}$, $[Na\subset As_4W_{40}O_{140}]^{27-}$ and $[K_8\subset P_8W_{48}O_{184}]^{40-}$.[13] Directed synthetic strategies have afforded these two classes of Ln-POM, with the lanthanoid center coordinating to the POM precursors in a predictable way, either occupying the single lacunary site in the sandwich complexes or replacing alkaline earth metals cations in the cryptand species. However, even with a restricted ligand environment, the predictability of the products has sometimes not been so straightforward. For instance the $[Ln(\alpha_2\text{-}P_2W_{17}O_{61})_2]^{17-}$, $[(H_2O)_3Eu(\alpha_2\text{-}P_2W_{17}O_{61})]_2^{14-}$, polymeric 1D $[\{Nd\text{-}(H_2O)_7\}_2\{Nd(H_2O)_3(\alpha_2\text{-}P_2W_{17}O_{61})\}_2]^{8-}$ and 2D network $[Nd_2(H_2O)_9(\alpha_2\text{-}P_2W_{17}O_{61})]^{4-}$ were all prepared from the same POM precursor.[20–22] A diverse range of more structurally complex Ln-POMs have also been reported from a variety of synthetic approaches.[13,14]

The compositional diversity of POMs that are suitable for reaction with lanthanoid salts provides a large library of possible precursors for exploration. Details relating to the formation and speciation of such inorganic ligands vary greatly depending on the central heteroatom, with the most extensively studied and well understood lacunary chemistry that of the tungstosilicates[23] and tungstophosphates.[24] These extensive studies have afforded invaluable information about the chemistry of heteropolytungstates with tetrahedral heteroatoms, while POMs with subvalent group 15 elements including arsenic(III), antimony(III) and bismuth(III), have also been investigated.[25–24] Reaction of these precursors with lanthanoid salts to form Ln-POMs further complicates their speciation chemistry due to the stereodirecting lone pair of electrons on the heteroatom preventing the formation of the plenary Keggin polyanions.

Our synthetic approach has mainly focussed on the use of the tungstoarsenate(III) POM precursors $\alpha\text{-}[AsW_9O_{33}]^{9-}$ and $[As_2W_{19}O_{67}(H_2O)]^{14-}$. Previously reported Ln-POMs synthesised from the $\alpha\text{-}[AsW_9O_{33}]^{9-}$ precursor include $[K\subset\{Eu(H_2O)_2(\alpha\text{-}AsW_9O_{33})\}_6]^{35-}$, $[Cs\subset\{Eu(H_2O)_2(\alpha\text{-}AsW_9O_{33})\}_4]^{23-}$ and $[Ho_5(H_2O)_{16}(OH)_2As_6W_{64}O_{220}]^{25}$ while $[YbAs_2W_{20}O_{68}(H_2O)_3]^{7-}$, $[Yb(H_2O)_2K(H_2O)_2As_2W_{19}O_{67}(H_2O)]^{10-}$ and $[La_2(H_2O)_6As_2W_{19}O_{67}(H_2O)]^{8-}$ have been obtained from reactions with $[As_2W_{19}O_{67}(H_2O)]^{14-}$.[30–34] In aqueous solution the $\alpha\text{-}[AsW_9O_{33}]^{9-}$ species likely exists in equilibrium with the $\beta\text{-}[AsW_9O_{33}]^{9-}$ isomer, as is evident

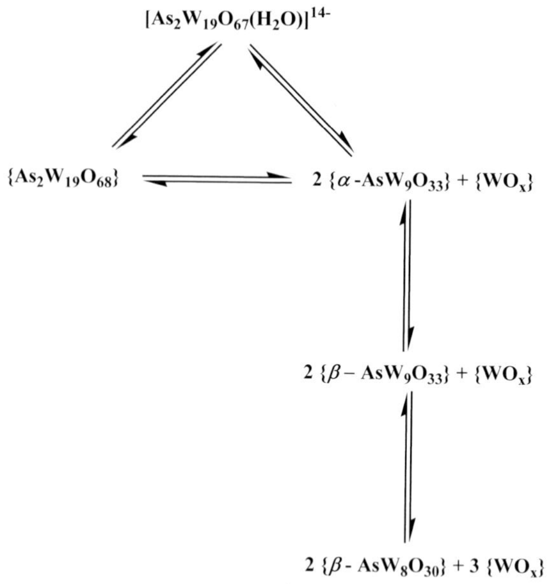

Scheme 1.

from ^{183}W NMR studies of the antimony(III) analogs.[28] Structurally, rotation of a single $\{W_3O_{13}\}$ triad by 60° results in the interconversion of the isomers, with the resulting terminal oxo ligands in the *fac*-WO_3 group of the β-$[SbW_9O_{33}]^{9-}$ POM more susceptible to protonation. It is anticipated that a similar situation arises when other subvalent group 15 elements are used as heteroatoms.[28] Throughout our investigations of tungstatoarsenates we have also observed a recurring $\{As_2W_{19}O_{68}\}$ structural motif in the presence of lanthanoid(III) ions that we hypothesize allows the stabilization of two $\{\alpha\text{-}AsW_9O_{33}\}$ fragments following condensation with an additional tungstate unit (Scheme 1).

The $\{As_2W_{19}O_{68}\}$ building block is more highly condensed than the $[As_2W_{19}O_{67}(H_2O)]^{14-}$ precursor (Fig. 1) and is evident in the majority of the Ln-POMs and Ln-org-POMs that we have obtained from both the α-$[AsW_9O_{33}]^{9-}$ and $[As_2W_{19}O_{67}(H_2O)]^{14-}$ precursors. The lacunary cavity in $\{As_2W_{19}O_{68}\}$ is very different in size and shape to those of either precursor and exemplifies the ability of polyanions that are sometimes considered rigid and inflexible, to accommodate the coordination of larger electrophiles such as lanthanoid(III) ions.

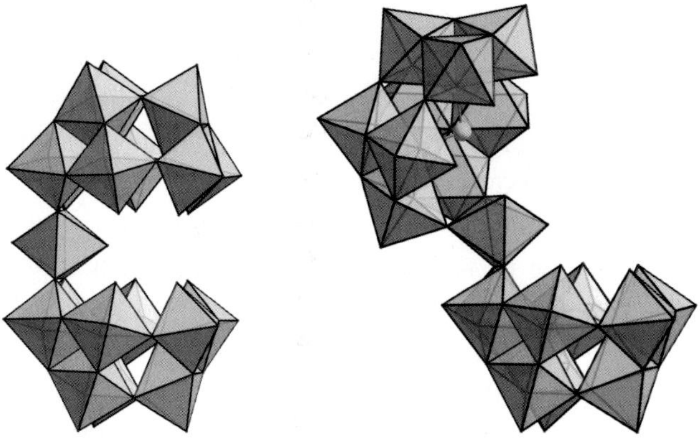

Fig. 1. The $[As_2W_{19}O_{67}(H_2O)]^{14-}$ precursor (left) and the $\{As_2W_{19}O_{68}\}$ building block formed *in situ* (right).

2.1. *Ln-POMs and Ln-org-POMs synthesized in the presence of acetate*

Our synthetic efforts in the area of Ln-POMs commenced with α-$[AsW_9O_{33}]^{9-}$ as POM precursor, resulting in the isolation of two families of novel high nuclearity Ln-org-POMs and another family of Ln-POMs: $[Ln_6As_6W_{65}O_{229}(OH)_4(H_2O)_{12}(OAc)_2]^{38-}$ with Ln = Gd (**1-Gd**) and Dy (**1-Dy**); $[Ln_{10}As_{10}W_{88}O_{308}(OH)_8(H_2O)_{28}(OAc)_4]^{40-}$ with Ln = Er (**2-Er**) and Yb (**2-Yb**); and $[Cs\subset Ln_6As_6W_{63}O_{218}(OH)_4(H_2O)_{14}]^{25-}$ with Ln = Eu (**3-Eu**), Gd (**3-Gd**), Tb (**3-Dy**), Ho (**3-Ho**) and Er (**3-Er**).[35,36] The synthesis of these compounds involved reaction of α-$[AsW_9O_{33}]^{9-}$ with lanthanoid(III) salts in sodium acetate/acetic acid buffer in a one-pot approach, followed by slow evaporation to produce X-ray diffraction quality single crystals and bulk samples for physical measurements. This synthetic procedure is one which has been used for many years by several research groups,[37–41] where the use of acetate buffer provides pH control and in some cases inclusion of acetate co-ligands.

Preparation of Ln-org-POMs **1** and **2** was achieved using essentially the same procedure, where the sodium salt of α-$[AsW_9O_{33}]^{9-}$ was reacted with the appropriate lanthanoid source in 1:3 molar ratio in 1 M acetate buffer at pH 4.7.[35] The reaction solutions were heated to 50°C, cooled to room temperature and a small amount of precipitate removed prior to evaporation. Inclusion of ammonium or cesium chloride as alternative cation sources was the only difference in the reaction conditions that yielded **1** and **2**, respectively. Complex **1** was isolated as the mixed sodium/ammonium salt, with **2** isolated as a mixed sodium/cesium salt.

The structure of complex **1** is asymmetric with C_1 point symmetry, although ignoring one of the acetate ligands results in approximate C_s symmetry (Fig. 2). The anion present in **1** is composed of two $\{As_2W_{19}O_{68}\}$ units (Fig. 1), with two additional β-$[AsW_9O_{33}]^{9-}$ fragments also present in the structure. Located between these POM fragments is a $\{Ln_6W_9O_{27}(OH)_4(H_2O)(OAc)_2\}$ core, which is composed of one $\{W_3O_{11}(OH)_2\}$, two $\{W_2O_9(OH)\}$ and two $\{WO_6\}$ units. Protonation of these types of structural units is not uncommon, however the location of protons on POM surfaces is not always possible due to their tendency to be statistically disordered over a number of sites with approximately equal basicity. The six lanthanoid centers in **1** display two coordination geometries, with five lanthanoid centers exhibiting 8-coordinate approximate square antiprismatic coordination geometries, while the sixth is a 9-coordinate distorted tricapped trigonal prism.

The structure of complex **2** is composed of four $\{As_2W_{19}O_{68}\}$ units each located on the corners of a $\{Yb_{10}As_2W_{12}O_{36}(OH)_8(H_2O)_{28}(OAc)_4\}$ central

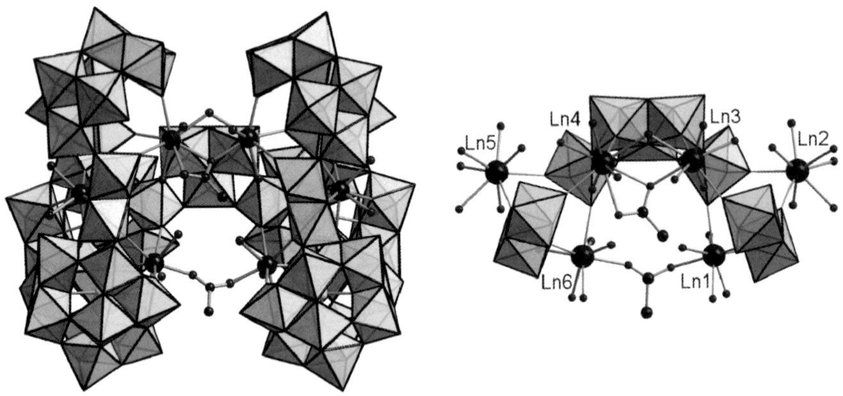

Fig. 2. Combined polyhedral/ball-and-stick representations of **1**. (Reproduced with permission from Hussain *et al.*, 2009a. Copyright © 2009, Royal Society of Chemistry.)

core (Fig. 3). This core comprises two edge-shared and acetate-bridged $\{W_2O_{10}\}$ groups, eight distorted monomeric $\{WO_6\}$ octahedra and two pyramidal $\{AsO_3\}$ groups connected by five crystallographically unique lanthanoid centers, all of which display 8-coordinate square antiprismatic coordination geometries. Bond valence sum calculations suggest single protonation of the eight μ_2-oxo ligands that bridge the lanthanoid ions to tungsten atoms of non-edge-shared $\{WO_6\}$ octahedral fragments. The presence of the $\{W_2O_{10}\}$, $\{WO_6\}$ and $\{AsO_3\}$ groups again indicates that fragmentation of the $[\alpha\text{-}AsW_9O_{33}]^{9-}$ starting material has occurred during the formation of **2**.

Slight changes to the reaction temperature and ionic strength employed in the synthesis of **2** afforded complex **3**, which was also isolated as a mixed sodium/cesium salt.[36] Thus the structural components that constitute **3** are very similar to those evident in **2**. Complex **3** is comprised of three $\{As_2W_{19}O_{68}\}$ units that are connected by a variety of Ln–O–W linkages. As in **1** and **2**, additional tungstate units are also present in **3**, in the form of two $\{W_2O_{10}\}$ units and two $\{WO_5\}$ units, which together with the six lanthanoid centers sites combine to form the $\{Cs\subset Ln_6W_6O_{14}(OH)_4(H_2O)_{14}\}$ core of the polyanion. A single cesium ion is located at the center of the polyanion and likely plays a structure directing role. Each of the

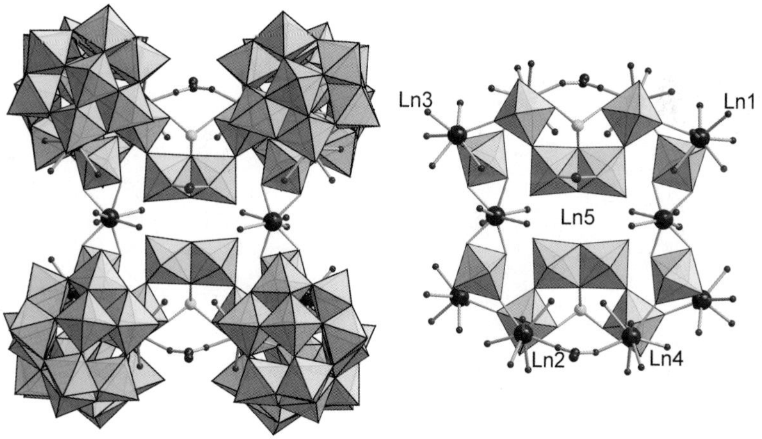

Fig. 3. Combined polyhedral/ball-and-stick representations of **2**. (Reproduced with permission from Ref. 35. Copyright © 2009, Royal Society of Chemistry.)

lanthanoids ions in **3** display distorted square antiprismatic coordination environments. In this particular structure, two lanthanoids are coordinated into each of the open pocket-like lacunary sites of the $\{As_2W_{19}O_{68}\}$ building blocks, with Ln···Ln separations of 5.8 Å.

Having prepared a number of compounds from the precursor α-$[AsW_9O_{33}]^{9-}$, we shifted our focus to the potentially more versatile $[As_2W_{19}O_{67}(H_2O)]^{14-}$ precursor.[42,47] The advantages that this polyanion offers over α-$[AsW_9O_{33}]^{9-}$ is the possibility of accessing the more condensed $\{As_2W_{19}O_{68}\}$ structural component (Fig. 1) by isomerization rather than *via* partial decomposition of α-$[AsW_9O_{33}]^{9-}$. Furthermore, on dissociation the $[As_2W_{19}O_{67}(H_2O)]^{14-}$ precursor yields two equivalents of α-$[AsW_9O_{33}]^{9-}$ and free tungstate, with the ensuing complicated equilibria, ultimately yielding a library of building blocks in any given reaction mixture (Scheme 1).

Several tungstoarsenates, $[As_6W_{65}O_{217}(H_2O)_7]^{26-}$ $[As_4W_{42}O_{143}(H_2O)_4]^{22-}$ and $[As_2W_{21}O_{69}(H_2O)]^{6-}$, have been synthesised from the $[As_2W_{19}O_{67}(H_2O)]^{14-}$ precursor with only minimal differences in synthetic conditions.[27,43,44] The resulting structures are closely related in terms of the constituent building blocks. One thing that is apparent from the reported syntheses, is that under the acidic conditions from which the compounds were obtained, the solution chemistry is complicated and involves several subtly balanced processes. In our work, the introduction of neodymium into a heated acidic reaction medium containing sodium chloride and $[As_2W_{19}O_{67}(H_2O)]^{14-}$, resulted in the isolation of a chain structure composed of $[Nd_3As_4W_{41}O_{141}(OH)$

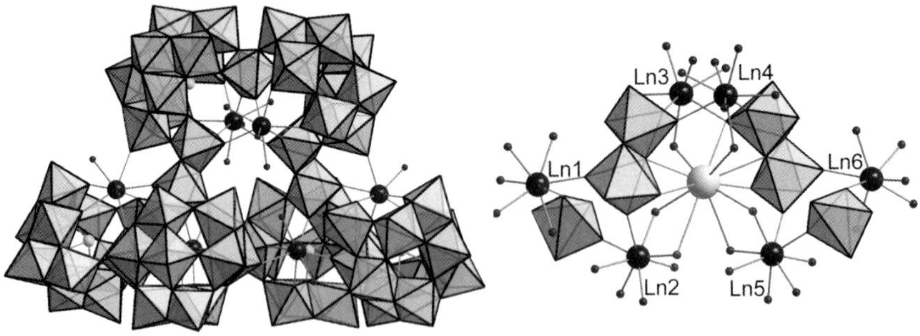

Fig. 4. Combined polyhedral/ball-and-stick representations of **3**. (Reproduced with permission from Ref. 36. Copyright © 2009, Royal Society of Chemistry.)

$(H_2O)_{10}]^{16-}$ (**4-Nd**) repeat units (Fig. 5) with C_1 point symmetry.[45] On dissecting the structure of $[Nd_3As_4W_{41}O_{141}(OH)(H_2O)_{10}]^{16-}$ further, three $\{\alpha\text{-}AsW_9O_{33}\}$ building blocks are apparent, together with a previously unreported $\{\beta\text{-}AsW_8O_{29}(OH)\}$ unit, as well as six additional tungstate units and three neodymium ions. From a mechanistic standpoint this provides further insight into the intricately balanced solution chemistry of tungstoarsenates. In order to obtain $\{\beta\text{-}AsW_8O_{29}(OH)\}$, the starting $[As_2W_{19}O_{67}(H_2O)]^{14-}$ precursor must firstly undergo dissociation, followed by isomerization of the resulting $\{\alpha\text{-}AsW_9O_{33}\}$ units to $\{\beta\text{-}AsW_9O_{33}\}$ that probably exist in equilibrium as mentioned above (Scheme 1). Some of this $\{\beta\text{-}AsW_9O_{33}\}$ species must then undergo a further loss of tungstate to form the $\{\beta\text{-}AsW_8O_{29}(OH)\}$ unit evident in **4**. The $\{\alpha\text{-}AsW_9O_{33}\}$ and $\{\beta\text{-}AsW_8O_{29}(OH)\}$ units in the structure of **4** are located on the periphery of a $\{Nd_3W_6O_{13}(H_2O)_{10}\}$ core. The tungsten metal centers link the various lacunary fragments in a manner similar to that seen in both $[As_2W_{21}O_{69}(H_2O)]^{6-}$ [46] and $\{As_2W_{19}O_{68}\}$, with the neodymium(III) ions seemingly providing further stability by forming links to multiple portions

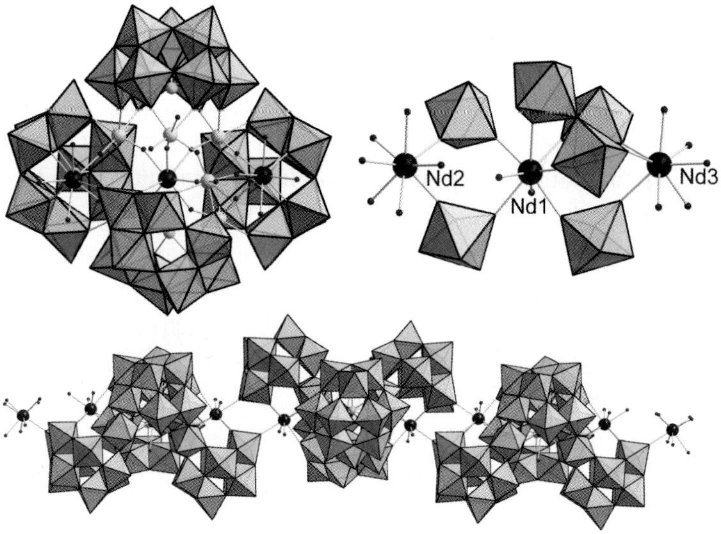

Fig. 5. Combined polyhedral/ball-and-stick representations of **4**. (Reproduced with permission from Ref. 45. Copyright © 2010, The American Chemical Society.)

of the tungstate assembly. At the core of the chain structure in **4** are the neodymium centers which act both as templates for the polyanion assembly and linkers to form the chains. Within **4** there are three crystallographically unique neodymium atoms, each displaying approximate square antiprismatic coordination geomety. Center Nd1 is located at the center of each tetrameric polyanion, with six of its eight bonds to oxo ligands of the additional tungstate units not belonging to the $\{\alpha\text{-}AsW_9O_{33}\}$ or $\{\beta\text{-}AsW_8O_{29}(OH)\}$ subunits. The remaining two coordination sites are filled by a terminal oxo ligand of the $\{\beta\text{-}AsW_8O_{29}(OH)\}$ unit, and a terminal aqua ligand. The remaining neodymium centers are involved in connecting the polyanions into chains, with four W–O–Nd linkages between neighboring clusters.

2.2. *Ln-org-POMs synthesized in the presence of glycine*

Continuing to build upon the use of $[As_2W_{19}O_{67}(H_2O)]^{14-}$ as our POM precursor of choice, we began to pursue the development of more elaborate chemistry involving the inclusion of carboxylate-containing species other than acetate, targeting Ln-org-POMs. Amino acids were an obvious choice of ligand and the simple glycine proved an excellent starting point, with glycine providing a buffering capacity akin to acetate. A key additional feature of glycine is the presence of a primary amine, which, when triply protonated, can facilitate the formation of electrostatic and hydrogen bonding interations with the polyanion surface, thus stabilizing its inclusion within the polyanion framework. While carboxylate-containing ligands are not so unusual in POM chemistry, amino acids have been little employed in this capacity, with $[Ce_4As_4W_{44}O_{151}(ala)_4(OH)_2(H_2O)_{10}]^{12-}$ representing a rare example of a structurally-characterized amino acid and lanthanoid containing polyoxotungstate.[47] Although not incorporated within the Ln-POM structure, the stereoselective interaction of amino acids with enantiomers of chiral Ln-POMs has been reported by Pope and co-workers,[48] while the Kortz group has successfully prepared a family of amino acid functionalized hexamolybdates.[27]

Reaction of a heated acidic aqueous 1 M sodium chloride solution containing $[As_2W_{19}O_{67}(H_2O)]^{14-}$, dysprosium nitrate and glycine yielded $[Dy_4As_2W_{22}O_{76}(H_2O)_{19}(gly)_2]^{2-}$ **(5-Dy)**.[45] Complex **5** has a chain structure,

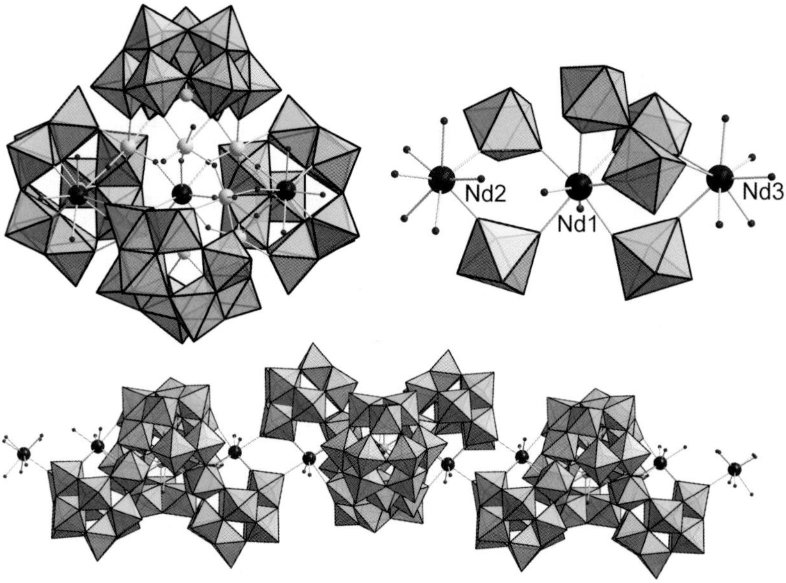

Fig. 6. Combined polyhedral/ball-and-stick representations of **5**. (Reproduced with permission from Ref. 45. Copyright © 2010, The American Chemical Society.)

with the repeat unit a $[Dy_4As_2W_{22}O_{76}(H_2O)_{19}(gly)_2]^{2-}$ polyanion that displays C_s point symmetry (Fig. 6). These polyanions are assembled from two $\{\alpha\text{-}AsW_9O_{33}\}$ subunits, that are then connected *via* two unprecedented $\{W_2O_5(gly)\}$ linking units and templated by a single sodium cation. The glycine molecules bridge the tungsten centers in a bidentate fashion with W–O bond lengths of 2.25(1) Å (Fig. 7). Due to the presence of the glycine molecules, the two $\{\alpha\text{-}AsW_9O_{33}\}$ units are crystallographically unique with N–H···O hydrogen bonding donation from the protonated amine to only one $\{\alpha\text{-}AsW_9O_{33}\}$. One hydrogen bond is observed between a tungsten-bound terminal oxo ligand, while two additional hydrogen bonds are located between the protonated amine and a neighborning polyanion unit. Overall the sodium templated polytungstate arrangement provides an open structure that accomodates four dysprosium centers. Three of the four dysprosium atoms are crystallographically unique with Dy1 and Dy3 displaying approximate square antiprismatic coordination geometries; meanwhile Dy2 has an approximately dodecahedral coordination

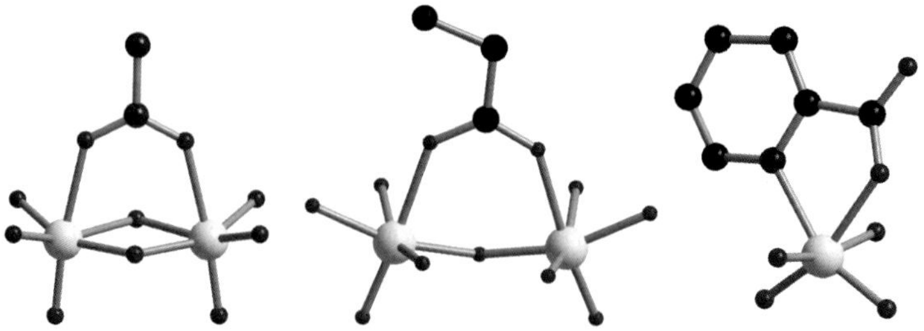

Fig. 7. Ball and stick representation of acetate, glycine and 2-picolinate binding modes to tungstates observed in compounds **5–9**.

geometry. Both Dy1 and Dy3 coordinate to the polyanion surface through four terminal oxo ligands belonging to the polyanion, meanwhile the two equivalent Dy2 centers only form two Dy–O–W linkages. The remaining coordination sites on each of the dysprosium centers are completed by terminal aqua ligands, except for one additional Dy–O–W connection, which is responsible for the chain structure of **5**. Each repeating POM unit in **5** is connected into chains through two crystallographically equivalent Dy–O–W linkages which occur through coordination of Dy3 to a terminal oxo ligand in the $\{W_3O_{13}\}$ cap of the $\{\alpha\text{-}AsW_9O_{33}\}$ unit.

The family of Ln-org-POM complexes $[Ln_4As_5W_{40}O_{144}(H_2O)_{10}(gly)_2]^{21-}$ with Ln = Gd (**6-Gd**), Tb (**6-Tb**), Dy (**6-Dy**) and Ho (**6-Ho**) was also synthesized from the $[As_2W_{19}O_{67}(H_2O)]^{14-}$ precursor and isolated as a mixed lithium/HDABCO salt where HDABCO is the monocationic form of 1,4-diazabicyclooctane.[49] Similar to **4** and **5**, **6** contains fragments originating from isomerization and decomposition of the tungstoarsenate precursor. The synthesis of **6** involved initial reaction followed by a very specific pH controlled recrystallization procedure incorporating lithium acetate. The organic compound DABCO was selected to provide countercations due to the pK_a values of 2.97 and 8.82 of its conjugate acids. These values are ideal for controlled protonation/deprotonation and therefore transformation from a dicationic to monocationic species at the pH of 5.7 used for recrystallization of **6** in the presence of lithium cations. Lithium acetate was included in the recrystalization medium to aid the dissolution of the complex, with the presence of lithium dramatically increasing the solubility of the polyanion as the pH is increased. This approach was based on the common practice of improving the solubility of polyoxotungstates for ^{183}W NMR spectroscopy by exchanging the countercations for lithium. Clearly, inclusion of lithium in the reaction mixture is a synthetic technique that can be exploited to maintain relatively insoluble POM building blocks in solution for long enough to access otherwise unobtainable structural architectures. Another interesting feature of the recrystallization procedure employed for **6** is that despite the large quantity of acetate in solution, the glycine molecules remains coordinated to the lanthanoid centers, with no acetate present in the final product. This observation suggests that the polyanion is most likely stable in solution, with the added stabilization of the intramolecular hydrogen bonding between the protonated amine of the glycine and the polyanion surface making ligand replacement unfavorable.

The structure of complex **6** is composed of two $\{As_2W_{19}O_{68}\}$ units, which are cross-linked through two symmetry equivalent W–O–Ln connections and an edge-shared di-oxo bridged ditungsten unit which is located on the molecular inversion center (Fig. 8). Each of the two unique lanthanoid centers is eight-coordinate. Lanthanoid center Ln1 coordinates to five terminal oxo sites of the POM framework, to two terminal aqua ligands and to a fully occupied oxo ligand, which has been refined as belonging to a partially occupied $\{As^{III}O_4\}$ unit with approximate "seesaw" coordination geometry (sp^3d hybridized to give a trigonal bipyramidal electron pair geometry with a stereoactive lone pair of

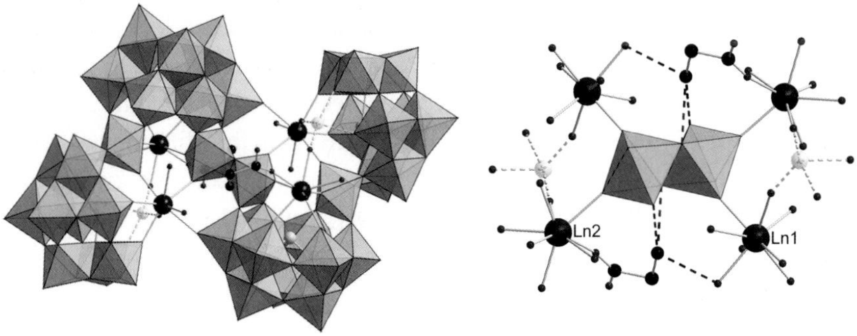

Fig. 8. Combined polyhedral/ball-and-stick representations of **6**. (Reproduced with permission from Ref. 49. Copyright © 2011, The American Chemical Society.)

electrons in an equatorial position). This oxygen position has therefore been interpreted as a μ_2-bridging oxo in the presence of arsenic at this site and as a terminal aqua ligand in the absence of the arsenic atom. A similar peripheral $\{As^{III}O_4\}$ unit has been observed previously in the polyoxomolybdate $[As_2Mo_8V_4AsO_{40}]^{5-}$.[50] The second crystallographically unique lanthanoid center, Ln2, coordinates to the POM through four Ln–O–W linkages, which fill one of the basal coordination planes of the rare earth metal. Additionally, Ln2 also coordinates to a partially occupied $\{As^{III}O_4\}$ unit, with terminal aqua ligands bound to the lanthanoid center when the arsenic site is vacant. The three remaining coordination sites are occupied by two terminal aqua ligands and a zwitterionic glycine through monodentate coordination to the carboxylate functional group. The pH of 5.7 at which **6** crystallizes, ensures the cationic nature of the glycine amine group ($pK_a = 9.78$). Both Ln1 and Ln2 display distorted square antiprismatic coordination geometries, giving rise to approximate D_{4d} local site symmetry. However, significant axial compression is observed for both lanthanoid centers and both coordination geometries deviate noticeably from strict D_{4d}. This is more pronounced for Ln2, which shows a substantial non co-planarity of the two square planes of the square antiprism.

2.3. *Ln-org-POMs synthesized in the presence of 2-picolinate*

The successful inclusion of glycine into Ln-org-POMs prompted us to consider other carboxylate-containing species as potential ligands. As 2-picolinate and its derivatives are well established as excellent ligands for lanthanoid centers that can act as organic chromophores, sensitizing lanthanoid-based luminescence, we were interested to incorporate 2-picolinate in Ln-org-POMs. Our initial synthetic efforts in this direction resulted in the isolation of two structural families of 2-picolinate containing Ln-org-POMs $[Tb_2As_2W_{19}O_{66}(H_2O)_2(pic)_4]^{10-}$ with Ln = (**7-Tb**) and $[Ln_8As_4W_{38}O_{132}(H_2O)_{22}(pic)_8]^{12-}$ with Ln = Tb (**8-Tb**) and Eu (**8-Eu**).[51,52] The compounds were prepared by careful selection and adjustment of the reaction parameters based upon the compatibility of the reagents. Selective crystallization of **7** or **8** was achieved by varying the lanthanoid salt employed as a starting material and the countercations available for crystallization. The use of terbium acetate affords **7-Tb**, while **8-Tb** is obtained when terbium nitrate is used and an excess of lithium ions is included in the reaction mixture. In this instance, the presence of lithium cations appears to promote the formation of **8-Tb**, probably as a consequence of the increased solubility of the polyanion building blocks compared to the typically less soluble potassium salts. Instances of similar structural dependence on cation have been reported previously for POMs. Thus through manipulation of counterion availability we were able to achieve the structural hierarchy observed for **7-Tb** and **8-Tb**, which has an important influence on the photophysical properties. Unfortunately, we were unable to obtain the lower nuclearity europium analog.

The structures of **7** and **8** are unprecedented and feature POM ligands bound to mononuclear 2-picolinate complexes of lanthanoid(III) ions, with additional tungstate linkers (Fig 9). Generation of the $\{\beta\text{-}AsW_8O_{30}\}$ species (which is the major structural component) from the $[As_2W_{19}O_{67}(H_2O)]^{14-}$ precursor requires a complex series of rearrangements (Scheme 1). These involve disassembly of the POM precursor, followed by isomerization of the resultant trilacunary $\{\alpha\text{-}AsW_9O_{33}\}$ species to the respective beta isomer and further tungstate loss, to give the $\{\beta\text{-}AsW_8O_{30}\}$ building block. Upon disassembly of the precursor, the tungstate released is likely chelated by the 2-picolinate, preventing further condensation (Fig. 7).

The molecular arrangement of complex **7** can be described as two $\{\beta\text{-}AsW_8O_{30}\}$ units, oriented at 180° with respect to each other (Fig. 9). This arrangement results in the formation of an asymmetric lacunary site, within which are encapsulated three $\{WO_2(pic)\}$ moieties and two terbium(III) centers in a sandwich-like configuration. The construction of **7** can be visualized to occur through the initial coordination of a terbium(III) cation (Tb1), which acts as a "hinge" on the backbone of the polyanion, bridging the $\{\beta\text{-}AsW_8O_{30}\}$ units. Opening of the "hinge" then facilitates insertion of two $\{WO_2(pic)\}$ species, with the 2-picolinate ligands flanking the belt region of the polyanion. Notably, coordination of the 2-picolinate ligands to the W(VI) centers observed in **7** and **8** represents the first report of a W(VI)-N coordinate bond within a POM (Fig. 7). The remaining coordination sites within the cavity of the polyanion host further $\{WO_2(pic)\}$ and $\{Tb(pic)(H_2O)\}$ moieties. The structure of complex

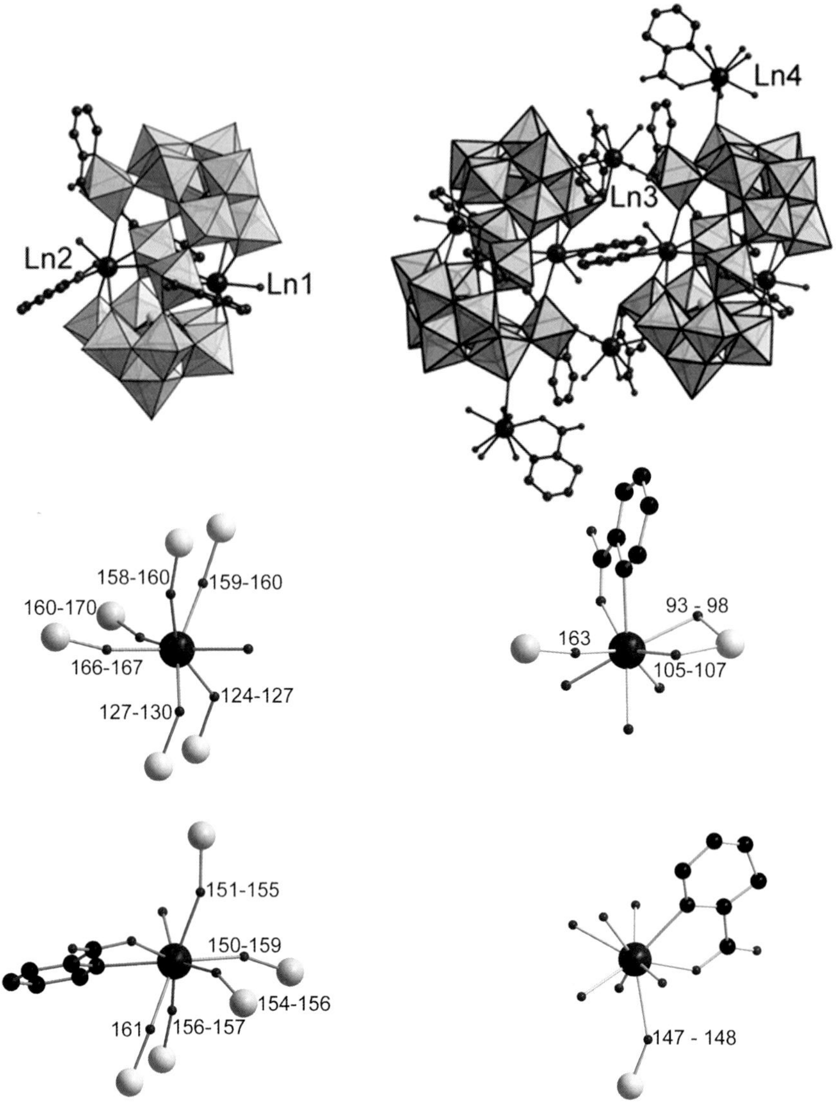

Fig. 9. Combined polyhedral/ball-and-stick representations of **7** (top left) and **8** (top right). (Reproduced with permission from Ref. 51. Copyright © 2010, Wiley.)

8 (Fig. 9, top right) can be considered to be based on two molecules of **7**, linked by two crystallographically equivalent {Ln(pic)$(H_2O)_3$} fragments. These sites possess distorted square antiprismatic coordination environments, with three Ln–O(W) bonds. Two of these bonds are involved in an edge-sharing link to the polyoxotungstate portion of the complex, with associated acute Ln–O–W bond angles of 93–98° and 105–107° (Fig. 9). Complex **8** also incorporates two further {Tb(pic)$(H_2O)_6$} units, which display 9-coordinate capped trigonal prismatic coordination geometries, with a single Ln–O(W) bond. Overall, the intricate assembly of the POM and {Ln(pic)} fragments in the structures of complexes **7** and **8** demonstrate the versatility of POMs to accommodate {LnL} (L = organic ligand) units in the generation of Ln-org-POMs.

A slight variation of the synthetic method that affords complex **8** gave rise to $[K_2Dy_2As_2W_{21}O_{72}(H_2O)(pic)_2]^{6-}$ (**9-Dy**).[53] Solid dysprosium nitrate, 2-picolinic acid and $[As_2W_{19}O_{67}(H_2O)]^{14-}$ were mixed in the stoichiometric ratio 2:4:1 at pH 3.4, resulting in single crystals of **9** after two days in a sealed vessel. The synthesis of **9** differs from that of **8** in only two particulars: the ionic strength is double and lithium ions are omitted. However, the structure of the two compounds are very different, with the 3D framework **9** containing {α-AsW_9O_{33}} building blocks instead of the {β-AsW_8O_{30}} units evident in the discrete molecular species **8**, in addition the 2-picolinate ligands in **9** do not coordinate to the lanthanoid centers, as is observed in **8**. It's notable that while **9** is formed within a few days, **8** crystallizes over a period of weeks and all efforts to speed up its synthesis actually result in a mixture of compounds that appears to contain **9**. It is likely that **9** is a kinetic product and **8** is a thermodynamic product of the same reaction, with the presence of solubilizing lithium ions and the lower ionic strength working together to keep the products in solution for long enough to allow the formation of **8**. The small differences in the synthesis of two such structurally diverse compounds exemplifies the complex equilibria that exists in POM solution chemistry and the important roles that ionic strength and cation identity play in determining the species that crystallizes from solution.

The structure of **9** can be considered as a 3D framework comprised of constituent {{Dy$(H_2O)_3$}$_2${$As_2W_{19}O_{68}$}{WO_2(pic)}$_2$} POM units and {K_4O_4} distorted cubane units (Fig. 10). A number of unique structural

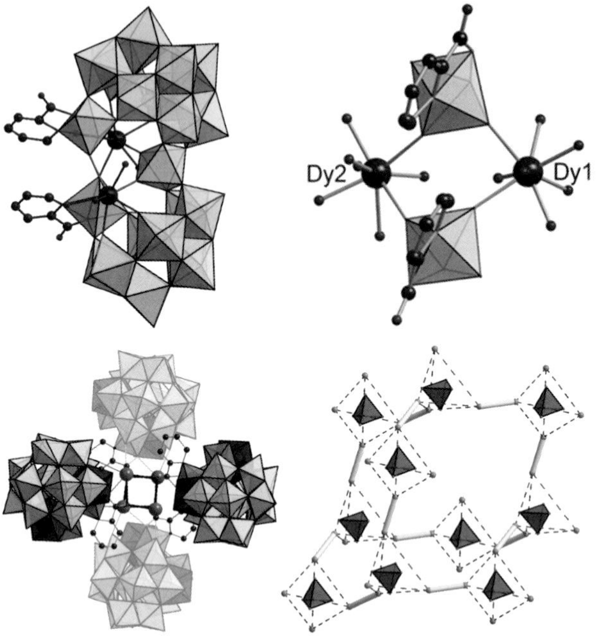

Fig. 10. Combined polyhedral/ball-and-stick representations of of **9**: the Dy-pic-POM building block (top), tetrahedral arrangement of Dy-pic-POM units around the templating potassium cubane (bottom left), and diamondoid network structure (bottom right). (Reproduced with permission from Ref. 53. Copyright © 2011, The Royal Society of Chemistry.)

features are evident in both building blocks. The two $\{WO_2(pic)\}$ moieties in each of the $\{Dy_2(H_2O)_6\{As_2W_{19}O_{68}\}\{WO_2(pic)\}_2\}$ POM building blocks of **9** are key to the resulting 3D structure. The tungsten center of each $\{WO_2(pic)\}$ fragment is coordinated to the pyridyl nitrogen and carboxylate oxygen donor atoms of a 2-picolinate ligand, forming a 5-membered chelate ring. These $\{WO_2(pic)\}$ fragments were observed previously in **7–8** and are generated *in situ* following partial scavenging of free tungstate that is released on dissociation of the $[As_2W_{19}O_{67}(H_2O)]^{14-}$ starting material into the constituent $\{\alpha\text{-}AsW_9O_{33}\}$ fragments.

Bound into the lacunary site of each $\{As_2W_{19}O_{68}\}$ subunit are two dysprosium(III) ions, which do not contribute to formation of the 3D framework, together with the two crucial $\{WO_2(pic)\}$ fragments (Fig. 7) located between the Dy centers. Incorporation on the two $\{WO_2(pic)\}$ units into the POM provides the route by which the POM building block can act as a 2-connecting unit to link the other building block of the 3D structure, the $\{K_4O_4\}$ distorted cubanes. Structurally, the potassium centers in the $\{K_4O_4\}$ units are each chelated by the carboxylate functional group of a picolinate ligand, with one of the oxygen atoms from each ligand occupying the corners of the cubane and the other linking the cubane to the POM through a K–O–W connection to the tungsten centers from the $\{WO_2(pic)\}$ fragments. The K–O(K/W) bonds are in the range 2.82–2.91 Å. Each $\{K_4O_4\}$ unit is connected to four independent POM units through the $\{WO_2(pic)\}$ tungsten centers, generating the tetrahedral arrangement of POMs around the central cubane template. The presence of analogous $\{K_4O_4\}$ distorted cubanes in extended framework materials is rare with only a few examples in the literature, none of which also contain POMs. Within the 3D structure of **9**, each $\{K_4O_4\}$ distorted cubane is a 4-connecting node, with the $\{\{Dy(H_2O)_3\}_2\{As_2W_{19}O_{68}\}\{WO_2(pic)\}_2\}$ POM units acting as 2-connecting linkers. This results in a diamondoid network topology with recognisable adamantane cages and interconnecting channels.

3. Magnetic Properties

Variable temperature magnetic susceptibility studies of compounds containing the Ln-POM and Ln-org-POM complexes **1-Gd**, **2-Yb**, **3-(Gd and Er), 6 (Gd, Tb, Dy and Ho), 7-Tb** and **8-Tb,** afforded X_mT versus T profiles that are typical for non-interacting lanthanoid(III) ions.[35,36,45,49] The gadolinium complexes **1-Gd**, **3-Gd** and **6-Gd** all display the spin-only magnetic behavior expected for gadolinium(III) ions ($4f^7$, $^8S_{7/2}$). Near perfect Curie-type susceptibility plots are observed with temperature-independent X_mT behavior down to very low temperatures. The observed Curie constants per gadolinium(III) ion are in the range 7.76–7.83 $cm^3\,K°\,mol^{-1}$, which are slightly lower than the calculated spin-only value of 7.88 $cm^3\,K°\,mol^{-1}$ because of the large spin-orbit interaction mixing other terms in to the $J = 7/2$ ground state.[54,55] Very small decreases in X_mT observed below 10 K are due to saturation and are not associated with antiferromagnetic exchange interactions between the gadolinium(III) centers.

At room temperature, compounds of the terbium(III)-based ($4f^8$, $^7F_6$) complexes **6-Tb**, **7-Tb** and **8-Tb** and, the dysprosium(III)-based ($4f^9$, $^6H_{15/2}$) complex **6-Dy**, the holmium(III)-based ($4f^{10}$, $^5I_8$) complex **6-Ho**, the erbium(III)-based ($4f^{11}$, $^4I_{15/2}$) complex **3-Er** and the ytterbium(III)-based ($4f^{13}$, $^2F_{7/2}$) complex **2-Yb** all reach $\chi_m T$ values close to the corresponding theoretical values calculated using the free ion approximation. For each of these complexes the observed increase in $\chi_m T$ upon heating from 2 K to room temperature is attributed to thermal population of higher energy multiplet states and is, again, of single-ion origin. In these systems, the ligand field causes a splitting of the J ground state into M_J sublevels and an energetic separation between the lowest and higher M_J states is due to both ligand field effects and spin-orbit coupling.

Access to pure bulk samples of multiple members of the isostructural family of complexes **6** allowed a detailed comparative investigation of the magnetic properties of these species.[49] In each complex, the two distinct lanthanoid sites both display approximately D_{4d}-symmetric, square antiprismatic coordination geometries. Modelling the magnetic characteristics of compounds **6-Tb, 6-Dy** and **6-Ho** involved the use of a single ion Hamiltonian that requires accounting for the interelectronic repulsion, spin-orbit-coupling, ligand field effects and Zeeman splitting induced by the external field. To reduce the number of independent ligand field fitting parameters, the two distinct Ln(III) sites were treated as a single average environment, the point symmetry of which was idealized to D_{4d}. In this ideal case, value of the ligand field parameters B^k_q with $q \neq 0$ (B^4_4 and B^6_4) would be exactly zero. Further, model calculations show that small values of these parameters, as expected for small deviations from ideal symmetry, have an insignificant effect on the magnetic data. Least squares fitting of the experimental data (Fig. 11) yielded the values for the ligand field parameters B^2_0, B^4_0 and B^6_0, and spin-orbit coupling constant ζ indicated in Table 1, with the negative signs of B^2_0, B^4_0 and the positive signs for B^6_0 corresponding to the point charge model for an axially-compressed square antiprismatic (D_{4d}) ligand field. The continuous trends in B^2_0, B^4_0 and B^6_0, going from terbium to dysprosium to holmium, are consistent with the increase in effective nuclear charge of these elements. It is notable that

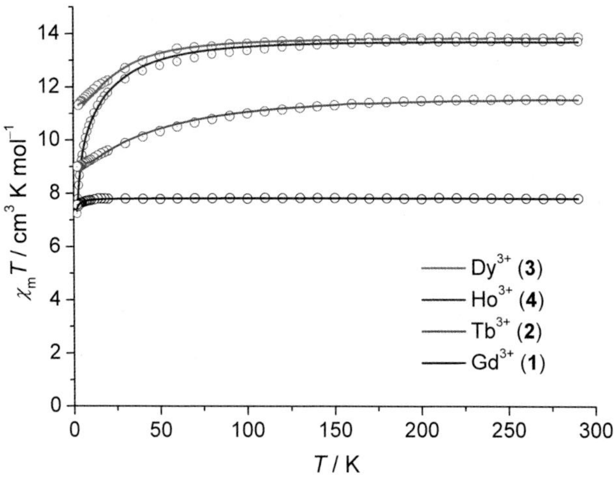

Fig. 11. Temperature dependence of $\chi_m T$ for compounds containing complex **6** of formula $(HDABCO)_8H_5Li_8$[**6**]$\cdot 25H_2O$ (Ln = Gd, Tb, Dy, and Ho), referenced to a single Ln center. The solid lines are least-squares fits as described in the text. (Reproduced with permission from Ref. 49. Copyright © 2011, The American Chemical Society.)

Table 1. Spin Hamiltonian parameters used for and derived from magnetochemical modeling of the $(HDABCO)_8H_5Li_8$[**6**]$\cdot$ $25H_2O$ family.

Parameters	6-Gd	6-Tb	6-Dy	6-Ho
Ln^{3+}	Gd^{3+}	Tb^{3+}	Dy^{3+}	Ho^{3+}
B^2_0/cm^{-1}	–	–25	–340	–480
B^4_0/cm^{-1}	–	–1510	–1425	–1375
B^6_0/cm^{-1}	–	155	205	225
ζ/cm^{-1}	1470	1705	1900	2160

both the signs of the three parameters and the trends observed are the same as those determined for the equivalent parameters in the $[Ln(\beta_2\text{-}SiW_{11}O_{39})_2]^{13-}$ family of compounds.[17] In addition, the trend in B^4_0 parameters for the family of complexes **6** is opposite to that observed for the $[Ln(W_5O_{18})]^{9-}$ family;[17] while the signs of, and trends in, the other

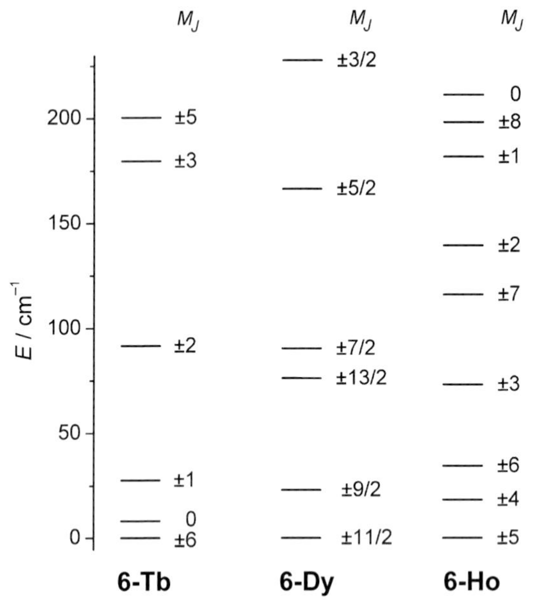

Fig. 12. Energy level diagrams of the ground-state multiplets for the $(HDABCO)_8H_5Li_8$[**6**]$\cdot 25H_2O$ family. Not shown are M_J sublevels above 230 cm^{-1} (for **6-Tb**: $M_J = \pm 4$ at 244 cm^{-1}; for **6-Dy**: $M_J = \pm 1/2$ at 261 cm^{-1} and $M_J = \pm 15/2$ at 345 cm^{-1}). (Reproduced with permission from Ref. 49. Copyright © 2011, The American Chemical Society.)

parameters are the same for the two structural families. In contrast the B^2_0 parameters are of opposite sign to those determined for the non-POM containing $[Pc_2Ln]^-$ family.[4] This seems to correlate directly with the negligible axial distortion evident in the $[Pc_2Ln]^-$ family versus the axial compression that is present in all three Ln-POM families of SMMs.

For **6-Tb, 6-Dy** and **6-Ho**, the energies of the M_J sublevels (Fig. 12) are determined by the ligand field parameters (Table 1) derived from the least squares fits. Notably, the exact splitting of the M_J sublevels for the terbium(III) sites in **6-Tb** is very sensitive to variations in $B^k{}_q$ values and the fits indicate the energetically lowest substates are $M_J = \pm 6$ (with a first excited $M_J = 0$ lying only 8 cm^{-1} higher in energy). The lowest substate level for the dysprosium(III) sites in **6-Dy** results as $M_J = \pm 11/2$ and for the holmium(III) sites in **6-Ho** it is $M_J = \pm 5$ (with first excited $M_J = \pm 11/2$ and $M_J = 0$ lying, respectively, 23 cm^{-1} and 18 cm^{-1} above in energy). Overall, the derived energy level diagrams for **6-Tb, 6-Dy** and **6-Ho** are broadly similar to those reported for the $[Ln(W_5O_{18})]^{9-}$ and $[Ln(\beta_2\text{-}SiW_{11}O_{39})_2]^{13-}$ structural families,[17] although the precise ordering of the M_J sublevels and the actual energies differ.

Frequency-dependent alternating current (AC) susceptibility measurements were performed on **6-Tb, 6-Dy** and **6-Ho**. While no significant out-of-phase susceptibility component (χ_m'') is observed down to 2.0 K for **6-Tb** and **6-Ho**, complex **6-dy** exhibits frequency-dependent in-phase and out-of-phase components characteristic of SMMs (Fig. 13). However, χ_m'' does not reach clearly defined maxima down to 2.0 K. In the absence of such maxima, the out-of-phase susceptibility data below 3.6 K were therefore fitted by a generalized Debye model, yielding an effective barrier $\Delta_{eff} = 3.9(1)$ K and an Arrhenius pre-exponential factor $\tau_0 = 1.9 \times 10^{-5}$ s. A relatively wide distribution of relaxation times is observed, indicating the presence of multiple relaxation mechanisms consistent with the presence of two crystallographically distinct dysprosium(III) ligand environments.

Of the Ln-org-POM family **6**, only the dysprosium-containing species, **6-Dy**, displays slow magnetic relaxation above 2.0 K, although it is possible that some of the other analogs may manifest slow magnetic relaxation at lower temperatures. It is informative to compare this behavior with that reported for the two mononuclear Ln-POM SMM families $[Ln(W_5O_{18})_2]^{9-}$ (Ln = Ho and Er) and $[Ln(\beta_2\text{-}SiW_{11}O_{39})_2]^{13-}$ (Ln = Dy, Ho, Er and Yb).[17] In the detailed study of these two structural families, the members that display slow magnetic relaxation all manifest Ising type magnetic anisotropy characterized by lowest energy sublevels with relatively large $|M_J|$ values, with the sublevels with small $|M_J|$ values lying at higher energy. This situation also exists for both **6-Dy** and **6-Ho** with a lowest energy sublevel with $M_J = \pm 11/2$ within the $J = 15/2$ ground state for **6-Dy**, whereas for **6-Ho** the lowest energy sublevel has $M_J = \pm 5$, within the $J = 8$ ground state. For the Tb-containing species **6-Tb**, although the $M_J = \pm 6$ is the lowest energy sublevel within the $J = 6$ ground state, only 8.1 cm^{-1} above lies the next highest energy level, characterized by $M_J = 0$. Of course the situation for **6-Tb**, **6-Dy** and **6-Ho** is complicated by the

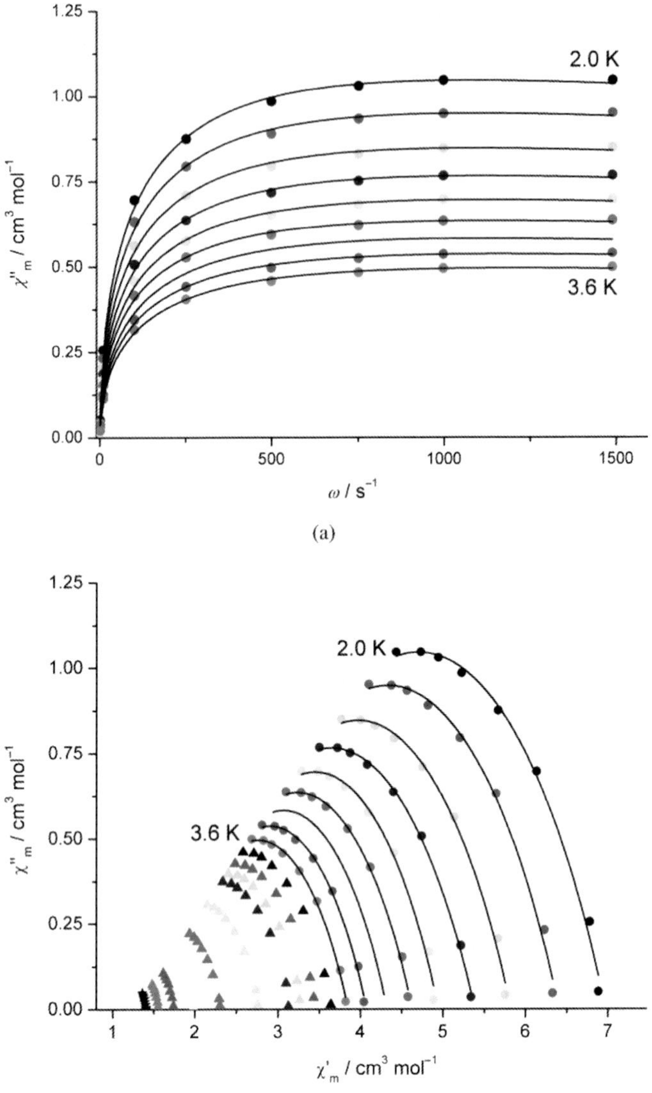

(b)

Fig. 13. Low-temperature AC susceptibility data for $(HDABCO)_8H_5Li_8$[**6-Dy**]$\cdot 25H_2O$. (a) Frequency dependence of the out-of-phase ac component χ_m'' from 2.0 to 3.6 K in intervals of 0.2 K. (b) "Cole-Cole" plot (χ_m'' vs. χ_m') from 2.0 to 10.0 K. Code: symbols = experimental data; black lines = least-square fits to a Debye model. Note that adequate fits are only obtained up to 3.6 K. (Reproduced with permission from Ref. 49. Copyright © 2011, The American Chemical Society.)

presence of two different lanthanoid environments, whereas only one such environment is present in $[Ln(W_5O_{18})_2]^{9-}$ and $[Ln(\beta_2\text{-}SiW_{11}O_{39})_2]^{13-}$. Thus the ligand field parameters and resulting energy levels derived can only represent an average for the two sites. In the study by the Coronado group the $[Ln(W_5O_{18})_2]^{9-}$ family of SMMs displayed slow magnetic relaxation at higher temperatures than the $[Ln(\beta_2\text{-}SiW_{11}O_{39})_2]^{13-}$ family, despite the similar pattern in the energies of the M_J substates for members of the two families containing the same lanthanoid ion.[17] From the careful analysis of the structural parameters associated with the local lanthanoid coordination environment it was apparent that although axially compressed, approximate D_{4d} local site symmetry was evident in both families, a deviation from strict axial symmetry existed in the $[Ln(\beta_2\text{-}SiW_{11}O_{39})_2]^{13-}$ family, which was most readily apparent from the deviation from coplanarity of the two coordinating planes of the square antiprism. This deviation from axial symmetry appears to give rise to faster magnetic relaxation, probably by mixing the different wavefunctions and enhancing the rate of quantum tunnelling through the energy barrier.

Members of the Ln-org-POM family **6** also display axially compressed distorted D_{4d} local site symmetry at the lanthanoid centers, with the deviation from strictly axial symmetry even greater than that observed for the $[Ln(\beta_2\text{-}SiW_{11}O_{39})_2]^{13-}$ family. This is particularly evident for one of the two inequivalent lanthanoid(III) sites. This deviation from strict axial symmetry clearly decreases the magnitude of the energy barrier for **6-Dy** estimated from AC susceptibility data, which is likely due to enhanced quantum tunneling associated with greater mixing of the wavefunctions. Given the relatively small separation between the low energy sublevels for complexes **6**, with the lowest energy levels separated by less than 25 cm^{-1} in each case, it is thus unsurprising that the energy barrier to magnetic relaxation determined for compound **6-Dy** is small. In any case, the observation of slow magnetic relaxation above 2.0 K for compound **6-Dy** and not for compound **6-Ho**, is consistent with a larger energy barrier for the dysprosium compound. Consistent with the studies from the Coronado group, our study of the magnetic properties of the family of complexes **6** suggests that a larger energy barrier to magnetization reversal will result for Ln-POMs where the local symmetry of the lanthanoid(III) ion deviates as little as possible from strictly axial.

4. Luminescence Properties

The POM ligands of Ln-POMs can act as inorganic "antennas" to sensitize the lanthanoid(III) luminescence. Photoexcitation of the O→M LMCT band of the POM ligand affords emission from the $^5D_0$ and $^5D_1$ excited states of europium(III) to the ^7F_J (J = 0–6) ground state multiplet for Eu-POM complexes and emission from the $^5D_4$ excited state of terbium(III) to the ^7F_J (J = 0–6) ground state multiplet for Tb-POM complexes (Fig. 14). In principle, ternary Ln-org-POMs can possess two different routes for sensitizing the lanthanoid emission, with both organic and inorganic ligands potentially acting as "antennas". Early studies of Eu-org-POMs reported sensitized Eu(III)-luminescence arising when: (i) a suspension of $[Eu(H_2O)_3(\alpha_2\text{-}P_2W_{17}O_{61})(Eu_2(H_2O)_7)_4]_4^{4-}$ in methanol was allowed to interact with a range of organic ligands and (ii) phenanthroline and bipyridine were added to

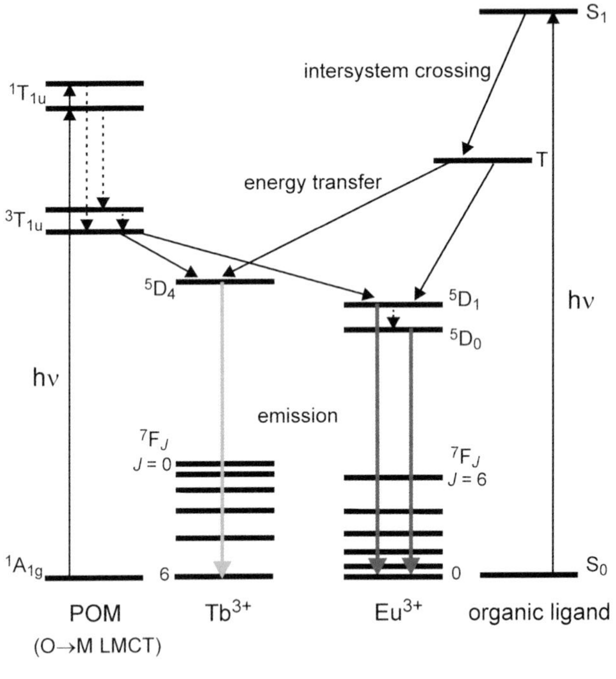

Fig. 14. Schematic energy level diagram of relaxation processes for the sensitized emission of Tb(III) and Eu(III) in Ln-org-POMs. (Reproduced with permission from Ref. 52. Copyright © 2012, The American Chemical Society.)

acetonitrile solutions of $[EuPW_{11}O_{39}]^{4-}$ and $[Eu(H_2O)_4P_2W_{17}O_{61}]^{7-}$, with the resulting luminescent species proposed to be ternary Eu-org-POM complexes of uncertain identity.[56–58] Subsequently Eu(III)- and Tb(III)-luminescence was observed for solid samples inferred to be $[M_4O_{12}Eu(picOH)_3]$ (M = Mo, W, HpicOH = 3-hydroxypicolinic acid) and $[Ln(W_5O_{18})_2(picOH)]$ (Ln = Eu and Tb), although again the luminescent materials were not structurally characterized.[59,60] The first structurally characterized luminescent Ln-org-POM was the polymeric $K_2Na[Eu_3(H_2O)_{14}(HNA)_3\{Eu(H_2O)P_5W_{30}O_{110}\}]\cdot 3H_2O$ (HNA= nicotinic acid.[61] However, in each of the cases listed above, the photophysical properties of the compounds were not studied in detail, allowing neither a site-specific determination of the luminescence properties of the different lanthanoid(III) centers, nor a delineation of the function of the organic ligand.

Our initial observation of intense green Tb(III)-based luminescence following irradiation of a solid sample of a compound of the octanuclear complex **8-Tb** with a laboratory UV lamp (λ_{ex} = 254 nm), was intriguing, given that a compound of the structurally related dinuclear complex **7-Tb** was not luminescent. It was immediately clear from this behavior that the two of the four inequivalent terbium sites in the octanculear complex that are not present in the dinuclear complex must be those responsible for the observed emission. In addition, similar irradiation of **8-Eu** afforded bright red Eu(III)-based emission, although unfortunately it was not possible to isolate a pure bulk sample of the europium analog of **7** for comparison. Detailed photophysical measurements performed on the three compounds involved irradiating the samples with two different wavelengths that corresponded to separate excitation of the organic 2-picolinate and the inorganic POM ligands. Measuring the emission at different temperatures has provided important insights into the luminescence properties of Ln-org-POMs, which were only possible because the Ln-org-POMs were structurally characterized.

By monitoring the room temperature Eu(III) emission of solid $Eu_{1.66}K_7$[**8-Eu**]$\cdot 54H_2O$ at 615 nm, the excitation spectrum was collected (Fig. 15(a)). This shows peak maxima at *ca.* 279 and 315 nm, which can be attributed to absorption by the coordinated organic 2-picolinate and inorganic POM ligands, respectively, resulting in subsequent Eu(III)-centered emission.[18,19,62–64]

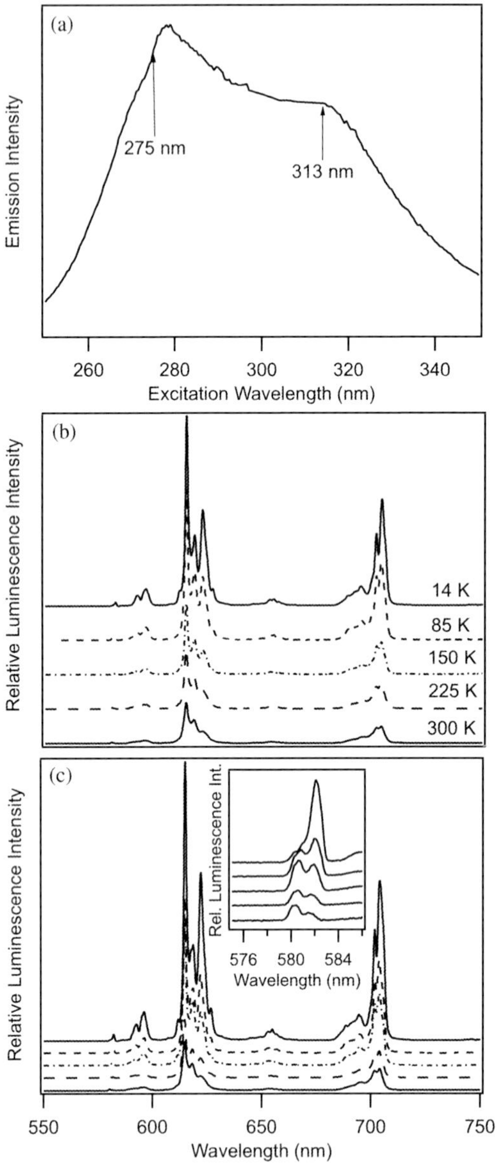

Fig. 15. (a) Excitation (λ_{emiss} = 615 nm), (b) temperature-dependent luminescence (λ_{ex} = 275 nm), and (c) temperature-dependent luminescence (λ_{ex} = 313 nm) spectra for solid $Eu_{1.66}K_7$[**8-Eu**]$\cdot 54H_2O$ with spectra shown at 14 K, 85 K, 150 K, 225 K and 300 K (top to bottom). Inset to c): temperature dependence of the highest-energy luminescence transitions. (Reproduced with permission from Ref. 52. Copyright © 2012, The American Chemical Society.)

The corresponding variable temperature emission spectra for solid $Eu_{1.66}K_7$[**8-Eu**]·$54H_2O$ are shown in Figs. 15(b) and (c). Spectra were measured with excitation wavelengths of 275 and 313 nm, which correspond to excitation through the 2-picolinate and POM ligands, respectively. The 14 K emission spectra of $Eu_{1.66}K_7$[**8-Eu**]·$54H_2O$ at both excitation wavelengths display characteristic transitions from the $^5D_0$ excited states of Eu(III) to the ^7F_J manifold. The apparent peak maxima are at *ca*. 582, 590, 620, 650 and 700 nm for the $J = 0, 1, 2, 3$ and 4 transitions, respectively. In europium(III) compounds the $^5D_0 \rightarrow ^7F_0$ transition is a singlet for each emitting site. Our best resolved spectrum at 14 K (λ_{ex} = 313 nm) shows a dominant peak at 582.0 nm with a width at half height of 1 nm, and a weaker maximum at 580.8 nm, higher in energy by 35 cm^{-1}. This small separation corresponds to the energy difference among the different europium emitting sites, indicating that the coordination environment is similar for the two or more emitting sites. The two peak maxima are also observed in the spectra measured with 275 nm excitation. In general for Eu(III)-luminescence, the magnetic-dipole allowed $^5D_0 \rightarrow ^7F_1$ transition is little affected by the coordination geometry, whereas the intensity of electric-dipole allowed $^5D_0 \rightarrow ^7F_2$ transition depends strongly on the coordination geometry and metal-ligand covalency. For compound $Eu_{1.66}K_7$[**8-Eu**]·$54H_2O$ the hypersensitive $^5D_0 \rightarrow ^7F_2$ transition is relatively very intense with $I(^5D_0 \rightarrow ^7F_2) / I(^5D_0 \rightarrow ^7F_1)$ intensity ratios of *ca*. 12 and 10 for λ_{ex} = 275 and 313 nm, respectively, suggesting low local symmetry at the emitting sites.

As the temperature is increased from 14 K to above 150 K the intensity of the weak, high-energy, peak of the $^5D_0 \rightarrow ^7F_0$ transition observed at 580.8 nm for $K_4Li_4H_4$[**8-Eu**]·$58H_2O$ increases compared to the more intense, lower energy peak at 582 nm for both excitation wavelengths, suggesting that the emission from the higher-energy site/s gains intensity through thermal population and energy transfer processes. At room temperature, the higher energy peak is dominant, a reversal of the intensity distribution at 14 K. Otherwise, the general trend is that the intensity of all transitions decreases as the temperature is increased up to 300 K, with very similar $I_{300\,K} / I_{14\,K}$ intensity ratios of *ca*. 5 and 6 for the $^5D_0 \rightarrow ^7F_1$ transitions for λ_{ex} = 275 and 313 nm, respectively. This is associated with a general increase in linewidth, but no significant shift in peak position, indicating the absence of major structural changes with temperature. The temperature dependence of the luminescence intensity for all excitation wavelengths is similar, possibly due to efficient energy transfer among the europium(III) sites, preventing the measurement of sensitization through POM absorption to the nearest europium(III) site.

Room temperature time resolved emission profiles were also collected for $Eu_{1.66}K_7$[**8-Eu**]·$54H_2O$ at 615 nm, with $\lambda_{ex} = 280$ nm. Multiexponential decay behavior was observed, consistent with the presence of four inequivalent Eu(III) sites. Fitting the decay profiles gave satisfactory agreement to a biexponential function, with lifetime values of $\tau_1 = 195 \pm 13$ s (19 %) and $\tau_2 = 393 \pm 3$ s (81%), which are consistent with expected values for hydrated complexes of europium(III).[65,66] The 80:20 ratio indicates a significant difference in the intensities of the two different emitting sites. As europium(III) is very sensitivite to quenching of the luminescence by coordinated aqua ligands, this suggests a different number of aqua ligands on the two sites.

The different spectroscopic experiments suggest the presence of at least two different emitting europium(III) centers in $Eu_{1.66}K_7$[**8-Eu**]·$54H_2O$ across the temperature range 14–300 K, regardless of whether the luminescence is sensitized by excitation through the organic 2-picolinate or inorganic POM ligands. The emission intensities display significant temperature dependence for both excitation wavelengths. Variable temperature spectroscopic studies of Eu-POMs have revealed the temperature dependence of the europium(III) emission intensity upon excitation into the O→M LMCT bands is always greater than under excitation of the f-f bands, due to thermal relaxation of the O→M LMCT states competing with energy transfer to the europium(III) ion. However, the extent of temperature dependence upon excitation into the O→M LMCT bands depends on both the Eu–O–M angles at the emitting site and the M–O–M angles within the POM ligand. Thermal deactivation of the O→M LMCT states is associated with d^1 hopping delocalization within the POM ligand, which depends on the configuration of both M–O–M and Eu–O–M linkages. Photoexcitation into the O→M LMCT bands allows the d^1 electron to hop to the Eu(III) site via the Eu–O–M linkage when the Eu–O-M angles are around 150°, through orbital mixing similar to the *dπ-pπ-dπ* mixing that occurs at corner-shared MO_6 octahedra with M–O–M bond angles of around 150°. Thus Eu-POMs with Eu–O–M angles of greater than 150° display a significant decrease in emission intensity with increasing temperature, while Eu-POMs with only edge-shared MO_6 octahedra and M–O–M angles of around 100° show little temperature dependence of the emission intensity due to localization of the d^1 electron on the MO_6 octahedra. A decrease of the intensity of Eu(III)-emission with increasing temperature has also been observed for complexes with only organic ligand sensitizers and ascribed to nonradiative relaxation from the $^5D_0$ state due to crossover to a low-lying LMCT state.[65] Typically, an increase in linewidth is also observed with temperature, particularly for less rigid structures.[64]

The structural complexity of **8-Eu** renders a precise assignment of the observed luminescence to specific europium sites impossible. Three of the four europium centers (Fig. 9), coordinate to picolinate ligands and are thus candidates as emitting sites following excitation through the organic chromophore, while potentially all four centers may display luminescence following excitation through the POM. Europium centers Eu1, Eu2, Eu3 and Eu4 have one, one, three and six aqua ligands, respectively, which can give rise to quenching of lanthanoid(III) emission, shortening luminescence lifetimes via non-radiative coupling with proximal OH oscillators.[66] They also have four, six, one and one Eu-O-W bridging angles greater than 145°, respectively (Fig. 9), which are associated with marked temperature dependence of the emission following excitation into the POM. Only Eu3 has two Eu-O-W bridging angles in the range 95–110°, due to edge-sharing, which are associated with little temperature dependence of the emission. The lack of luminescence of terbium-containing **7-Tb** indicates that the emitting centers in **8-Tb** are Tb3 and Tb4 and it is likely the equivalent Eu3 and Eu4 sites in isostructural **8-Eu** are also emitting, at least following excitation through the 2-picolinate ligand. The shorter luminescence lifetime is assigned to Eu4, with six aqua ligands to quench the luminescence, while the longer lifetime is assigned to Eu3, which only has three aqua ligands.

By monitoring the room temperature Tb(III)-emission of $K_4Li_4H_4$ [**8-Tb**]$\cdot 58H_2O$ at 490 nm, the excitation spectrum was also collected, showing a peak maximum at *ca.* 276 nm and a smaller shoulder at slightly lower wavelength, both of which we attribute to absorption by the coordinated 2-picolinate ligands, resulting in subsequent Tb(III)-centered emission (Fig. 16(a)). The absence of any other excitation peaks in the UV region indicates that in contrast to the europium analog, sensitization of the terbium(III) ions at room temperature proceeds solely via the organic chromophores, and does not involve POM-centered LMCT states. This is in accordance with results from Yamase *et al.*, who have shown that room temperature sensitization via the latter pathway is not efficient for terbium(III), due to radiationless deactivation involving Tb(IV)-W(V) charge transfer (CT) states.[18,19]

The 14 K emission spectra of $K_4Li_4H_4$[**8-Tb**]·$58H_2O$ at excitation wavelengths of 275 and 313 nm both display characteristic transitions from the $^5D_4$ excited state of terbium(III) to ground state ^7F_J manifold (Figs. 16(b) and 16(c)), consistent with excitation through both 2-picolinate and POM ligands at this temperature. The apparent peak maxima are at ca. 492, 544, 583, 622, 650, 670 and 680 nm for the $J = 6, 5, 4, 3, 2, 1$ and 0 transitions, respectively. For terbium(III) the hypersensitive electric-dipole allowed transitition is $^5D_4 \rightarrow {}^7F_6$, which can be compared with the environment-insensitive magnetic-dipole allowed $^5D_4 \rightarrow {}^7F_5$. For $K_4Li_4H_4$[**8-Tb**]·$58H_2O$ at 14 K, $I(^5D_4 \rightarrow {}^7F_5) / I(^5D_4 \rightarrow {}^7F_6)$ intensity ratios of *ca.* 0.05 are determined for both $\lambda_{ex} = 275$ and 313 nm, again suggesting low local symmetry at the emitting sites.

As the temperature is increased from 14 K to 300 K, the intensity of all transitions again decreases for both excitation wavelengths. However the change with temperature is much more marked for $\lambda_{ex} = 313$ versus $\lambda_{ex} = 275$ nm, as illustrated in Fig. 16. The integrated luminescence intensity decreases by a factor of 30 between 14 K and 300 K for the spectra measured with an excitation wavelength of 313 nm, resulting in very weak luminescence at room temperature, but only by a factor of 2 for the spectra measured with an excitation wavelength of 275 nm over the same temperature range. This is consistent with the absence of a peak due to POM absorption in the 300 K excitation spectrum and indicates much greater temperature sensitivity for excitation through the POM than compared with excitation through the 2-picolinate ligand, as has been observed previously.[63] The decrease in intensity is again accompanied by an increase in linewidth, but no significant shift in wavelength.

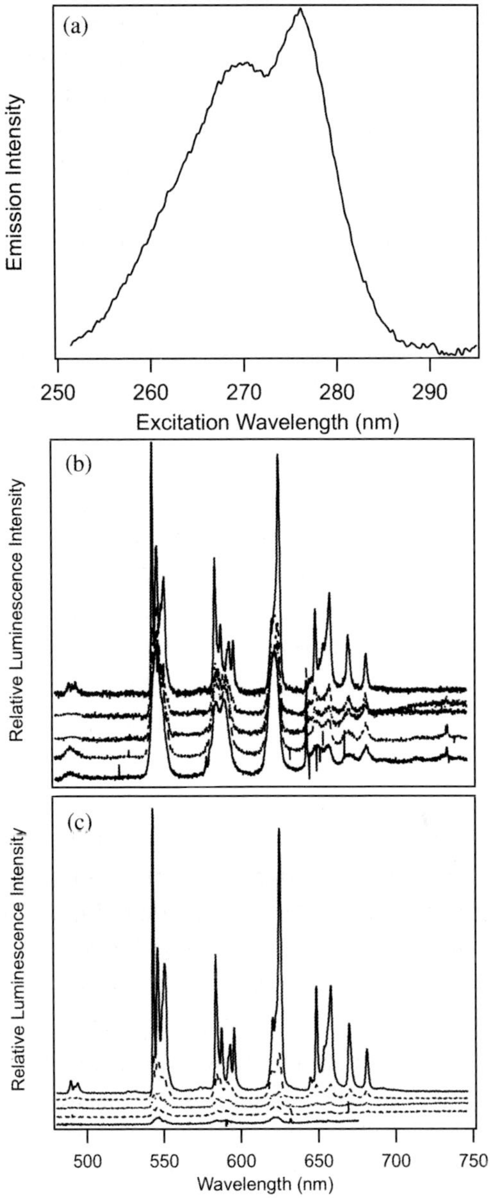

Fig. 16. (a) Excitation (λ_{emiss} = 490 nm), (b) temperature-dependent luminescence (λ_{ex} = 275 nm), and (c) temperature-dependent luminescence spectra (λ_{ex}= 313 nm) for solid $K_4Li_4H_4$[**8-Tb**]·$58H_2O$ with spectra shown at 14 K, 85 K, 150 K, 225 K and 300 K (top to bottom). (Reproduced with permission from Ref. 52. Copyright © 2012, The American Chemical Society.)

Room temperature time resolved emission profiles were collected for $K_4Li_4H_4$[**8-Tb**]·58H_2O at 490 nm, with λ_{ex} = 280 nm. A multiexponential decay behavior was again observed, which could be fit satisfactorily to a biexponential function, with lifetime values of τ_1 = 58 ± 2 s (51%) and τ_2 = 274 ± 10 s (49%), consistent with expected values for terbium(III) complexes with 2-picolinate and aqua ligands.[67] The approximately equal contributions for the two emitting sites suggests that the metal-centered emission intensities are similar for the two sites, in contrast to $Eu_{1.66}K_7$[**8-Eu**]·54H_2O. This is consistent with the greater sensitivity to quenching of the luminescence by coordinated aqua ligands for europium(III) compared with terbium(III), since the smaller energy gap for the lowest energy transition for europium(III) renders the vibrational coupling to ν(OH) modes more efficient than for terbium(III).

The variable temperature emission spectra obtained for solid $H_{0.5}K_{8.5}$Na[**7**]·30H_2O with excitation wavelengths of 275 and 313 nm show much weaker luminescence than that observed for compound $K_4Li_4H_4$[**8-Tb**]·58H_2O. The spectra obtained at 14 K have features similar to those obtained for $K_4Li_4H_4$[**8-Tb**]·58H_2O, with the expected terbium(III) transitions. As the temperature is increased with λ_{ex} = 275 nm, the intensities of all the transitions decrease as per the other compounds, however at 225 K and above peaks are weak and broadened, consistent with no visible luminescence at room temperature. The temperature-dependent behavior with λ_{ex} = 313 nm is a little more complicated, with all of the transitions decreasing in intensity as the temperature is increased from 14 K to 225 K.

As no sensitized Tb(III)-emission is detected for **7-Tb** at room temperature, we are able to deduce the contribution to the observed luminescence from each metal center in **8-Tb**. The lack of room temperature luminescence for **7-Tb** suggests that metal centers Tb1 and Tb2, which are present in both **7-Tb** and **8-Tb**, do not contribute to the luminescence observed for **8-Tb**. The center Tb1 has no 2-picolinate ligands, and since the excitation spectrum of **8-Tb** has shown that sensitization proceeds only via the organic chromophore at room temperature, and not via the POM, we conclude this metal center is not sensitized and does not contribute to the luminescence. A similar conclusion can be made for Tb2, although a 2-picolinate ligand is coordinated. Analysis of bond angles and coordination geometries for each of the Tb metal centers (Fig. 9) sheds some light on this effect. Previous work from Yamase has shown that the efficiency of emission is strongly temperature dependent in POM complexes which contain Ln–O–W bond angles of *ca.* 150°.[18,63] The presence of five such (corner-sharing) bond angles to Tb2 (151–161°) in both **7-Tb** and **8-Tb** facilitates efficient quenching of the emission, through involvement of the Tb(IV)-W(V) CT states.

For compound **8-Tb**, we therefore assign the observed biexponential decay components to Tb3 and Tb4. For Tb3, which has three Tb-O(W) bonds, two of these are involved in an edge-shared linkage with smaller Tb-O-W bond angles of 107° and 98°, which will be less efficient for quenching the Tb(III)-luminescence. Similarly, for Tb4, there is a single Tb-O-W bond with an angle of 147°, but this is evidently insufficient to completely quench the emission sensitized by the 2-picolinate ligand. Based on the room temperature luminescence lifetimes, we assign the long and short lived components to Tb3 and Tb4 respectively, due to the presence of three and six water ligands that can quench the luminescence.

Our synthesis of bulk samples of structurally characterized ternary Ln-org-POMs has allowed for the first time site-specific determination of the luminescence properties for the different terbium(III) and europium(III) centers and a delineation of the luminescence-sensitization through both the organic and inorganic ligands. At room temperature both organic 2-picolinate and inorganic POM ligands serve to sensitize the Eu(III)-luminescence in complex **8-Eu,** following excitation with UV light. Although the luminescence is more intense at 14 K, there is little difference in the temperature dependence observed for sensitization through both types of antenna. In contrast, at room temperature the Tb(III)-luminescence in the terbium analog **8-Tb** is only sensitized through the organic ligand. However at lower temperature the POM becomes an effective inorganic antenna for the Tb(III)-luminescence.

5. Concluding Remarks

The tungstoarsenates(III) α-$[AsW_9O_{33}]^{9-}$ and $[As_2W_{19}O_{67}(H_2O)]^{14-}$ have proved invaluable precursors for the formation of new Ln-POMs and Ln-org-POMs. The presence of the stereoactive lone pairs on the arsenic(III) heteroatoms is structure-directing, affording POM building blocks with open cavities and/or sites ideal for coordination to lanthanoid centers. The diversity of POM fragments evident in the final compounds is a testament to the versatility of the POM precursors and the complexity of the solution chemistry. Incorporation of carboxylate-containing organic ligands with additional functional groups into the reaction solution gives rise to structurally novel Ln-org-POMs. In these species, binding of the other organic functional groups to the lanthanoid center and/or POM framework is often essential to the integrity of the structural architectures obtained. There is considerable potential to extend the studies described here by investigating (i) other POM precursors with subvalent heteroatoms; and (ii) alternative examples of the wide variety of carboxylate containing-organic ligands.

The Ln-org-POMs synthesized in this manner include the first polylanthanoid SMM with inorganic ligands, with magnetic studies indicating the importance of high axial symmetry at the lanthanoid site for optimizing SMM behavior. We have also obtained the first structurally characterized Ln-org-POMs with Ln(III)-luminescence sensitized by both organic and inorganic ligands. The observed difference in the temperature dependence of the sensitization of the terbium(III) luminescence through the inorganic and organic antennas clearly illustrates the importance of distinguishing sensitization pathways in molecular and nanoscale luminophores and may be relevant for future applications of Ln-org-POMs in phosphor thermometry. Finally, the challenge remains to combine the luminescence and SMM properties to achieve luminescent Ln-org-POM SMMs. Such bifunctional molecular materials may facilitate the imaging, and/or addressing, of individual magnetic molecules on surfaces and in devices for applications in high density data storage or quantum computing.

Acknowledgements

The Australian Research Council and University of Melbourne are thanked for financial support. Invaluable contributions to this work from Chloë Miller and Firasat Hussain in our group are gratefully acknowledged, as are contributions from our collaborators Robert Gable, Evan Moore, Paul Kögerler, Manfred Speldrich, Christian Reber, Valérie Baslon and Lorenzo Sorace.

References

1. Rinehart J. D. and Long J. R. *Chem Sci* **2** (2011) 2078–2085.
2. Gatteschi D., Sessoli R. and Villain J. *Molecular Nanomagnets*, Oxford University Press, Oxford. (2006).
3. Winpenny R. *Molecular Cluster Magnets,* World Scientific, Singapore (2012).
4. Ishikawa N., *et al.*, *Inorg. Chem* **42** (2003) 2440–2446.
5. Ishikawa N., *et al.*, *J Am Chem Soc* **125** (2003) 8694–8695.
6. Ishikawa N., *et al.*, *Angew Chem Int Ed* **44** (2005) 2–5.
7. Sessoli R. and Powell A. K. *Coord Chem Rev* **40** (2009) 2328–2341.
8. Sorace L., Benelli C. and Gatteschi D. *Chem Soc Rev* **40** (2011) 3092–3104.
9. Bünzli J.-C. G. and Piguet C. *Chem Soc Rev* **34** (2005) 1048–1077.
10. Moore E. G., Samuel A. P. S. and Raymond K. N. *Acc Chem Res* **42** (2009) 542–52.
11. Binnemans K. *Chem Rev* **109** (2009) 4283–4374.
12. Carlos L. D., *et al.,Chem Soc Rev* **40** (2011) 536–549.
13. Pope M. T. *Handbook on the Physics and Chemistry of Rare Earths* **38** (2008) 337–381.
14. Bassil B. S and Kortz U. *Z. Anorg Allg Chem* **636** (2010) 2222–2231.
15. Dolbecq A., *et al.*, *Chem Rev* **110** (2010) 6009–6048.
16. AlDamen M. A., *et al.*, *J Am Chem Soc* **130** (2008) 8874–8875.
17. AlDamen M. A., *et al.*, *Inorg Chem* **48** (2009) 3467–3479.
18. Yamase T. *Chem Rev* **98** (1998) 307–326.
19. Yamase T. *Handb Phys Chem Rare Earths* **39** (2009) 297–354.
20. Zhang C., *et al.*, Inorg Chem **44** (2009) 3569–3578.
21. Qunhui L., *et al.*, *Inorg Chem* **41**, (2002) 6112–6117.
22. Ying L., *et al.*, *Inorg Chem* **45** (2006) 2055–2060.
23. Tézé A., *et al.*, *Inorg Synth* **27** (1990) 85–96.
24. Contant R., Klemperer W. G. and Yaghi O. *Inorg Synth* **27** (1990) 104–111.
25. Jeannin Y., *C R Acad Sci Paris, Serie II*, (1999) 161–165.
26. Müller A., *et al.*, *Angew Chem Int Ed Engl* **35** (1996) 171–173.
27. Kortz U., *et al.*, *Angew Chem Int Ed* **40** (2001b) 3384–3386.
28. Bösing M., *et al.*, *Chem Eur J* **3** (1997) 1232–1237.
29. Kortz U., *et al.*, *Angew Chem Int Ed* **41** (2002) 4070–4073.
30. Yamase T., Naruke H. and Sasaki Y. *J Chem Soc Dalton Trans* (1990) 1687–1696.

31. Fukaya K. and Yamase T. *Angew Chem Int Ed* **42** (2003) 654–658.
32. Merca A., *et al.*, *J Cluster Sci* **18** (2007) 711–719.
33. Lefebvre F., *et al. Inorg Chim Acta* **73** (1983) 173–178.
34. Ismail A. H., *et al.*, *Z Naturforsch* **65b** (2010) 383–389.
35. Hussain F., *et al.*, *Chem Commun* (2009) 328–330.
36. Hussain F., *et al.*, *Dalton Trans* (2009) 4423–4425.
37. Wassermann K., *et al.*, *Inorg Chem* **35** (1996) 3273–3279.
38. Kortz U., *et al.*, *Inorg Chem* **40** (2001a) 4742–4749.
39. Kortz U. *J Clust Sci* **14** (2003) 205–214.
40. Godin B., *et al.*, *Angew Chem Int Ed* **44** (2005) 3072–3075.
41. Botar B., Kögerler, P. and Hill, C. L. *Inorg Chem* **46** (2007) 5398–5403.
42. Tourné C., *et al.*, *C R Acad Sci Ser III* **277** (1973) 643–645.
43. Chang S., *et al.*, *J Mol Struct* **875** (2008) 462–467.
44. Chang S., *et al.*, *Aust J Chem* **63** (2010) 680–686.
45. Ritchie C. and Boskovic C. *Cryst Growth Des* **10** (2010a) 488–491.
46. Jeannin Y. and Martin-Frère J. *J Am Chem Soc* **103** (1981) 1664–1667.
47. Chen W., *et al.*, *Dalton Trans* (2007) 4293–4301.
48. Sadakane M., Dickman M. H. and Pope M. T., *Inorg Chem* **40** (2001) 2715–2719.
49. Ritchie C., *et al.*, *Inorg Chem* **50** (2011a) 7004–7014.
50. Müller A., *et al.*, *Anorg Allg Chem* **620** (1994) 599.
51. Ritchie C., *et al.*, *Angew Chemie Int Ed* **49** (2010b) 7702–7705.
52. Ritchie C., *et al.*, *Inorg Chem* **51** (2012) 1142–1151.
53. Ritchie C., Miller C. E. and Boskovic C. *Dalton Trans* **40** (2011b) 12037–12039.
54. Sytsma J., *et al.*, *Phys Rev* B **52** (1995) 12668–12676.
55. Shuskus A. J. *Phys Rev* **127** (1962) 2022–2024.
56. Zhang C., *et al.*, *C R Chimie* **8** (2005b) 1035–1044.
57. Boglio C., *et al.*, *Inorg Chem* **45** (2006) 1389–1398.
58. Burton-Pye B. P. and Francesconi L. C. *Dalton Trans* **40** (2011) 4421–4433.
59. Rodrigues M. J. E., *et al.*, *Materials Science Forum* **514–516** (2006) 1305–1312.
60. Granadeiro C. M., *et al.*, *J Alloys Compounds* **451** (2008) 422–425.
61. Lu Y., *et al.*, *Inorg Chim Acta* **360** (2007) 2063–2070.
62. Yamase T., *et al.*, *J Phys Chem A* **101** (1997) 5046–5053.
63. Ozeki T. and Yamase T. *J Alloys Compounds* **192** (1993) 28–29.
64. Zolin V. F., *et al.*, *J Alloys Cmpds* **380** (2004) 279–284.

65. Berry M. T. May P. S. and Xu H., *J Phys Chem* **100** (1996) 9216–9222.
66. Beeby A., *et al.*, *J Chem Soc Perkin Trans 2* **3** (1999) 493–503.
Watson W. M., *et al.*, *Inorg Chem* **14** (1975) 2675–2680.
67. Rosen D. L. and Niles S. *Appl Spec* **55** (2001) 208–216.

Chapter 7

POLYOXOTHIOMETALATES POTM

FRANCIS SÉCHERESSE and EMMANUEL CADOT
Institut Lavoisier de Versailles, UMR 8180, Université de Versailles St Quentin 78035 Versailles, France

1. Introduction

The transition elements M = V (V), Nb (V), Ta (V), Mo (VI), W (VI) at their highest degree of oxidation form with oxygen large polynuclear compounds commonly named polyoxoanions or polyoxometalates (POMs).[1,2] This term refers to both isopolyanions $\{M_xO_y\}^{n-}$, generally obtained by the self-condensation of a single transition metal cation and heteropolyanions $[X_aM_xO_y]^{m-}$ which contain one or more additional heteroatoms X. Regarding the large number of possibilities that exist for the metal and the heteroatom, many heteropolyoxometalates have been identified. The huge structural diversity that characterizes the chemistry of polyoxometalates makes this field an exciting domain for inorganic chemistry, the Keggin anion being probably the most famous representative. Extremely prolific over the past 150 years, this chemistry is still an important and fruitful subject of scientific exploration.[3,4] Polyanions have many applications in diverse areas such as catalysis, medicine, electrochemistry, chemical analysis and recently nanochemistry, and the interest in this type of compound has continued growing in the recent decades. A widely developed approach to design new POMs consists in the functionalization of a vacant archetypal POM precursor in order to generate large polyanionic structures with targeted reactivity and new properties.[5,6] Moreover, the position of polyoxoanions at the border of classical and supramolecular chemistries makes them excellent candidates to access this last modern area.

Polyoxoanions are known since Berzelius[7] and the first polyoxoanions containing organic groups have been reported in the early twentieth century. The structural characterizations of $(NH_4)_6[(HCO)_2Mo_8O_{28}].2H_2O$, described by Miolati in 1908, and $[(R_2AsO2)\ Mo_4O_{13}H]_{2-}$ prepared by Rosenheim and Belicki in 1913, probably marked the emergence of the coordination chemistry of polyoxometalates.[8,9] The inclusion of organic ligands in the structures of polyoxoanions allowed their functionalization and activation with respect to various reactions.[10] However, the major economic center that represents catalysis remained for a long time the main driving *force* of this research. Oxides of transition metals are known to be good heterogeneous catalysts, and diplay a wide range of reactivity towards organic substrates.[11] They are employed in reactions such as oxidation of methanol, selective oxidation of hydrocarbon chains or olefin metathesis. Viewed as analogues of soluble oxides, polyoxometalates represent good surface models for the study of the mechanisms of catalytic reactions.[12] The substitution of oxygen atoms of the skeleton is the most convenient route to introduce organic ligands in the polyoxometalate framework. Indeed, based on electronic analogy, the oxo ligand has been replaced by HO^-, HN_2^-, N_3 ligands.[13] However, if they all display a π-donor character sufficient to stabilize the highest oxidation states of the metal, their respective tendency to form multiple bonds is different. In this perspective, the imido derivatives of polyoxometalates revealed of real interest in recent years.[14] Despite the isolobal analogy that can be made, the substitution of an oxo ligand by a nitrido ligand has allowed so far the isolation of only one nitridopolyoxometalate $[Mo_6O_{18}N]^{3-}$[15] However, many imido derivatives have been reported. The structures of the type $(n\text{-}Bu_4N)_2\ [Mo_6O_{18}\ (NR)]$ have been determined deriving directly of the parent anion $[Mo_6O_{19}]^{2-}$. Derivatives of $[Mo_6O_{19-x}\ (NAr)_x]^{2-}$, $x \leq 6$, have been isolated opening the polyfonctionalization of oxometallic species.[17] In contrast to these results, attempts to substitute oxygen atoms by alkoxo ligands evidenced the different reactivity of these ligands with respect to polyoxometalates. In particular ethoxy ligands revealed more efficient to adopt a bridging coordination mode between two metal atoms than to substitute terminal oxygen atoms of the POM structure as expected.

In order to further modify the reactivity of polyoxometalates, research has been conducted on the reductive substitution of ligands.[18] The reductive nitrosylation of oxo compounds usually takes place by reaction with hydroxylamine precursors leading to derived species containing the unit $\{Mo(NO)\}^{3+}$ their structures resembling those commonly depicted as oxo compounds. Nitrosyl derivatives can be formally described as products of the substitution of a $\{Mo^{VI}O_4\}^{4+}$ fragment of a polyoxometalate by one $\{Mo^{II}(NO)\}^{3+}$ unit. Derivatives of the Lindqvist structure $[M_5O_{18}(M'^{II}NO)]^{3-}$, (M, M′ = Mo, W) have been isolated and fully characterized.[19] The methylation of some of them was observed while no reaction occurs with the analogous oxo $[Mo_6O_{19}]^{2-}$ dianion.[20] Thus, the activation of polyanions by insertion of nitrosyl ligand is an example that encourages to anticipate a lot of new reactions for these new compounds.

The formation of an isopolyanion results from the connection of MO_6 octahedra by sharing edges and/or faces. The domain of possible architectures of the heteropolyoxometalates family has been greatly expanded, but the structural chemistry of polyoxometalates is limited to a few types known as Keggin, Lindqvist and Dawson types and few others. The discovery of new structural types requires to have in hands new structural precursors, or to develop new connections between usual available motives. The use of reduced metal centers allowed to generate new architectures, and many polyoxomolybdates have been constructed by assembling the bimetallic unit $\{Mo_2O_8\}$.[6,21] This basic structural unit is characterized by the presence of a Mo^V–Mo^V metal-metal bond and consists of two edge-sharing octahedra. A combination of these fragments by vertices or edges has been used to build the structures shown in Fig. 1, and the introduction of organic ligands also participated to extend the field of structural chemistry of polyoxometalates.

Thus, a variety of new geometries became accessible *via* the association of octahedral, tetrahedral or pyramidal units associated by vertices, edges or faces.

A multitude of new compounds has been obtained by using simple organic ligands such as alkoxy ligands, carbonate, carboxylate and squarate, organophosphonates.[6] In such systems, compounds with high nuclearity are frequently encountered, the number of assembled metal centers varying from 2 to 22 in the case of mixed valence clusters of vanadium. However, the most common structures encountered contain six metal atoms and many of them are derived from the structure of the spherical anion $[M_6O_{19}]^{2-}$. The

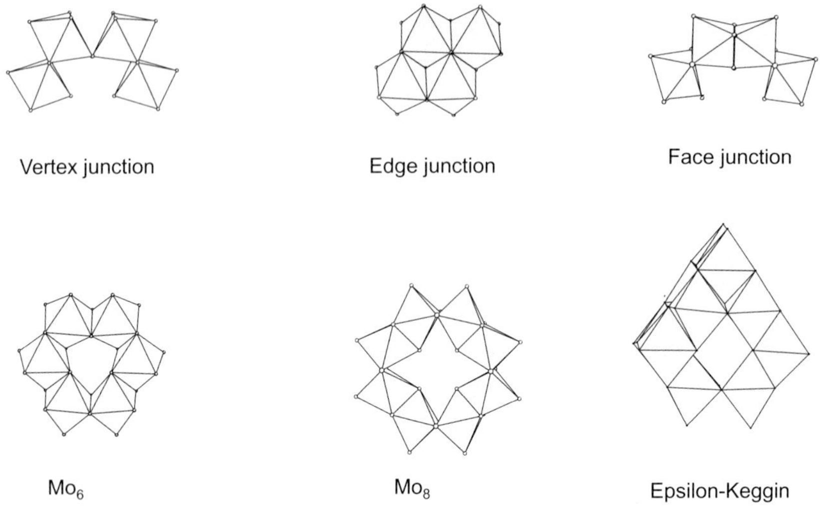

Fig. 1. Examples of structures derived from the dinuclear $\{Mo_2O_8\}$ basic unit.

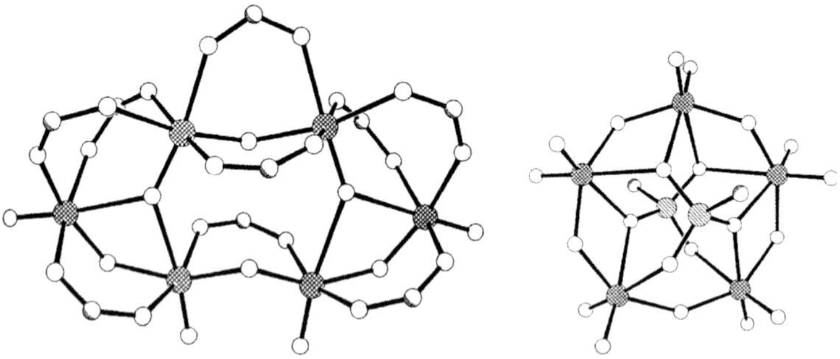

Fig. 2. Frameworks of $[V_6O_{10}(PhCO_2)_9]$ and $[Mo_5O_{15}(PhPO_3)_2]^{4-}$, respectively.

geometry of the mixed-valence cluster $[V_6O_{10}(PhCO_2)_9]$ is however a typical example of a new structural mode. Indeed, it contains two different units, two mononuclear moieties $\{(VO)O(RCO_2)_2\}$ and a tetranuclear $\{(VO)_4O_3\ (RCO_2)_3\}$ group linked through four bridging benzoate ligands.[22] The two cluster frameworks are shown in Fig. 2 (left). The metal octahedra of the first are connected only by sharing a vertex, while those of the tetranuclear unit alternately share an edge and a vertex. The different pentamolybdates of general formula $[Mo_5O_{15}(RPO_3)_2]$ n-(R = Me, Ph-: n = 4; $C_2H_4NH_3$–,–$CH_2N(CH_2)_4O$: n = 2)[23] exhibit a common architecture of another kind. The structure of the anion $[Mo_5O_{15}(PhPO_3)_2]^{4-}$ consists of five $\{MoO_6\}$ octahedra bound by a vertex or an edge. This arrangement generates a cyclic structure in which each face is topped by a tetrahedral phosphonate group as represented in Fig. 2 (right).

This paragraph, deliberately limited to these two examples, is not representative of the huge variety of the structures identified in this area but it is a good illustration of the possibilities offered by the introduction of organic components within oxometallic architectures.

Isopolyanions, heteropolyanions and more generally polyoxometalate derivatives are the members of an extensive family of compounds which differ in their chemical composition and structure. The recent developments in the chemistry of polyoxometalates are very exciting and only the more recent have been detailed here. Although this chemistry has greatly developed over the past decades, the process of formation of the soluble species remains enigmatic and often uses the notion of self-assembly to bring a justification. The control of the growth process of these structures is not yet clearly established and the design of clusters of predicted nuclearity and architecture remains a risky target.

Depending on the size of the objects it manipulates, the chemist very often differentiates chemistry in two domains, one is the domain of molecular compounds and at the opposite the domain of solids and 3D-structures. At the boundary of these two areas many compounds have been recently designed so called nanoclusters and giant molecules.[24] A very fruitful strategy for the chemistry of giant polyoxometalates in solution consists to generate species that can be condensed by using metal-oxo building blocks. The polytungstates obtained can generally be described as assemblies of archetypal POMs, generally Keggin anion, mono or polyvacant. They result of the nucleophilic reaction of these structures with electrophiles such as transition metal cations, lanthanides and actinides, $\{Mo_2O_2S_2\}^{2+}$ or $\{Mo_3S_4(H_2O)_9\}^{4+}$ cationic fragments. For example, a giant tungsten-based POM containing up to 148 metal centers has been obtained. The structure of this anion $[Ce_{16}As_{12}W_{148}O_{524}(H_2O)_{36}]^{76-}$ composed of twelve trivacant Keggin anion centers assembled by $\{CeW_5\}$ tungsten groups is shown in Fig. 3.[25]

Fig. 3. Representation of the molecular structure of $[Ce_{16}As_{12}W_{148}O_{524}(H_2O)_{36}]^{76-}$.

To design structures of very different types, the same strategy can be applied but starting from different species that do not derive from Keggin or Dawson anions. To achieve high degrees of condensation, the charge density of the expected product must remain constant during the growth to ensure its solubility. This is generally issued by substituting a metal center by another of lower oxidation state. The condensed species, thus more negatively charged, need to be stable in solution to be connected by cations.

The mixed-valence polyoxomolybdates that contain a large number of metal atoms are generally composed of the same oxometallic units, differently assembled, but also by other species. The works of Müller *et al.* are exemplary in this field.[26] Wheel-shaped compounds or sphere containing up to 354 molybdenum atoms have been obtained. Their structures can be described as an assembling of the same unit $\{Mo_{11}\} \equiv \{(Mo)(Mo_5)(Mo_5)\}$. The structures of the giant clusters $\{Mo_{11}\}_{14} = [Mo_{154}(NO)_{14}O_{420}(OH)_8(H_2O)_{70}]^{(25\pm5)-}$ and $\{Mo_{11}\}_{12} = [Mo_{132}O_{372}(H_2O)_{72}(CH_3COO)_{30}]^{42-}$ are represented in coordination polyhedra in Fig. 4.

The principles that permit to organize the molecular architecture of a compound are the subjects of many investigations. Randomly, concepts and

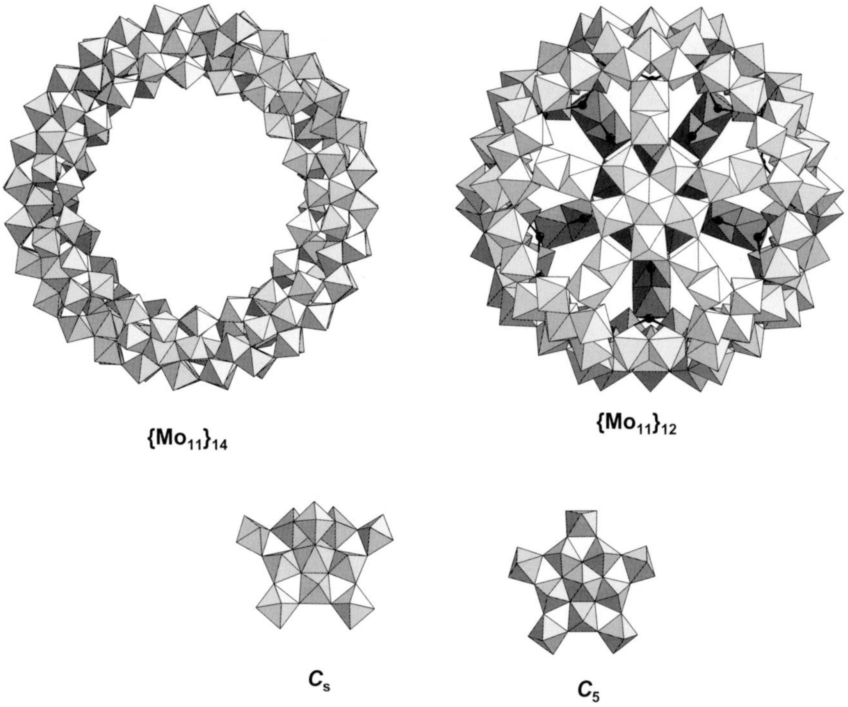

Fig. 4. Representations of the molecular structures of $[Mo_{154}(NO)_{14}O_{420}(OH)_8(H_2O)_{70}]^{(25\pm5)-}$ and $[Mo_{132}O_{372}(H_2O)_{72}(CH_3COO)_{30}]^{42-}$ with their corresponding constitutional building units.

definitions necessary for the understanding of the design of original compounds emerged, and among them, the term "template" points out the effect or the entity that controls the assembly of basic units in a wide inorganic architecture. Under this influence, far from the so-called classical architecture of polyoxometalates, clusters of amazing basket-shaped ring or shell have been isolated.[6] These new structures reflect the templating effect directed by species as different as halides, carboxylates, oxoanions, alkylammonium or alkali metal cations, water molecules and neutral organic molecules. These directing agents are generally encapsulated in the frameworks they control, maintained by covalent or noncovalent (electrostatic, hydrogen bonding, Van der Waals forces) interactions. In solution, the templates regulate the mobility of the basic units, usually free, and direct them through their characteristic size, shape and charge towards an ordered assembly.

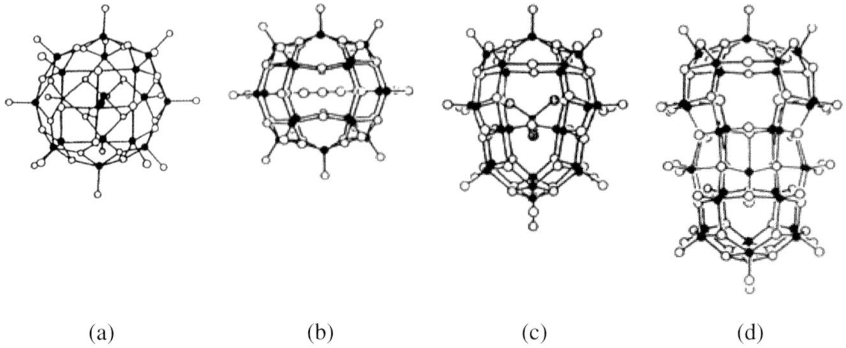

Fig. 5. Representations of the molecular structures of (a) $[V_{18}O_{42}]^{12-}$, (b) $[H_2V_{18}O_{44}N_3]^{5-}$, (c) $[HV_{22}O_{54}(ClO_4)]^{6-}$, (d) $[V_{34}O_{82}]^{10-}$.

Historically, oxopolyvanadates of the V^{IV} or mixed valence V^{IV}/V^{V} were recognized as the first family of compounds whose architectures were induced by template effect.[27] In the presence of halide anions, nitrate or azido anions, hollow spherical structures were obtained, closely resembling that of the anion $[V_{18}O_{42}]^{12-}$ which encloses a water molecule. The influence of these encapsulated species modifies the geometrical characteristics associated to the structures: symmetry close to D_{4d} is observed with halide ions, while the elongated sphere containing azido ions adopts a D_{2h} symmetry.

The molecular structure of the anion $[HV_{22}O_{54}(ClO_4)]^{6-}$ encloses the perchlorate ion. That of $[V_{34}O_{82}]^{10-}$ consists of a cage of thirty vanadium atoms arranged around the core $(VO)_4O_4$. Representations of these structures are collected in Fig. 5. The formation of such structures, in addition to the template effect produced by the anion, is made possible by the pyramidal coordination adopted by the vanadium. Such modes of organization are less observed with molybdenum or tungsten which clearly prefer octahedral coordination. It is remarkable to note that no covalent bonding exists between the template and the molecular backbone in these structures. The position of the anions in the cavity is determined by weak interactions with the atoms of the inorganic framework.

These complexes called "host-guest", for their dual membership in inorganic chemistry and supramolecular chemistry, can establish a bridge between these two areas.[28] The structural influence of a templating agent on the assembly of inorganic units permits to imagine an infinite number of possible architectures. Borrowed to supramolecular chemistry, the term template should be applied strictly to refer to weak interactions between the agent and the structure. However, its definition has been expanded in inorganic chemistry to describe any directing effect whatever is the bonding scheme. Its common use, sometimes unclear, helps to make a little more tenuous the boundaries of coordination chemistry and supramolecular chemistry.

The insertion in a molecular structure of soft donating centers such as sulfur atoms is expected to bring new properties to a POM because sulfur, in addition to his own properties of softness and polarization, presents an intense redox activity, different from that of oxygen. If thio- and oxometalates exhibit some similarities, they represent in fact two clearly separated fields governed by their own concepts, challenges and methods of investigation.

1.1. *Thiometalates*

V, Nb, Ta, Mo, W and Re transition elements regarding their high oxidation state form tetrahedral species whose intense colors very earlier intrigued Berzelius.[29] In contrast to the chemistry of polyoxometalates, thiometalates chemistry remained for a long time an under-explored domain which encountered these last thirty years a true development.[30] Today, the interest for thiometalates is duc to their implication in catalysis and bioinorganic chemistry because of the central role of MoS_2 and WS_2 as catalysts for hydrogenation reactions and in hydrodesulfurization processes, deoxygenation and denitrogenation of oil fractions.[31] In bioinorganic chemistry, their interest is directly connected to the important role they play in molybdoenzymes.[32] A great diversity of sulfur-containing ligands is available for the synthesis of thiometalates.[33] For most of the 3d-metals, the stability of the species formed with S-ligands depends on the polarizability of the free electron pair of sulfur which decreases following the sequence $S^{2-} > RS^{-} > SR_2$. The most extensively used are thioether ligands, cyclic thioethers, dithiolene ligands, dithiolates and thiolates. However, initially motivated by the modeling of the active sites of molybdoenzymes, the chemistry of sulfido and disulfido complexes remains by far the most developed family of thiometalic compounds.

In addition, metal-sulfide clusters should play an important role as key-components in some crucial challenging applications using new, cheap and *ecocompatible* processes for the energy conversion. Recently, the derivation of molybdenum sulfides as dimers, trimers and their cationic derivatives can act as powerful and viable materials for electron and photoelectron based catalytic hydrogen evolution.[34]

The introduction of sulfur in polyoxometalate frameworks is expected to significantly modify the structural and electronic properties of the POM in order to tentatively develop complementary properties at the molecular level, such as the electrocatalytic function of the {M-S} core combined with the electron storage capacity of the polyoxometalate. The study of acidification of the basic thioanions $[MS_4]^{2-}$ (M=Mo, W), revealed that the condensation process is much more complex in the case of thiometallates than oxometalates since it is systematically accompanied with intramolecular redox reactions.[35] However, the heating of a solution of $(NH_4)_2WS_4$ in acetonitrile allowed to isolate the tungsten anion $[W_3S_9]^{2-}$. In aqueous media, its partial hydrolysis led to the corresponding mixed-valence species $[W_3OS_8(H_2O)]^{2-}$.[36] In acetonitrile or DMF, the acidification process of $[P(C_6H_5)_4]_2[WS_4]$ is slow enough to permit the control of the formation and the crystallization of the tetrameric anion $[W_4S_{12}]^{2-}$.[37] Conversely, the proton attack of $[MoS_4]^{2-}$ proceeds very fast and invariably produces amorphous and polymeric substances globally identified as "MoS_3". However, the formation of reduced binuclear or trinuclear Mo^V or Mo^{IV} species can be obtained from aqueous solutions of ammonium sulfur, using polyoxomolybdate salts as precursors ($Na_2MoO_4 \cdot 2H_2O$ or $(NH_4)_6Mo_7O_{24} \cdot 4H_2O$) and elemental sulfur as external oxidant. Depending on the experimental conditions, several species can be obtained differing in their nuclearity, oxidation state of the Mo centers, their coordination number and the attached terminal or bridging ligands (oxo, sulfido or disulfido groups). The formation pathway of these species is represented in Fig. 6.[38]

Different attempts to functionalize polyoxometalates by direct addition of sulfide or hydrogenosulfide failed because the facile and rapid reduction of the POM metal provokes its rapid degradation into thio-species. However, replacement of oxygen by sulphur has proved possible in less reducible polyoxometalates. Klemperer *et al.* opened the way with the

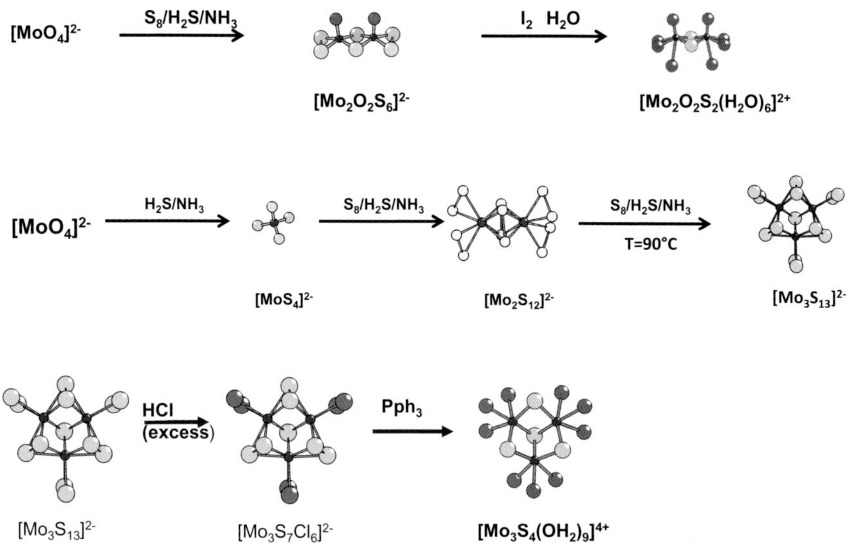

Fig. 6. Formation pathway of thiometalates derived from the $[MoO_4]^{2-}$ precursor (S in yellow, O in red). **See Color Plate 23**.

mixed Nb(Ta)-W Linqvist $[W_5MO_{19}]^{3-}$ (M=Nb^V or Ta^V) succeeding to replace a terminal $\{M{=}O\}^{3+}$ group by a M=S group to form the first thio-POM $[W_5MSO_{18}]^{3-}$.[48] The methodology has been successfully transposed to the Keggin $[PW_{11}MO_{40}]^{4-}$ allowing to isolate and characterize the first sulphur-containing Keggin derivative $[PW_{11}MSO_{39}]^{4-}$.[49] But, because of the poor hydrolytic stability of the M=S bond, this synthetic route imposes to use only non aqueous solvent which obviously limits the interest of the method. For this reason Sécheresse and Cadot chose a completely different approach, the challenge being to develop polyoxothiometalates (POTM) in water.

2. Polyoxothiometalates POTM

2.1. *The choice of a precursor*

The design of sophisticated clusters by the step by step building block approach requires specific precursors since the properties of the resulting cluster closely depend on that of the precursors. A "*good precursor*" is

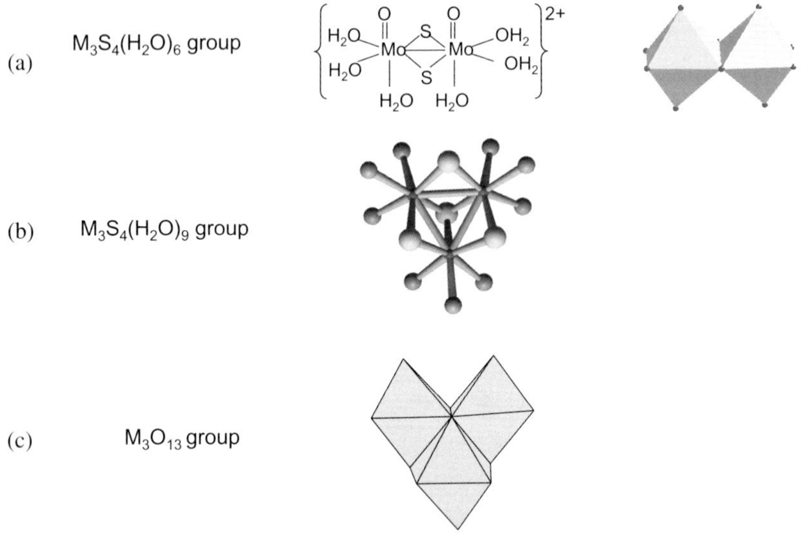

Fig. 7. Representation of the $[Mo_2O_2S_2(H_2O)_6]^{2+}$ and $[Mo_3S_4(H_2O)_9]^{4+}$ thio-precursor used in self condensation process (a) and in coordination to vacant POMs (a and b.) In grey M^{VI} and in yellow M^{V} metal cations. **See Color Plate 24**.

expected to display a reasonable thermodynamic stability along the condensation process, an adapted electronic structure is also required, and its initial geometry has to be chosen with respect to that of the targeted product.

2.1.1. *$[Mo_2O_2S_2]^{2+}$ a promising precursor*

According to previous works in Mo/O/S chemistry developed in Versailles, the dication $\{MoO_2S_2\}^{2+}$ has been chosen for matching these requirements, and its synthesis scheme is represented in Fig. 6 and its geometry in Fig. 7(a).

The two equivalent octahedral Mo^{V} (d^1) centers are mutually linked by a double S-bridge, each Mo atom being bonded to a terminal oxygen atom through a short multiple bond (1.64 Å). The two Mo=O fragments occupy two equivalent axial *syn* positions. The two d^1electrons of the Mo atoms are coupled to form a metal-metal bond. The Mo^{V}–Mo^{V} distance of 2.8 Å is very short compared to Mo^{VI}–Mo^{VI} analogues which are about 3.5 Å. This metal-metal bond widely contributes to the stability of the dication which remains stable in HCl 12M. In aqueous solution, three water molecules complete the coordination of each molybdenum atom to form the aqua dication $[Mo_2O_2S_2(H_2O)_6]^{2+}$. The three water molecules are not equivalent since the Mo-OH$_2$ bond *trans* to the Mo=O double bond (axial position) is weaker than the two others (equatorial positions) as revealed bond lengths and reactivity. Each Mo-atom is highly charged (+5) that induces strong polarizing effects to the terminal water ligands located in the xy plane which are polarized enough to be ionized.

2.1.2. *$[Mo_3S_4]^{4+}$ a possible building block for the coordination to vacant POMs*

The chemistry of molybdenum-sulfur clusters is extremely rich when compared to that of transition metal ions. These compounds are involved in non-biological catalysis a good example being the hydrodesulfurization (HDS) reaction,[39] while in biological systems, Mo-S clusters are present in the Fe/Mo cofactor of the nitrogenase enzymes.[40]

Mo-S clusters exhibit also an activity in the reduction of proton in Hydrogen Evolution Reaction (HER)[41] providing alternatives to the usual models based on the electrocatalytic activity of metal-associated to hydrogenases.[42] The electrocatalytic activity of the $\{Mo_3S_4\}$ cluster was first reported by Shibahara[43] and recently reinvestigated for the design of modified electrode.[44] $\{Mo_3S_4\}$, by its incomplete coordination and oxidation state of the Mo-centers offers rich possibilities in coordination chemistry and redox applications.[45]

The aqua-cationic species $[Mo_3S_4(H_2O)_9]^{4+}$ which presents a strong analogy with the basal $\{W_3O_{13}\}$ oxo triad, see Figs. 7(b, c) can be incorporated[46] within the vacancy of a POM framework to access mixed heterometalic saturated polyoxothiometalates.[47] In first approximation the geometry of $[Mo_3S_4(H_2O)_9]^{4+}$ can be described as a cube with a metal missing corner; three distorted Mo-polyhedra are fused together *via* one vertex (μ_3-X) and two edges (μ_2-X). The three molybdenum atoms have the formal +IV oxidation state (d^2 electronic structure) and the Mo$_3$ triangular core is ensured by 3 metal-metal bonds (Mo-Mo are distances about 2.8 Å). In this situation, the coordination at each Mo-atom is achieved by 3 water ligands. These water molecules are labile that gives the possibility to do chemistry at the periphery of the pseudo-cube and formally replace a W_3O_{13} group in POM architectures.

2.2. *Self-condensation of the $\{Mo_2O_2S_2\}^{2+}$ precursor*

Coucouvanis first prepared the $\{Mo_2O_2S_2\}^{2+}$ core by selective oxidation of the S_4^{2-} and S_2^{2-} terminal polysulfido ligands of the $[Mo_2O_2S_8]^{2-}$ anion by iodine in DMF.[38a] Six molecules of DMF complete the coordination sphere of the central $\{Mo_2O_2S_2\}$ core. The synthesis has been adapted to aqueous medium: the starting material is the $[Mo_2S_6O_2]^{2-}$ anion, the final product is the expected $\{Mo_2O_2S_2(H_2O)_6\}2^+$, see Fig. 6.

Indeed, inter-building block connections via an olation process (Mo-OH-Mo bonds) result from the addition of a base to a solution of the $\{Mo_2\}$ precursor in the presence or not of an assembling group.

$$2n\ M\text{-}OH_2 + nOH^- \rightarrow (M\text{-}OH\text{-}M)n + 2n\ H_2O$$

A yellow powder is obtained by direct addition of potassium hydroxide to a mixture of the dithiocation and potassium iodide about *p*H 2–3. After crystallization in water, the neutral dodecameric wheel, $[Mo_{12}O_{12}S_{12}(OH)_{12}(H_2O)_6]$, noted $\{Mo_{12}\}$, has been isolated.[50] The structure is that of a Mo_{12} ring formed of six $\{Mo_2O_2S_2\}$ fragments, linearly connected by double hydroxo bridges as represented in Fig. 8.

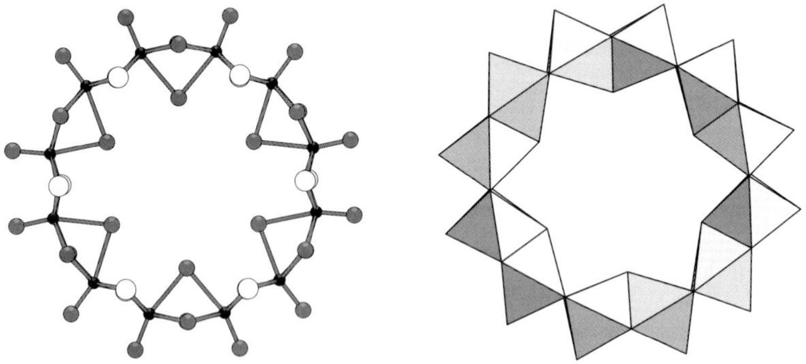

Fig. 8. Molecular structure of $[Mo_{12}O_{12}S_{12}(OH)_{12}(H_2O)_6]$ in ball and stick (Mo in blue, S in yellow, O in red) and polyhedral representations (Mo^V as yellow polyhedra). **See Color Plate 24**.

The cyclic backbone $\{Mo_{12}O_{12}S_{12}(OH)_{12}\}$ is neutral and delimits a central open cavity of about 11 Å, lined by six water molecules. The inner water molecules are labile enough to be exchanged for specific anions, suggesting the cavity exhibits a *cationic* character imposed by the twelve Mo(V) atoms. This situation is quite different from that encountered in organic chemistry with crown ether compounds where the cavity is lined by *donor* oxygen atoms giving the cavity an *anionic* character. Only the four equatorial aquo ligands of the initial dithiocation are involved in the olation reaction. The four equatorial aquo ligands are ionized below pH 2, provoking the oligomerization process, while the two axial water molecules are unaffected in the formation of the Mo_{12}-ring.

As the size of the soluble species is increased, the problem of the solvent-compound interactions becomes crucial. Indeed, to avoid the self-aggregation or spontaneous precipitation from solution, it is necessary that these large molecules adapt their size and shape in solution to minimize their interactions, a complementary way being to tune their contact surface with the solvent to idealize the entropy value of solvent-interaction. The molecular surfaces that mostly achieve these two ideas are that of balls and rings.

2.3. *Condensations* via *inorganic assembling groups*

2.3.1. *Halides as templates*

The recrystallization of $\{Mo_{12}\}$ in DMF containing tetrabutylammonium iodide or tetraethylammonium chloride leads to $Cat_2[X_2Mo_{10}S_{10}O_{10}(OH)_{10}(H_2O)_5]$ with Cat = Nbu_4 and X = I, or Cat = Net_4 and X=Br, respectively,[51] described as a $\{Mo_{10}\}$ wheel encapsulating two anions as represented in Fig. 9.

The two molecular structures are similarly built, the neutral decanuclear ring exhibiting the same connection scheme as the parent Mo_{12}-ring. The two halide ions are symmetrically located on both sides of the mean plane defined by the ten Mo atoms. The distances between the two halides and the five-oxygen atom of the ring are short enough to suggest the stability of the supramolecular arrangement is ensured by an H-bonding network involving the protons of the inner water molecule. In water, successive equilibria based on halide exchanges permit to obtain the iodide

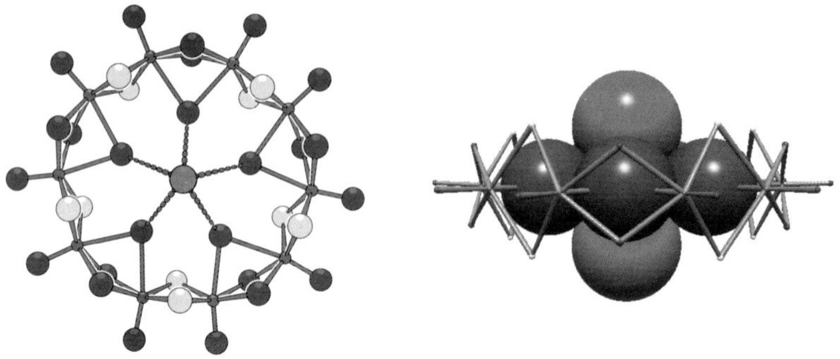

Fig. 9. Representations of $[X_2Mo_{10}S_{10}O_{10}(OH)_{10}(H_2O)_5]^{2-}$ in ball and stick and in space filling models (Mo in blue, S in yellow, O in red and X in purple). **See Color Plate 25**.

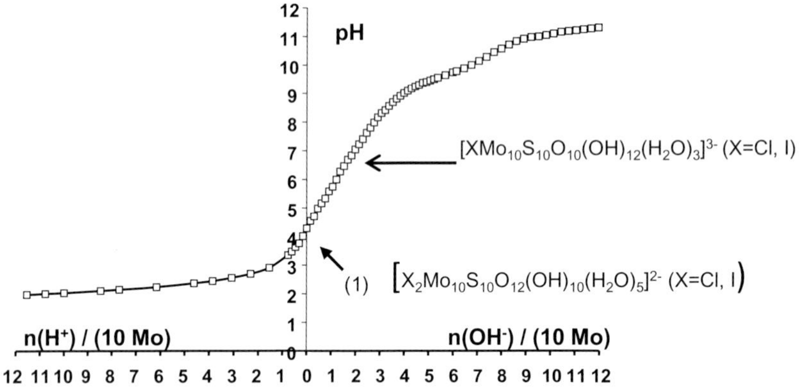

Fig. 10. Potentiometric titration of $\{X_2\text{-}Mo_{10}\}$ in the presence of halides.

free $\{Mo_{12}\}$ cluster. In contrast, to maintain the $\{Mo_{10}\}$ backbone, the presence of halide ions (Cl^- or I^-) or the use of an aprotic solvent is required. This point has been confirmed by varying the anion.

Potentiometric titrations of the neutral $\{I_2\text{-}Mo_{10}\}$ wheel (Point (1) in Fig.10) in the presence of halides show the successive ionizations of the five inner water molecules.[52] Within *p*H 6–6.5, the $[XMo_{10}S_{10}O_{10}(OH)_{12}(H_2O)_3]^{3-}$ (X=Cl, I) anion has been characterized. The molecular structure resembles that of the $[Cl_2Mo_{10}S_{10}O_{10}(OH)_{10}(H_2O)_5]^{2-}$parent except that two inner aquo ligands have been replaced by two hydroxo groups. Surprisingly, at the same pH, the $[Cs_2Mo_8O_8S_8(OH)_8(H_2O)]$ octamolybdenum ring has been isolated but only in the presence of cesium as counteraction.

In the solid state the $\{Mo_8\}$ ring directly interacts with two Cs^+ cations located on the two opposite sides of the ring and two hydroxo ions are trapped in the cavity.The presence of the two hydroxo-groups within the cavity is very important since it reveals it is possible to change the polarity of the cavity from cationic, as usually observed for these rings, to anionic which opens the way to react the cavity with cations. In DMF, variable temperature ^1H NMR and ^{133}Cs NMR experiments carried out on $[Cs_2Mo_8O_8S_8(OH)_{10}(H_2O)]$ confirmed that supramolecular interactions between the cesium cations and the anionic ring still exist in solution.[53]

Halides, iodides, sulfates and sulfonates which have a comparable anionic size can be viewed as templates susceptible to impose the nuclearity of the wheel which can swing from 10 to 12.

2.3.2. *Examples of other weak bases acting as templating groups*

Sulfate and sulfonate anions[54] also stabilize the $\{Mo_{10}\}$ ring *via* the same central set of hydrogen bondings, and in the case of ethylsulfonate, the two organic groups point outward the ring plane in opposite directions to minimize repulsions in the solid state as represented in Fig. 11.

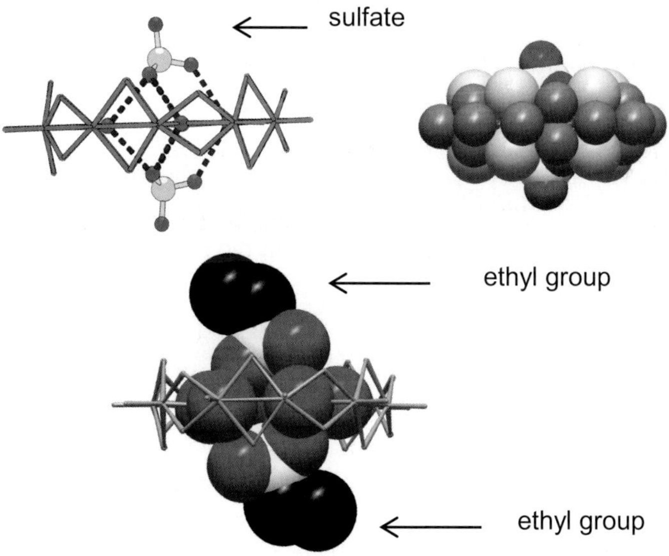

Fig. 11. Mo_{10}-ring stabilized by sulphate and ethylsulfonate groups (O in red). **See Color Plate 25**.

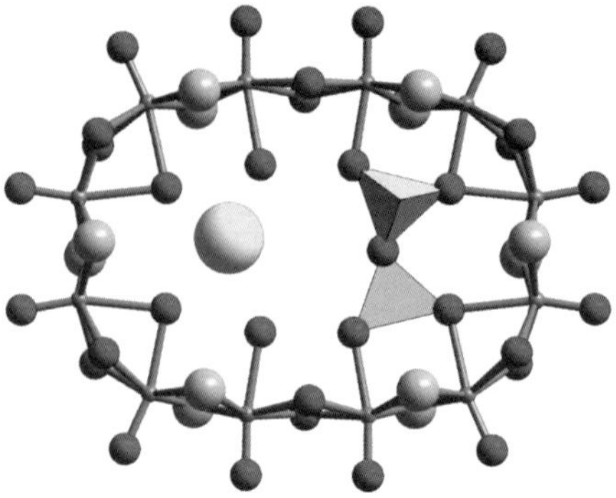

Fig. 12. View of the molecular structure of $[Cl(P_2O_7)Mo_{12}O_{12}S_{12}(OH)_{12}(H_2O)_4]^{5-}$ (Mo in blue, S in yellow, O in red and Cl in green). **See Color Plate 26.**

2.3.2.1. Dynamics associated to phosphate template effect

Thus, the facile ability of the cyclic rings to self-rearrange supported by the possibility of such systems to bind soft bases open exciting perspectives in the field of host-guest chemistry. To quantitatively study the stability of the $\{Mo\}_{10}$ rings in solution a chemical probe has been introduced in the ring. For that, the mixed species $[X(P_2O_7)Mo_{12}O_{12}S_{12}(OH)_{12}(H_2O)_4]^{5-}$ X=Cl, Br, I, represented in Fig.12 have been synthesized and structurally characterized before they were studied by ^{31}P NMR.[55]

The P_2O_7 group is σ-bonded to the Mo centres while the halide templates weakly interact with the four water molecules located within the hemicycle of the $\{Mo_{12}\}$ ring giving rise to the following exchange equilibrium characterized, as expected, by a weak thermodynamic constant K = 0.63±0.03 which decreases in the rank Cl –Br –I, consistent with the polarization properties of the anions.

$$[X(P_2O_7)Mo_{12}O_{12}S_{12}(OH)_{12}(H_2O)_4]^{5-} \rightarrow [(P_2O_7)Mo_{12}O_{12}S_{12}(OH)_{12}(H_2O)_4]^{4-} + X^-$$

In fact, such an exchange cannot be reduced to the simple and reversible in/out motion of the halide of the ring. ^{31}P standard and DOSY NMR measurements revealed powerful tools to study the dynamic of the system at the time scale of NMR.

The mechanism of exchange is much more sophisticated than reflected by the naive writing of the exchange reaction. The halide anion leaves the cavity together with two water molecules generating two under-coordinated Mo-centers that let the diphosphate group tangle in the relaxed cavity.

The key point which deserves to be examined now is whether the anion-templating effect is strong and tunable enough to permit the design of host-guest geometries with predictable shape and size and if this approach can be generalized.

2.3.2.2. Rings based on phosphate anions

Phosphates are interesting species since they are soluble in water, are good chelating groups and display active acid-base properties together with a small size and tetrahedral shape (spherical in solution). Two types of synthesis have been considered: one consists in replacing inner labile groups by stronger coordinating groups in preformed wheels (the previous $[X(P_2O_7)Mo_{12}O_{12}S_{12}(OH)_{12}(H_2O)_4]^{5-}$ seems obviously a good precursor), the other is the direct acid-base condensation of $\{Mo_2O_2S_2(H_2O)_6\}^{2+}$ in the presence of assembling groups of selected shape and size.

As expected, the halide anion in the previous $[X(P_2O_7)Mo_{12}O_{12}S_{12}(OH)_{12}(H_2O)_4]^{5-}$, can be selectively removed for a $[HPO_4]^{2-}$ group to give the mixed $\{P_2O_7\}/\{HPO_4\}$ containing groups $K_6[(HPO_4)(P_2O_7)Mo_{12}S_{12}O_{12}(OH)_{12}(H_2O)_2]\bullet 19H_2O$.[55] As represented in Fig. 13, the X^- elimination is accompanied by a significant change of the ring shape which adopts a "pear-like" conformation.

The reaction of $[Mo_2O_2S_2(H_2O)_6]^{2+}$ with a large excess of phosphate led to the $[(HPO_4)_4Mo_6O_6S_6(OH)_6]^{5-}$ hexanuclear wheel encapsulating a central phosphato group surrounded by three additional peripheral phosphates while in stoichiometric conditions, the $[(HPO_4)Mo_{10}O_{10}S_{10}(OH)_{10}(H_2O)_3]^{2-}$ and $[(HPO_4)_2Mo_{12}O_{12}S_{12}(OH)_{12}(H_2O)_2]^{4-}$ have been separately isolated owing to the domain of concentration chosen.[56] With $\{P_3O_{10}\}$ ten $[Mo_2O_2S_2(H_2O)_6]^{2+}$ self arrange in the heart-shaped architecture $[(P_3O_{10})Mo_{10}O_{10}S_{10}(OH)_{10}]^{5-}$. In Fig.14, are given, some representations of phosphate templated molybdenum wheels illustrating the diversity of molecular structures that can be deliberately designed.

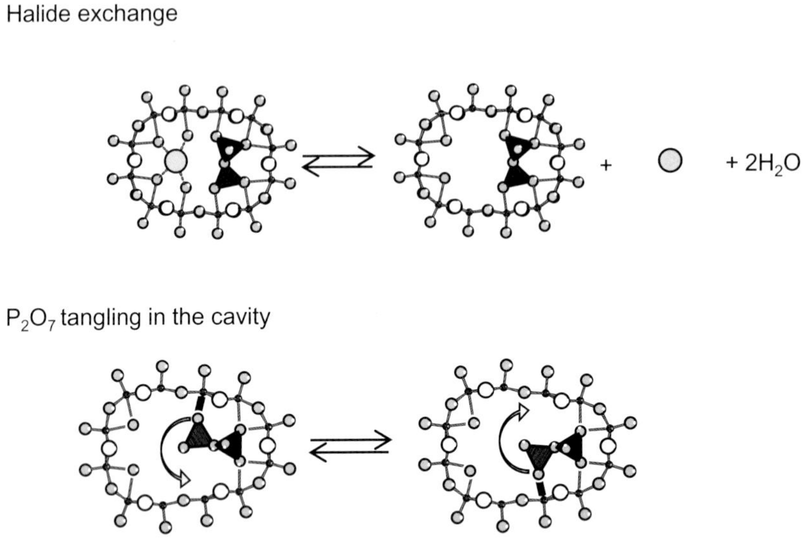

Fig. 13. Illustration of the P_2O_7 tangling effect after elimination of the encapsulated halide. Copyright American Chemical Society.

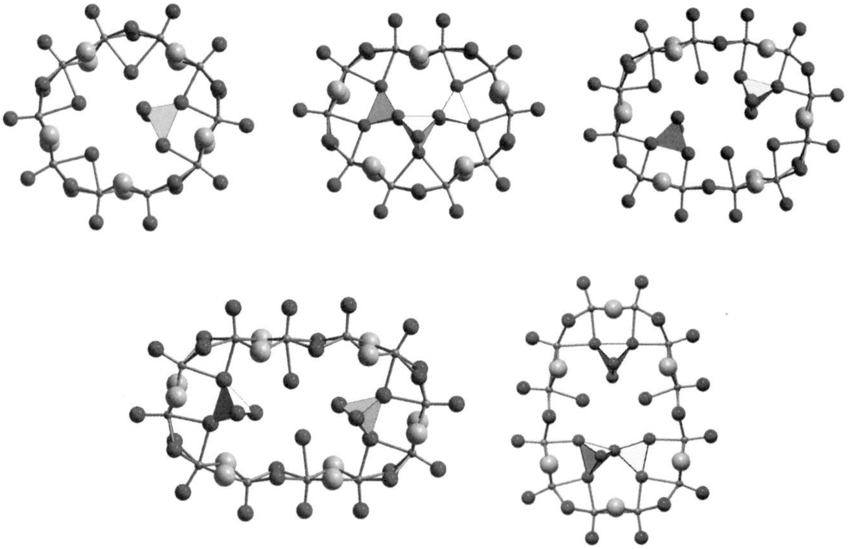

Fig. 14. Ball and stick representations of: top, $[(HPO_4)Mo_{10}O_{10}S_{10}(OH)_{10}(H_2O)_3]^{2-}$, $[(P_3O_{10})_2Mo_{10}O_{10}S_{10}(OH)_{10}]^{5-}$, $[(HPO_4)_2Mo_{12}O_{12}S_{12}(OH)_{12}(OH)_{2]}^{6-}$, and below, $[(HPO_4)_2Mo_{12}O_{12}S_{12}(OH)_{12}(H_2O)_2]^{6-}$ and $[(HPO_4)(P_2O_7)Mo_{12}S_{12}O_{12}(OH)_{12}(H_2O)_2]^{6-}$, (Mo in blue, S in yellow, O in red and phosphate and diphosphate guests in polyhedral models). **See Color Plate 26**.

2.4. *POTM derived from organic templates*

2.4.1. *Ring design with carboxylate and dicarboxylates*

The overall geometry of the POTM issued from the condensation of the $[Mo_2O_2S_2(H_2O)_6]^{2+}$ precursor seems imposed by the template effect of the directing anions present in solution. With phosphate ions, the size and shape of the molybdenum skeleton self adapts to the size and number of anions which have been inserted in the ring as summarized in Fig.14. As the reactivity and size of the acetate ion is quite comparable to those of phosphates they logically have been used as templating groups.

The structure of the triacetato anion $[Mo_{12}O_{12}S_{12}(OH)_{12}(CH_3COO)_3]^{3-}$ presentented in Fig.15 highlights the adaptation the $\{MO_{12}\}$ framework relatively to the constraints brought by the presence of three acetates in the core of the ring.[57] The cycle contains six $\{Mo_2O_2S_2\}$ units connected by two hydroxo bridges. The three acetate groups force the cycle to adopt the idealized D_{3h} symmetry of an equilateral triangle, with edges of 11 Å. An oxygen atom of each acetate ion bridges two dinuclear units while the second completes the coordination sphere of a molybdenum atom. As the result of the "pince effect" imposed by the acetates, the metal frame is constrained in three points characterized by more closed Mo-Mo-Mo angles of 135° and importantly three molybdenum atoms have lost a labile water molecule to adopt a pyramidal geometry.

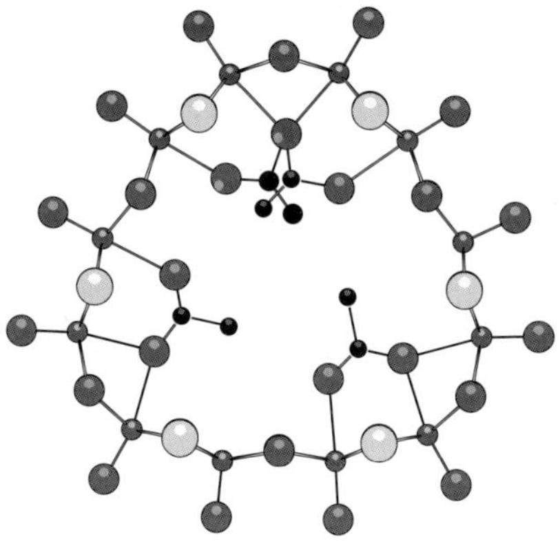

Fig. 15. Representation of the triangular shaped $[Mo_{12}O_{12}S_{12}(OH)_{12}(CH_3COO)_3]^{3-}$ (Mo in blue, S in yellow, O in red and C in black). **See Color Plate 27.**

All the templating groups presented up to now are comparable for being spherical in solution, the only difference coming from their different charges. Organic dicarboxylates present the advantages of containing two terminal chelating carboxylate groups separated by an organic chain which can be easily varied in length and chemical nature. That gives the opportunity to build families of template compounds that permits to draw tendencies and propose generalizations and also to establish a link between organic and inorganic chemistry.

This last point is interesting for bringing methods of characterization specific to organic chemistry (mass spectroscopy, NMR methods, and orhers). In Fig. 16 are given representations of the carboxylates that have

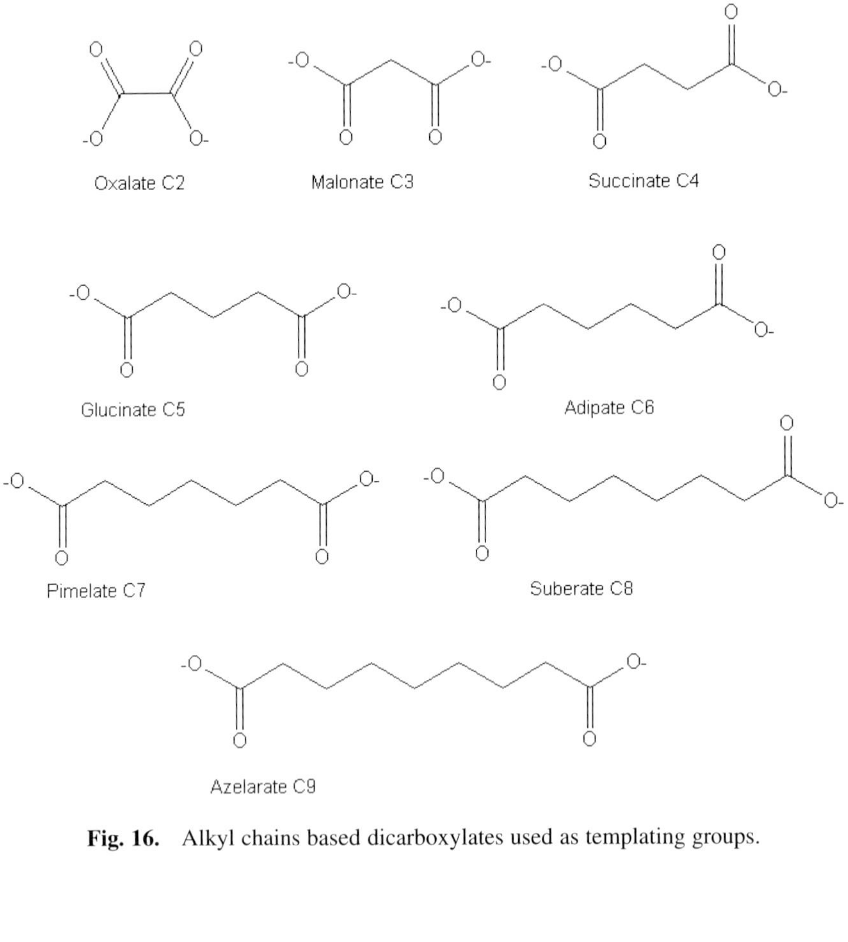

Fig. 16. Alkyl chains based dicarboxylates used as templating groups.

been used as templates for the design of organic-inorganic cyclic POTM of variable shapes and sizes.

The syntheses of these species have been achieved by using two different methods. Either the dicarboxylate anion has been directly introduced into an aqueous solution of the $[Mo_2O_2S_2\ (H_2O)_6]^{2+}$ precursor, either it has been generated *in situ* from the corresponding dicarboxylic acid by neutralisation. In both cases, the pH of the mixture has been adjusted to 3–4 by addition of an alkaline solution. The resulting products have been quantitatively isolated by precipitation and crystallization as salts of alkali metal cations.

In a first approach three dicarboxylic acids with chains of variable lengths have been tested, the oxalic acid (2 carbon atoms), the glutaric acid (5 carbon atoms), and the pimelic acid (7 carbon atoms), respectively.[58] With oxalic acid the octanuclear $[Mo_8O_8S_8\ (OH)_8(C_2O_4)]^{2-}$ wheel, noted $\{Mo_8Ox\}^{2-}$ hereafter, has been obtained.

It can be noted that octanuclear $\{Mo_8Ox\}$ ring skeletons are not unknown in the field of polyoxometalate chemistry. Four dinuclear $\{Mo_2O_2S_2\}$ units are connected by pairs through two hydroxo-bridges which generates alternately short Mo-Mo intrafragment distances of 2.8 (6)Å) and long interfragment ones of 3.2 Å. Each oxygen atom of the oxalate anion bridges two neighboring Mo centers via two Mo-O bonds of equivalent lengths (2.470 (3) –2.509 (3) Å). The eight molybdenum atoms have octahedral environment that means their coordination sphere is complete. The circular enchainment of the $\{Mo_2O_2S_2\}$ building blocks is achieved by face sharing octahedra as represented in Fig.17.

With glutaric acid, the decanuclear $[Mo_{10}O_{10}S_{10}\ (OH)_{10}(C_5H_6O_4)]^{2-}$ noted $\{Mo_{10}Glu\}^{2-}$ has been obtained. The molecular structure of $\{Mo_{10}Glu\}^{2-}$ shown in Fig. 17 reveals the distorted conformation of the ten Mo-ring yet reported for the $\{X_2Mo_{10}\}$ anions described in Sec. 2.3.1. The stress caused by the presence of the glutarate anion in the center of the cavity is responsible of the difference observed in the inter-units Mo-Mo-Mo angle values in the range 133°–152°. The four oxygen atoms of the organic anion ensure the connection with the metal skeleton by formation of five equivalent Mo-O bonds of 2.291 (4)–2.335 (5) Å. The elliptical shape of the cavity imposed by the glutarate anion makes that five Mo-centers are under-coordinated

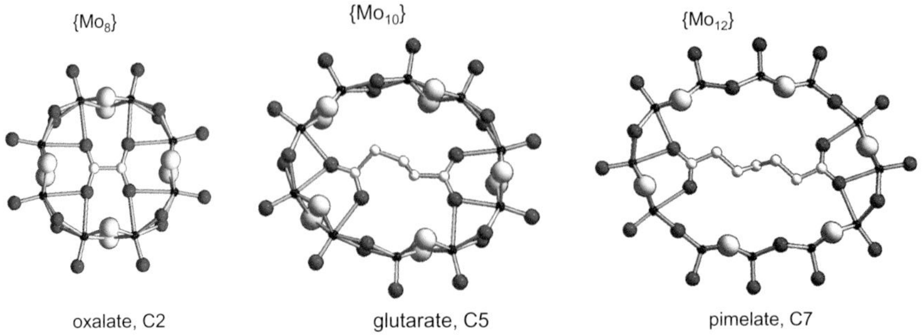

Fig. 17. Molecular structures of $\{Mo_8Ox\}^{2-}$, $\{Mo_{10}Glu\}^{2-}$, and $\{Mo_{12}Pim\}^{2-}$ showing the templating effect on the size and shape of the host-guest compounds (Mo in blue, S in yellow, O in red and C in black). **See Color Plate 27**.

adopting a square pyramidal geometry. Thus the overall architecture of $\{Mo_{10}Glu\}^{2-}$ appears composed of octahedra and pyramids edge and face sharing. The molecular structure of $[Mo_{12}O_{12}S_{12}(OH)_{12}(C_7H_{10}O_4)]^{2-}$ noted $\{Mo_{12}Pim\}^{2-}$ is given in Fig.17. The overall structure of the ring closely resembles that obtained with glutaric acid except a twelwe membered ring is obtained. Six $\{Mo_2O_2S_2\}$ units are connected by pairs by two hydroxo bridges ensuring the neutral circular $Mo_{12}O_{12}S_{12}(OH)_{12}$ backbone. The Mo_{12} ring is hugely deformed by the presence of the hindering anion in the cavity in such a way that six molybdenum atoms are obliged to adopt a pyramidal coordination.

The octahedral coordination of the six remaining Mo centers is completed by the oxygen atoms of the pimelate anion which displays two coordination modes, either single bonded or bridging. At each end of the alkyl chain, an oxygen atom is bonded to a metal center (Mo-O = 2.341 Å) while the second bridges two adjacent atoms of molybdenum through two non-equivalent Mo-O bonds of 2.359 (9) Å and 2.440 (9) Å, respectively. The cyclic $\{Mo_{10}Ox\}^{2-}$ molecular structure is perfectly symmetric because the oxalate dianion guest is perfectly symmetric (C_2 axis). That means that based on the same principles, the use of more hindering symmetric guests would permit to design larger host-guest wheels.

Trimesate (3 carboxylate groups, C_3 axis) led to the perfectly symmetric Mo_{12}-wheel $[Mo_{12}O_{12}S_{12}Trim_3]^{3-}$ represented in Fig. 18, the main

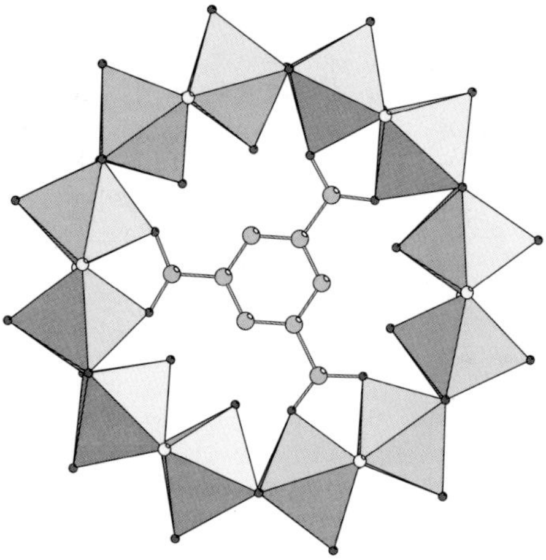

Fig. 18. Representation of the perfectly circular $[Mo_{12}O_{12}S_{12}Trim]^{3-}$ (Mo in blue, S in yellow, O in red and C in black). **See Color Plate 28**.

difference with $\{Mo_{10}Ox\}^{2-}$ and $\{Mo_{12}terp\}^{2-}$ being that in these wheels 6 Mo-atoms have a saturated coordination (octahedral environment) the other six being in a pyramidal geometry due to the hindrance of the carboxylate guest. The terephtalate ring $\{Mo_{12}terp\}^{2-}$ is symmetrically comparable to the oxalate wheel represented in Fig. 17,[59] but as terp is more hindering, 8 Mo-atoms have a pyramidal coordination, see Fig. 20.

It is remarkable to mention that the design of such host-guest wheels is made possible thanks to the lability of the water molecules located *trans* to the M = O bonds which are easily removed owing to the size of the hindering template. The templating effect of dicarboxylates has been clearly demonstrated by the formation of related 8, 10 and 12 Mo-membered rings. The flexibility and free rotations of the aliphatic chains limit the growing of the wheel size to 12 Mo atoms. As represented in Figs. 17 and 19, on increasing the number of carbon atoms, the alkyl chain is distorted inside the cavity, a limit length is reached that does not permit the extension of the template effect. Changing the chemical nature of the guest for more rigid groups is one of the possible solutions to increase the size of the wheel. The under-coordination of Mo atoms in the host-guest product induced by the distortion of the ring gives the wheels amazing properties of dynamic that are described in Sec. 2.4.3.

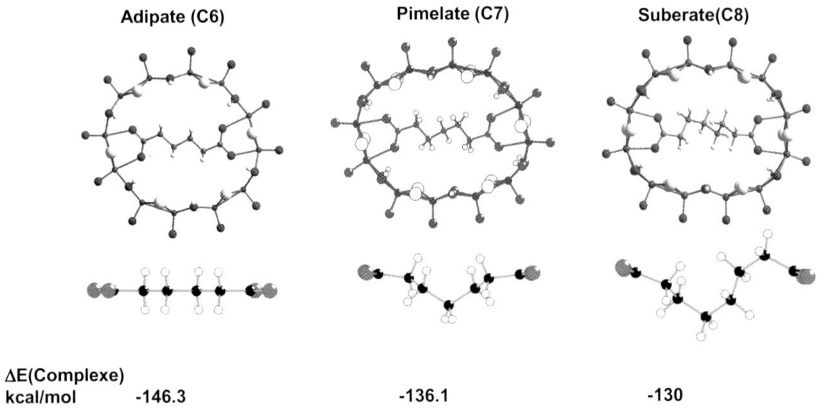

Fig. 19. Top, view of the distorsion generated on increasing the length of the inner alkyl chain (Mo in blue, S in yellow, O in red and C in black), below in black, view of the optimized DFT geometries of the alkyl chains with the corresponding energies. **See Color Plate 28**.

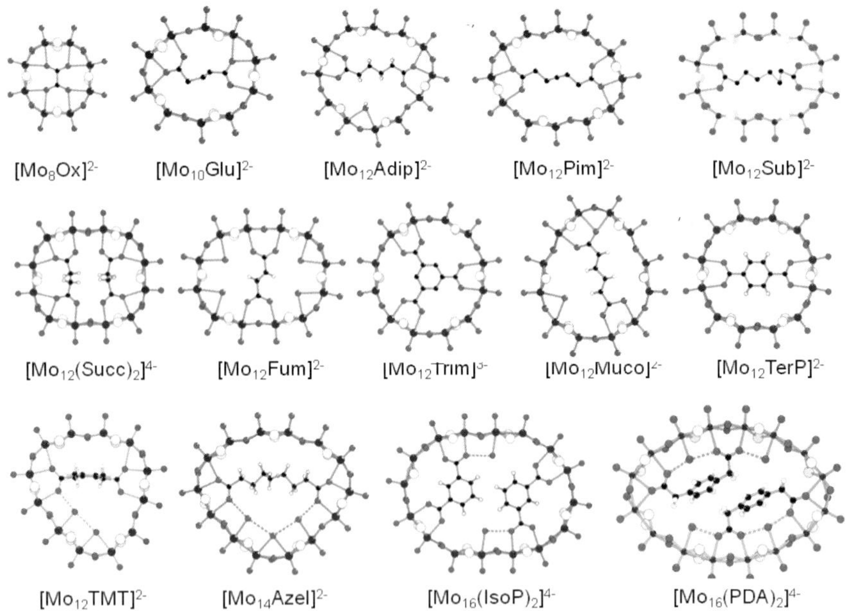

Fig. 20. Host-guests compounds obtained with flexible and/or rigid dicarboxylate anions (Mo in blue, S in yellow, O in red and C in black).(Copyright © Wiley 2011.) **See Color Plate 29**.

2.4.2. *Adaptability and stability of host-guest rings*

2.4.2.1. Example of C6, C7, and C8 based rings

The self-condensation of the $[Mo_2O_2S_2]$ building block around C6 (adipate), C7 (pimelate) or C8 (suberate) dicaboxylate chains leads exclusively to the $\{Mo_{12}\}$ rings[58,60,61] represented in Fig. 19. The relative stability of these $\{Mo_{12}\}$ rings has been determined by the study of the ligand exchange reaction (1).

The concentrations of the species present in solution, $[Mo_{12}\text{-Sub}]^{2-}$, L_2^{2-}, $[Mo_{12}\text{-}L_2]^{2-}$ and Sub^{2-} ions have been obtained by ^{1}H NMR and used to determine the equilibrium constant K_{rel} (2) which permitted to establish the order of stability $[Mo_{12}Adip]^{2-} > [Mo_{12}Pim]^{2-} > [Mo_{12}Sub]^{2-}$. The relative stabilities of the three $\{Mo_{12}\}$ wheels have been faced to DFT calculations in order to identify the structural and/or thermodynamic parameters influencing their stability. The most unexpected DFT results is that the energetic cost between circular and elliptical conformations is only ~3 kJ mol^{-1} suggesting that the flexible inorganic ring does not participate to a large extent in the host-guest stability. In fact, DFT calculations established that the main energetic factor contributing to the host-guest stability is governed by the distortion of the alkyl chain, unfolded for the C6 alkyl chain and highly S-distorted for the longer C8 one (see Fig. 19). It results that for a given size, the stability of the host-guest system is decreased as the length of the alkyl chain is increased.

This type of study has been extended to other $\{Mo_{12}\}$ host-guest systems containing *rigid* dicarboxylates, see Fig. 20, such as $Muco^{2-}$ (Muconate), TMT^{2-}(Tetra-Methyl-Terephtalate), $Trim^{2-}$ (Trimesate), and $TerP^{2-}$(Terephtalate).[60] The resulting stability scale follows the order $[Mo_{12}\text{–Trim}]^{3-} >> [Mo_{12}\text{-TMT}]^{2-} > [Mo_{12}\text{-Muco}]^{2-} > [Mo_{12}\text{-Adip}]^{2-} \sim [Mo_{12}\text{-TerP}]^{2-} > [Mo_{12}Pim]^{2-} > [Mo_{12}Sub]^{2-}$ suggesting that host-guest assemblies are better stabilized by rigid groups.[60c]

$$[Mo_{12} - Sub]^{2-} + L_2^{2-} \rightleftarrows [Mo_{12} - L_2]^{2-} + Sub^{2-} \qquad (1)$$

$$K_{rel} = \frac{[Mo_{12} - L_2]\cdot[Sub]}{[Mo_{12} - Sub]\cdot[L_2]} \qquad (2)$$

In addition, DFT calculations carried out on optimized geometries revealed that water molecules play an important role in the stabilization of the host-guest assembly. These water molecules are located inside the ring as inner coordinating groups and outside as first solvation shell. They form an extended H-bond network which significantly contributes to about 12–17 kJ.mol^{-2} to the host-guest stability.

The DFT resulting stability ranking, $[Mo_{12}\text{-Trim}]^{3-}$>>$H_2O@[Mo_{12}\text{-TMT}]^{2-}$>$H_2O@[Mo_{12}\text{-Adip}]^{2-}$>$H_2O@[Mo_{12}\text{-Muco}]^{2-}$>$[Mo_{12}\text{-TerP}]^{2-}$>$[Mo_{12}\text{-Pim}]^{2-}$>$[Mo_{12}\text{-Sub}]^{2-}$, matches quite well with the experimental one with only one inversion between $[Mo_{12}\text{-Adip}]^{2-}$ and $[Mo_{12}\text{-Muco}]^{2-}$.[60c] The most stable $\{Mo_{12}\}$ system is obtained with trimesate as guest, probably related to strong entropic effects. The stability of the wheels mainly depends on the nature of the guest, rigid or flexible, but also on the presence of encapsulated water molecules engaged in inner and outer hydrogen bonds. $[Mo_{16}(PDA)_2(H_2O)_4]^{4-}$ is a representative example for which four inner water molecules are embedded within two inner H-bond networks involving the adjacent carboxylate groups (represented by dotted lines in Fig. 20).

These two H-bond networks are reinforced by two CH-π interactions between the two inner PDA^{2-} ligands.[61] NMR studies supported by DFT calculations points out the prominent role of the water molecules trapped in the ring and that of the H-bond network which participates to the host-guest molecular stability and to the structure of the conformers stable in solution.

2.4.2.2. Intrinsic stability

The stability of $[Mo_{12}Sub]^{2-}$ has been studied in water by ^{1}H NMR. The ^{1}H NMR spectrum of $[Mo_{12}Sub]^{2-}$ in D_2O consists of three broad signals (unresolved triplets) located in the low frequency region of the spectrum, characteristic of an encapsulated suberate (Sub^{2-}) anion.

A variable temperature ^{1}H NMR experiment has been carried out in D_2O between 275 K and 305 K (see Fig. 21). As the temperature is decreased, signals of growing intensities assigned to uncoordinated suberate anions are observed.

The presence of free suberate anions in solution suggests the $[Mo_{12}Sub]^{2-}$ host-guest decomposes according to the following equilibrium.

$$[Mo_{12}\text{-Sub}]^{2-} + 6\,H_2O \rightleftarrows [Mo_{12}(H_2O)_6] + Sub^{2-},\ K_{eq} \qquad (3)$$

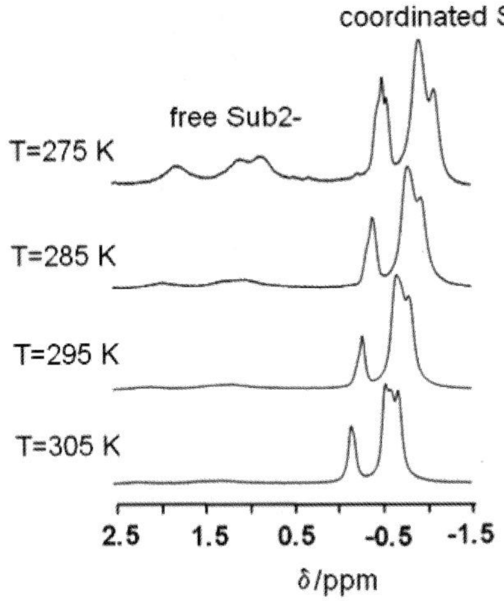

Fig. 21. Variable ^{1}H NMR experiment showing the increasing concentration of released Sub^{2-} on cooling the solution.

The integrated signals of uncoordinated Sub^{2-} and $[Mo_{12}\text{-}Sub]^{2-}$ permit to evaluate the respective concentrations in solution of the species so that the standard thermodynamic parameters $\Delta_R H^*$ and $\Delta_R S^*$ can be easily derived leading to the values $\Delta_R H^*$=-60.2 kJ mol^{-1} and $\Delta_R S^*$=−297 J mol^{-1} K^{-1}, which indicates that the encapsulation process of the template is mainly driven by entropy.[60c]

A similar study has been previously carried out in D_2O on $[Mo_{12}(succ)_2]^{4-}$ (Fig. 22). ^{1}H NMR variable temperature experiments evidenced that free $Succ^{2-}$ anions were observed as the temperature was decreased.

Two successive equilibria have been postulated and formulated by Equations 4 and 5. The variations of the concentrations of $[Mo_{12}(Succ)_2]^{4-}$, $[Mo_{12}\text{-}Succ]^{2-}$ and $[Succ]^{2-}$, respectively as a function of temperature are represented in Fig. 22.

$$[Mo_{12}(Succ)_2]^{4-} \rightleftarrows [Mo_{12}Suc)] + Succ^{2-} \tag{4}$$

$$[Mo_{12}Succ]^{2-} \rightleftarrows [Mo_{12}] + Succ^{2-} \tag{5}$$

The thermodynamic parameters associated to the two reactions have been obtained from a Van't Hoff's treatment of the data. Both the enthalpy

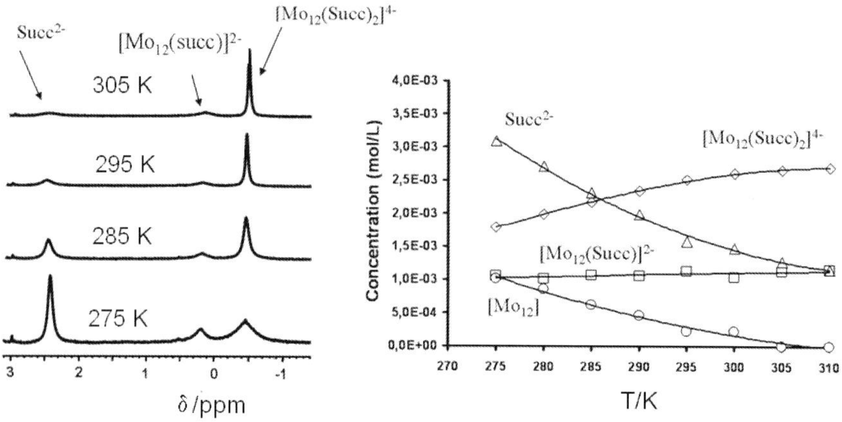

Fig. 22. Evidence of released $Succ^{2-}$ in solutions of $[Mo_{12}(succ)_2]^{4-}$. Quantitative treatment of ^1H NMR spectra.

and entropy variations are negative, $\Delta_R H^*(4) = -29$ kJ.mol^{-1} and $\Delta_R S^*(4) = -158$ J.K^{-1}.mol^{-1} for Eq. 3; $\Delta_R H^*(5) = -55.7$ kJ.mol^{-1} and $\Delta_R S(5) = -251$ J.K^{-1}.mol^{-1} for Eq. 4. These values support the conclusions stated for $[Mo_{12}Sub]^{2-}$ and confirm the encapsulation of dicarboxylates within the Mo-rings is mainly entropy dependent.

2.4.3. *Example of dynamic associated to dicarboxylate templating*

Some informations concerning the parameters that govern host-guest interactions have been obtained from ^1H NMR experiments in D_2O. For $\{Mo_{10}Glu\}^{2-}$ and $\{Mo_{12}Pim\}^{2-}$ only resonances of the coordinated anion are observed which indicates that, unlike the phosphate groups of $\{PMo_{10}\}^{2-}$ and $\{P_2Mo_{12}\}^{4-}$ in fast exchange with free phosphate ions, glutarate and pimelate groups are tightly blocked within the cavity. However, the width of the peaks confirms that dynamic processes take place in solution.

The identification of the species involved in dynamics has been realized with solutions of $Li_2\{Mo_8Ox\}$. $18.5H_2O$, $Li_2\{Mo_{10}Glu\}$. $11H_2O$ and $Li_2\{Mo_{12}Pim\}\cdot 22\ H_2O$ in CD_3CN which have been studied by ^1H NMR experiments at variable temperatures, see Fig. 23. In each spectrum three groups of signals corresponding to the three types of hydrogen atoms present in the wheel are observed. The peaks attributed to the CH_2 groups

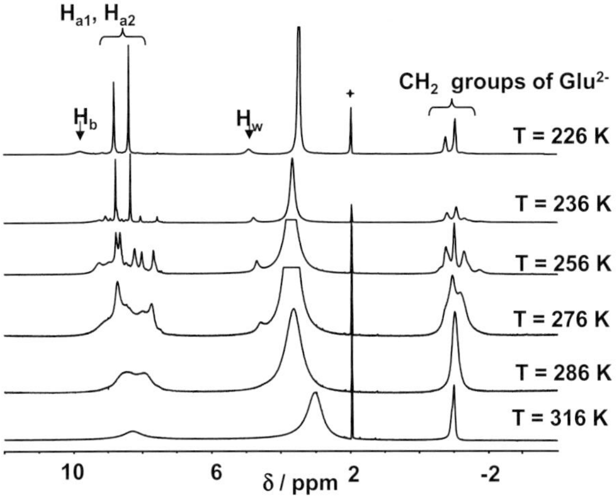

Fig. 23. ^{1}H NMR characterization of $\{Mo_{10}Glu\}^{2-}$ in CD_3CN via the chemical shifts of the CH_2 groups of the alkyl chain and OH bridges of the Mo-ring.

of the alkyl chain are observed between 0.0 and −1.0 ppm, with chemical shifts close to those recorded in D_2O. Peaks attributed to the water are located between 4 and 2.0 ppm, and the resonances of the OH bridging groups are observed at higher frequencies with chemical shift values in the range 7–10 ppm. The observed relative intensity ratio of 0.6 corresponds to that of the 6H of the $\{C_5H_6O_4\}$ glutarate group and 10H of the bridging hydroxo group displayed by the XRD structure.

As the temperature is decreased significant changes occur in the shape of the spectra affecting both the CH_2 and OH groups. The signal attributed to hydroxo groups is broadened at 316 K for gradually splitting into several peaks. At 256 K, 10 resonances are resolved. In addition, 7 sharp peaks (CH_2 groups) emerge from the broad signal of the glutarate present at 316 K. Below 256 K, the shape of the spectrum is significantly simplified by reduction of the number of peaks observed in the two regions. At 226 K, three signals attributed to the hydroxo groups are observed at 9.78, 8.78 and 8.36 ppm, respectively, together with two signals of the CH_2 groups at −0.78 and −1.06 ppm of relative intensities of 1/2/2 and 1/2, respectively.

An additional broad peak is observed below 276 K. It emerges gradually from the water peak, between 3 and 4 ppm, to reach the value of 4.91 ppm at 226 K.

Table 1. Chemical shifts and relative intensities for $\{Mo_{10}Glu\}^{2-}$ at variable temperatures.

	Chemical shifts (ppm); relative intensities	
T,K	H of the Mo_{10} ring	H of the CH_2 groups of Glu^{2-}
316	8.25;5	−1.11;1
		−0.98;2
226	9.78;1	−0.78;1
	8.78;2	−1.06;2
	8.36;2	
	4.91;2	

This observation permits to assign the peak of relative intensity of 1 to an aquo ligand located inside the cavity in fast exchange with free water molecules. This type of exchange is made possible since 4 Mo-atoms of the cycle exhibit a lack of coordination for being pentacoordinated. Assignments of the signals and chemical shifts and relative intensities for $\{Mo_{10}Glu\}^{2-}$ at 316 K and 226 K are given in Table 1.

2.4.4. *DOSY NMR methods applied to host-guest rings*

Diffusion NMR methods are a powerful tool for characterizing dynamics in solution. They have been successfully applied to resolve mixtures, to characterize biological systems and more recently to supramolecular chemistry concerns.[62–69]

The main advantage of Diffusion Ordered Spectroscopy (DOSY) is it gives access quasi-directly to the self-diffusion coefficient D_x of a molecule. This coefficient is correlated to the size and the molecular weight of the molecule. Polyoxothiomolybdates are inorganic-organic hybrids in which the organic part is a nice probe for ^{1}H NMR experiments. In addition, their structural diversity permits to investigate series of compounds which is ideal to establish comparisons, to draw tendencies and make common approximations.

The Einstein–Smoluchowski–Stokes diffusion theory has been successfully applied to oxothiomolybdate rings series.[60c,61] The self-diffusion coefficient D_x of a compound x in solution is proportional to $(v_x{\cdot}M_x)^{-1/3}$, where M_x is the molecular weight and v_x the specific partial volume of the species. As the $[Mo_{12}Trim]^{3-}$ anion is very stable in solution, it has been chosen as reference to calculate the diffusion coefficients D_x of various wheels (reference: D_r; M_r = 2139 g.mol^{-1} for the unsolvated compound). As the wheels in an oxothiomolybdate series exhibit a quite similar cyclic arrangement, it can be assumed that their specific partial volumes are almost the same ($v_r \approx v_x$) along the series. This approximation permits to easily calculate the D_r/D_x ratio:

$$D_r\,\alpha(v_r \cdot M_r)^{-1/3} \qquad \frac{D_r}{D_x} = \sqrt[3]{\frac{M_x}{M_r}}$$
$$D_r\,\alpha(v_x \cdot M_x)^{-1/3}$$

In D_2O and in DMSO, the experimental molecular weights obtained for $\{Mo_2O_2S_2\}$-based rings are consistent with those expected.[60c,61,70] The Stokes–Einstein relation permits to calculate the diffusion coefficient for spherical objects (relation 1) where k is the Boltzmann constant, T the temperature, η the solvent viscosity and R_H the hydrodynamic radius. A series of $\{Mo_2O_2S_2\}$-based rings ranging from Mo_{10} to Mo_{23} has been prepared and completed with two references, i- the two 12-Crown-4 and 18-Crown-6 crown-ethers selected for their low molecular weights, ii- the Müller's Keplerate $[Mo_{72}Mo_{60}O_{372}(H_2O)_{72}(CH_3COO)_{30}]^{42-}$[71], chosen for its high molecular weight. In the series, the various wheels exhibit comparable spherical shapes and because of the fast rotational tumbling, they behave as dynamical hollow spheres with molecular weights ranging from 200 to 20000 g.mol^{-1}.

$$D = \frac{kT}{6\pi\eta R_H} \quad (1) \qquad R_H = \sqrt[3]{\frac{3M}{4pr\mathrm{N_A}}} \quad (2) \qquad \mathrm{Log}\,D = A - 1/3\,\mathrm{Log\,M} \quad (3)$$

For a spherical particle, the hydrodynamic radius R_H may be calculated by using relation 2, where M is the molecular weight, ρ the density of the species and N_A, the Avogadro number. The relation between the self-diffusion coefficients D and the molecular weights M is easily obtained from relation 3 where A is a constant depending of the solvent and temperature.

Table 2. Molecular weights and diffusion coefficients of $\{Mo_2O_2S_2\}$-based rings obtained from ^1H DOSY experiments with organic and inorganic references, crown-ether and Kleperates resprectively.

Compounds	**M_w(anion) (g/mol)**	**D ($\mu m^2.s^{-1}$) in D_2O**
12-crown-4	176.2	782
18-crown-6	264.3	623
$[Mo_3S_4(Hnta)_3]^{2-}$	983.5	389
$[Mo_{10}Glu]^{2-}$	1740.2	395
$[Mo_{12}Muco]^{2-}$	2090.3	342
$[Mo_{12}TerP]^{2-}$	2096.3	358
$[Mo_{12}TMT]^{2-}$	2188.4	335
$[Mo_{12}Trim]^{3-}$	2139.3	315
$[Mo_{16}(IsoP)_2]^{4-}$	2940.5	295
$[Mo_{16}(PDA)_2]^{4-}$	3032.6	279
$[Mo_{18}(Mo_3S_4(nta)_3)]^{5-}$	4040.8	259
$\{[Mo_{18}(Mo_3S_4(nta)_3)][Mo_2O_2S_2(nta)_2]\}^{9-}$	4836.0	240
$[Mo_{132}O_{372}(CH_3COO)_{30}(H_2O)_{72}]^{42-}$	21684.4	144

The values of the diffusion coefficients of the wheels obtained for the series are given in Table 2.[67] They decrease as the molecular weight is increased, and the experimental variation of log$D$ as a function of log$M$ is linear with a slope of −0.338, in agreement with the value expected in 3.[60c]

These results illustrate that cyclic POTM can be considered as dynamical hollow spheres following the Stokes–Einstein rule, and that their molecular weights can be estimated from facile DOSY experiments. NMR experiments revealed to be also an excellent tool to study the dynamics of POTM in solution together with their thermodynamic stability.

3. Polyoxothiometalates Derived from the Archetypal Keggin Structure

3.1. *Functionalization of polyvacant polyoxometalates*

Besides the self-condensation of the preformed $\{M_2S_2O_2\}^{2+}$ dication, another elegant route to get polyoxothiometalates consists in grafting the

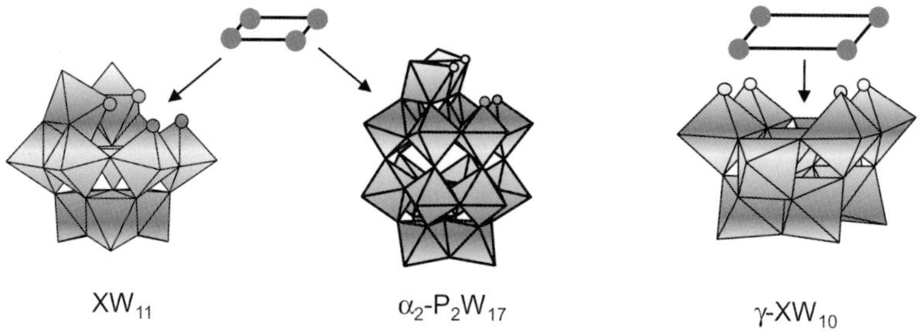

Fig. 24. Polyhedra representations of mono- and divacant Keggin and Dawson POMs highlighting the shape and size of the vacancy border.

$\{M_2S_2O_2\}^{2+}$ cationic precursor, M = Mo or W, to preformed lacunary polyoxometalates. Different examples are presented in this paragraph dealing with the monovacant Keggin α-$[PW_{11}O_{39}]^{7-}$ and Dawson $[P_2W_{17}O_{61}]^{10-}$ and the divacant Keggin $[\gamma\text{-}SiW_{10}O_{36}]^{8-}$ which are presented in Fig. 24.

As historically the divacant $[\gamma\text{-}SiW_{10}O_{36}]^{8-}$ polyoxoanion has been the first to be functionalized, it will be firstly presented.

3.1.1. *Example of* $[\gamma\text{-}SiW_{10}O_{36}]^{8-}$

The first reported polyoxothioanion has been obtained by direct addition of the $\{M_2O_2S_2\}^{2+}$ group (M = Mo or W) to the divacant $[\gamma\text{-}SiW_{10}O_{36}]^{8-}$[72] The structure of the resulting $[\gamma\text{-}SiW_{10}M_2S_2O_{38}]^{6-}$ anions clearly shows that the lacuna of $[\gamma\text{-}SiW_{10}O_{36}]^{8-}$ has been filled by the $\{M_2O_2S_2\}^{2+}$ fragment resulting in the formation of four M-O-W bridges as represented in Fig. 25. The phosphato analogue $[\gamma\text{-}PW_{10}M_2S_2O_{38}]^{5-}$ has been obtained by the same way.[73]

The observed M^V–M^V distances in the $\{M_2O_2S_2\}$ fragments, 2.832(1) and 2.815(1) for M = Mo and M = W), respectively, confirm that the metal-metal bond has been retained after grafting the dinuclear fragment into the POM structures. These anions are stable in aqueous solution, as confirmed UV-vis specroscopy, electrochemistry and ^{183}W and ^{31}P multinuclear NMR experiments. ^{183}W NMR spectra of the tungsten anion $[\gamma\text{-}XW_{12}S_2O_{38}]^{6-}$ (X = P^V or Si^{IV}) display in the −100/−200 ppm usual chemical shift range the resonances characteristic of the W^{VI} atoms in octahedral oxo environment, while that of the two equivalent W^V nuclei of the $\{W_2O_2S_2\}$ core are

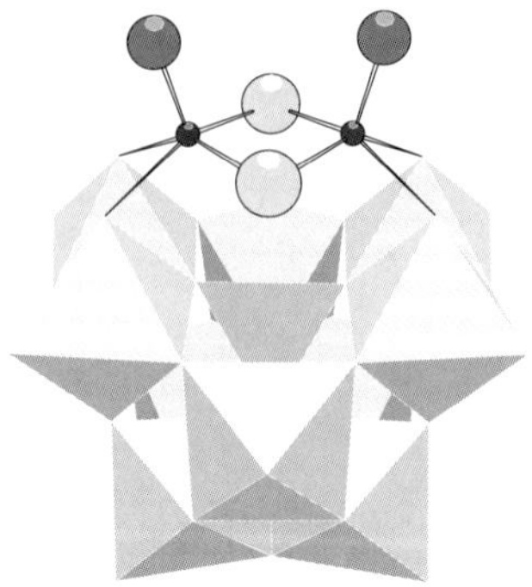

Fig. 25. Polyhedra representation of the molecular structure of $[\gamma\text{-SiW}_{10}M_2S_2O_{38}]^{6-}$, M = Mo, W.

observed at +1078 ppm and +1041 ppm for X = Si and X = P, respectively. Such important chemical shifts are characteristic of reduced tungsten atoms in a sulphur environment[74] confirming that the W-W metal-metal bond is still present in the POTM. The brown color of their solutions is attributed to the strong localization of the *d*-electrons within the metal-metal bond since the solutions of the isoelectronic $[\gamma\text{-SiW}_{12}O_{40}]^{6-}$, previously reported by Tézé *et al.*[75] are blue, a color characteristic of the delocalization of the two *d*-electrons over the POM structure and at the origin of the intervalence charge transfer (IVCT) observed at 1280 nm.

The strong stabilization of the metal-metal bond due to the presence of the two bridging sulfur atoms has been confirmed by DFT calculations carried out on the reduced $[\gamma\text{-SiW}_{10}M_2E_2O_{38}]^{6-}$, M = Mo or W and E = O or S,[76] showing the significant increase of the gap between the HOMO localized on the $\{M_2O_2S_2\}$ fragment and the LUMO mainly composed of the *d* shell orbitals of the W^{VI} atoms, see Fig. 26.

3.1.2. *Examples of the monovacant Keggin α–$[PW_{11}O_{39}]^{7-}$ and Dawson $[P_2W_{17}O_{61}]^{10-}$*

The direct addition of the $\{Mo_2O_2S_2\}^{2+}$ oxothiocation to the monovacant Keggin $\alpha\text{-}[PW_{11}O_{39}]^{7-}$ and Dawson $[P_2W_{17}O_{61}]^{10-}$ anions leads to the sandwich-like compounds $[(PW_{11}O_{39})_2(Mo_4S_4O_4(H_2O)_2)]^{10-}$ and $[(P_2W_{17}O_{61})_2(Mo_4S_4O_4(H_2O)_2)]^{16-}$, respectively.[77,78] The two anions present a quite similar geometry in which the POM subunits behave as bis-bidentate ligands

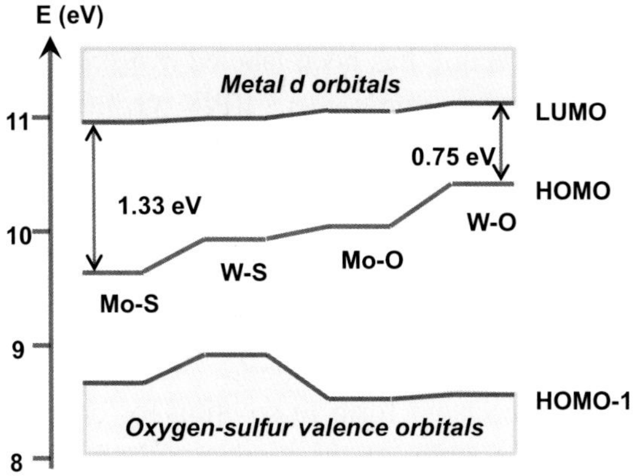

Fig. 26. Orbital diagram resulting from DFT calculations showing the evolution of the HOMO-LUMO gap on replacing sufur by oxygen in the $\{M_2O_2S_2\}$ fragment of $[\gamma\text{-}SiW_{10}M_2E_2O_{38}]^{6-}$, E = O, S, M = Mo, W.[76]

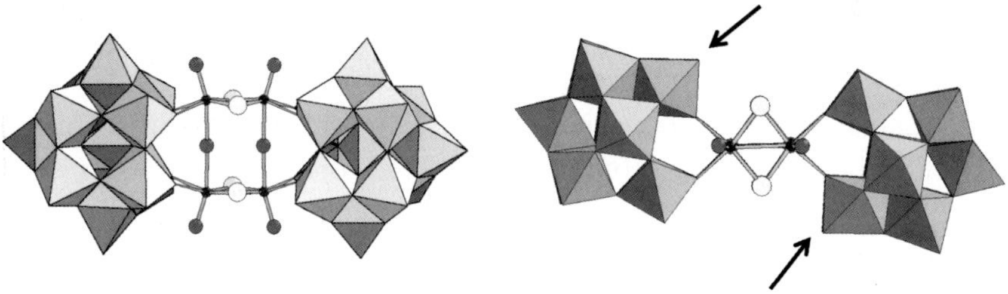

(a) (b)

Fig. 27. (a) View of the transoid isomer $[(PW_{11}O_{39})_2(Mo_4S_4O_4(OH_2)_2)]^{10-}$; (b) the two equivalent octaedra maked with arrows are 180° staggered relatively to the two $\{PW_{11}\}$. The central cluster $\{H_4Mo_4S_4O_6\}$ is represented in ball and stick model, (Mo in blue, S in yellow, O in red), a coordination polyhedra representation is used for the two $\{PW_{11}\}$. **See Color Plate 29**.

constraining the two $\{Mo_2O_2S_2\}$ to interact through a quasi-linear double Mo-O-Mo bridge. As represented in Fig. 27, the four oxygen atoms delimiting the vacancy of $[(PW_{11}O_{39})_2(Mo_4S_4O_4(H_2O)_2)]^{10-}$ are bonded to two $\{Mo_2O_2S_2\}$ fragments, the two POM subunits adopting a 180°-staggered disposition.

This disposition is not so rare since a similar association has been previously reported by R. G. Finke *et al.* for the structure of the $\{(P_2W_{17}Ru)_2\text{-}O\}$ dimer.[79]

Pope proposed the terminology of *transoid* for such an arrangement; thus, the molecular structure has been labeled *transoid*, which obviously suggests the possibility of a *cisoid* isomer.[80] $[(PW_{11}O_{39})_2(Mo_4S_4O_4(OH_2)_2)]^{10-}$ and $[(P_2W_{17}O_{61})_2(Mo_4S_4O_4(H_2O)_2)]^{16-}$ not only present a similar molecular arrangement based on the possibility of *transoid-cisoid* isomerism but also behave similarly in solution.

It can be noticed that a *cisoid-transoid* isomerism has been recently reported by Kortz *et al.* concerning square-planar Pd^{2+} cations trapped between monovacant POMs.[81]

The presence of the cisoid form in solution for both compounds has been deduced from the study of the kinetic conversion of the *transoid* isomer by ^{31}P and ^{183}W NMR methods. The results presented here interest $[(P_2W_{17}O_{61})_2(Mo_4S_4O_4(H_2O)_2)]^{16-}$. A representation of the postulated structural transformation is given in Fig. 28.

At the equilibrium a proportion of 30:70 (*transoid*:*cisoid*) is obtained, this value being independent of the temperature as expected for process governed by the entropy. Consequently the activation enthalpy is found to be the same for the two opposite reactions, $\Delta H^{\neq}_{1} = \Delta H^{\neq}_{-1} = 68.6$ kJ.mol^{-1}, while the the activation entropy, $\Delta S^{\neq}(1) = -27.7$ J.K^{-1}.mol^{-1} and $\Delta S^{\neq}(-1) = -35.5$ J.K^{-1}mol^{-1} are different. The $\Delta H^{\neq}$ value determined for the *transoid-cisoid* isomerization process is significantly lower than that observed by Anderson and Hill for the classical β to α Dawson isomerization of $[P_2W_{18}O_{62}]^{6-}$, $\Delta H^{\neq} = 83$ kJ.mol^{-1}.[82] This has been attributed to the weakening of the Mo-O bonds within the heterogeneous Mo-O-W bridges

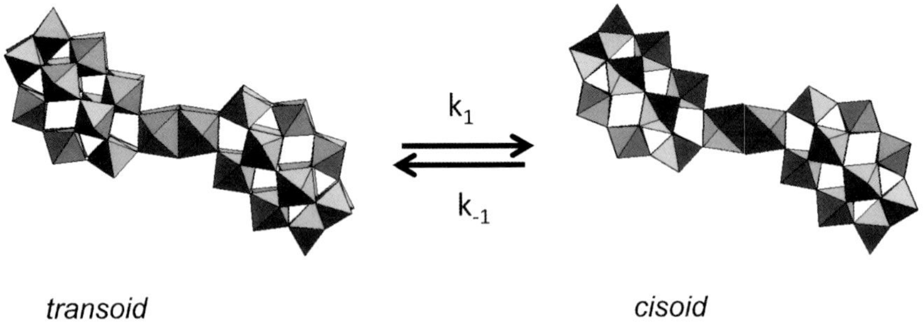

Fig. 28. Representation of the structural change resulting from the *transoid-cisoid* isomerization of $[(P_2W_{17}O_{61})_2(Mo_4S_4O_4(H_2O)_2)]^{16-}$ in solution (Mo^{VI} in grey, Mo^{V} in yellow). **See Color Plate 30.**

compared to the W-O-W junctions in the pure Dawson polyanion. The $\{M_2O_2S_2\}$ fragment is really an efficient linker permitting to assemble divacant POM subunits and generate inorganic supramolecular assemblies. These sandwich-like assemblies exhibit a high flexibility due to the specific coordinating properties of the Mo^V that provokes the weakening of the Mo-O-W bridges on replacing W^{VI} in the parent structure.

3.1.3 *Trivacant polyoxometalates as ligands*

Trivacant POMs have the specificity to display structural and chemical differences derived from the location and the size of the vacancies. Thus, trivacant Keggin and Dawson POMs are usually classified as A- or B-type isomers. Their representations are given in Fig. 29 with the shape and size of the respective lacuna. The hexagonal border of the vacancy is designed on the top of each type of isomer, the red points representing the terminal nucleophilic oxygen atoms. Three A- or B-type trivacant POMs have been chosen to be reacted with the $[M_2O_2S_2(H_2O)_6]^{2+}$ oxocation, namely $[A\text{-}\alpha\text{-}PW_9O_{34}]^{9-}$, $[B\text{-}P_2W_{15}O_{56}]^{12-}$, and $[B\text{-}\alpha\text{-}AsW_9O_{33}]^{9-}$ respectively.

Both types of isomers exhibit a crown of six terminal nucleophilic oxygen atoms. The perimeter of the crown in the type-A isomer is about 21 Å while it appears significantly reduced to about 17 Å in the B-type. In $[B\text{-}P_2W_{15}O_{56}]^{12-}$, a terminal P-O group is pointing towards the center of the crown while in $[B\text{-}\alpha\text{-}AsW_9O_{33}]^{9-}$ the center of the crown is occupied by the lone pair electrons of the central As^{III} ion. These structural differences

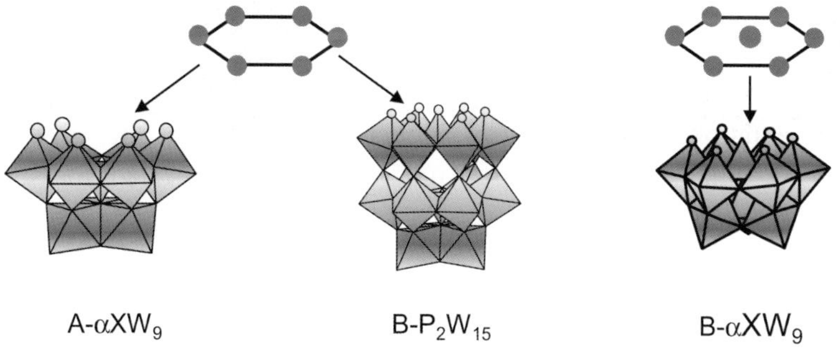

Fig. 29. Representations of the molecular structures of A- and B-type isomers of trivacant POMs with the shape of the associated vacancy border.

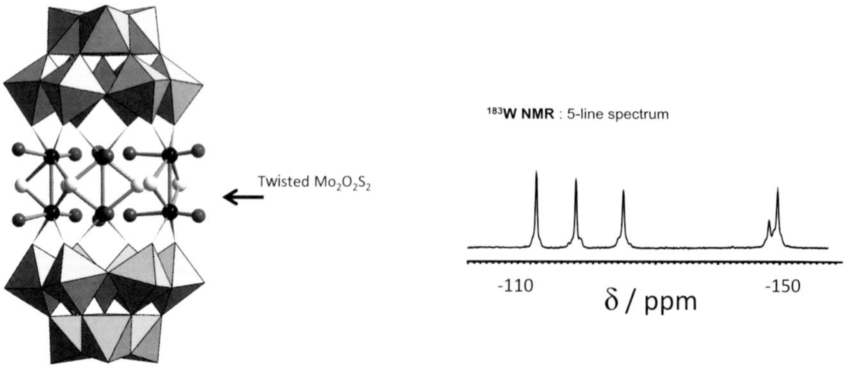

Fig. 30. Representation of $[(PW_9O_{34})_2(Mo_2O_2S_2(H_2O)_2)_3]^{12-}$ highlighting the 180° rotation of a $\{Mo_2O_2S_2(H_2O)_2\}$ group.^{183} W NMR spectrum showing the C_{2v} symmetry.

are at the origin of specific coordination properties with metal cations. Different possibilities of coordination of the ditopic $\{M_2O_2S_2\}$ fragment with the oxygen atoms located at the border of the vacancy of the trivacant POMs exist and modular architectures such as dimer, tetramer and hexamer are expected.

3.1.3.1. Reaction of $\{M_2O_2S_2\}$ with $[A\text{-}\alpha\text{-}PW_9O_{34}]^{9-}$

Three oxocations $[M_2O_2S_2(H_2O)_6]^{2+}$ M = Mo, W, directly coordinate to the six oxygen atoms of the border of the vacancy of the trivacant $[A\text{-}\alpha\text{-}PW_9O_{34}]^{9-}$. The resulting $[(PW_9O_{34})_2(Mo_2O_2S_2(H_2O)_2)_3]^{12-}$ is represented in Fig. 30: two $\{PW_9\}$ units are connected by three parallel $\{Mo_2O_2S_2(H_2O)_2\}$ groups acting as pillars for the two Keggin moieties. One of the $\{Mo_2O_2S_2(H_2O)_2\}$ fragment is twisted by 180° relatively to the two others.

The rotation by 180° of a $\{Mo_2O_2S_2\}$ linker lowers the idealized symmetry of the molecule from D_{3h} to C_{2v}. ^{183}W NMR experiments confirm this orientation is maintained in aqueous solution since the ^{183}W NMR spectrum obtained is composed of five lines as expected for such a compound in C_{2v} symmetry.

3.1.3.2. Reaction of $\{M_2O_2S_2\}$ with $[B\text{-}P_2W_{15}O_{56}]^{12-}$

The reactivity of $[B\text{-}P_2W_{15}O_{56}]^{12-}$ is obviously different from that of $[A\text{-}\alpha\text{-}PW_9O_{34}]^{9-}$ because of the presence in the center of the hexagonal vacancy of the additional oxygen atom of the central P = O group (see Fig. 29) which drastically affects the stereochemistry of the condensation processes. Instead of the dimeric $[(PW_9O_{34})_2(Mo_2O_2S_2(H_2O)_2)_3]^{12-}$, a spectacular planar tetrameric arrangement is obtained.[83]

The molecular structure of the tetra- anion $[(H_2P_2W_{15}O_{56})_4\{Mo_2O_2S_2(H_2O)_2\}_4\{Mo_4S_4O_4(OH)_2(H_2O)\}_2]^{28-}$ represented in Fig. 31 can be viewed as the association of two dimers resulting of the association of two $\{P_2W_{15}\}$ moieties with two $\{Mo_2O_2S_2(H_2O)_2\}$ linkers connected to two central $\{Mo_4S_4O_4(OH)_4(H_2O)\}$ groups. The four $\{Mo_2S_2O_2\}$ groups are mutually connected in a symmetric ribbon-like $\{Mo_8\}$ shape decorated by four pendent $\{P_2W_{15}\}$ POMs. In this structure, the $[Mo_2O_2S_2]$-building blocks are mutually bonded via hydroxobridges as in the wheels previously described[84] in Sec. 2.2. The tetrameric molecular structure is maintained in aqueous solution as confirmed by ^{31}P and ^{183}W NMR experiments: two equal ^{31}P resonances and fifteen equal ^{183}W resonances are observed as expected for the global C_{2h} symmetry of the anion.

Fig. 31. Polyhedra representation of $[(H_2P_2W_{15}O_{56})_4\{Mo_2O_2S_2(H_2O)_2\}_4\{Mo_4S_4O_4(OH)_2(H_2O)\}_2]^{28-}$ (Mo^{VI} in grey polyhedra, Mo^{V} in yellow polyhedra). **See Color Plate 30.**

3.1.3.3. Reaction of $\{M_2O_2S_2\}$ with $[A\text{-}\alpha\text{-}AsW_9O_{33}]^{9-}$

In the trivacant $[A\text{-}\alpha\text{-}AsW_9O_{33}]^{9-}$ the nucleophilic oxygen atoms are located at the vertices of the hexagonal vacancy without any other atom in the center of the hexagon, but $[A\text{-}\alpha\text{-}AsW_9O_{33}]^{9-}$ differs from the previous vacant POMs examined by the presence of the As^{III} lone pair located in the center of the vacancy and which widely opens it. The $[A\text{-}\alpha\text{-}AsW_9O_{33}]^{9-}$/ $[Mo_2O_2S_2(H_2O)_4]^{2+}$ system is extremely dependent on the ionic strength and the nature of the alkali cations used to crystallize the final products. Thus, three functionalized POMs have been isolated and fully characterized in solution and in the solid state. Surprisingly the three sophisticated clusters exhibit the same POM_2L_3 ($L=Mo_2O_2S_2$) global stoichiometry $\{POM_2L_3\}^{12-}$, $\{POM_4L_6\}^{12-}$ and $\{POM_6L_9\}^{36-}$, respectively. Each compound can be selectively crystallized from aqueous solutions by choosing the nature and the concentration of the alkali cations (Li^+, Na^+, K^+, Rb^+ and Cs^+) necessary for the crystallization as shown in Fig. 32.

$\{POM_2L_3\}^{12-}$. $[(AsW_9O_{33})_2(Mo_2O_2S_2)_3]^{12-}=(\{POM_2L_3\}^{12-})$ is obtained in the presence of large cations such as Rb^+ and Cs^+ in concentrated medium (1 M) of LiCl and on heating the solution to 75°C. That tends to prove the tetrameric configuration is probably, in those conditions, the thermodynamic stable product. The molecular structure of $\{POM_2L_3\}^{12-}$, is the simplest arrangement of two POM units linked by three $\{Mo_2S_2O_2\}$ pillars as yet observed for $[(PW_9O_{34})_2(Mo_2O_2S_2(H_2O)_2)_3]^{12-}$ described in Sec. 3.1.3.1.

$\{POM_4L_6\}^{24-}$. $\{POM_4L_6\}^{12-}$ is directly crystallized from solutions containing Rb^+ and Cs^+ cations. The molecular structure of $\{POM_4L_6\}^{12-}$ consists of four $\{AsW_9O_{33}\}$ POMs occupying the vertices of a tetrahedron while the six $(Mo_2O_2S_2)$ linkers form the six edges.

The 12 Mo-atoms are in octahedral environment, the sixth position of the octahedron being occupied by a water molecule.

$\{POM_6L_9\}^{36-}$. The hexameric anion is obtained only in the presence of Na^+, K^+ or Li^+ cations (see Fig. 32) from solutions with a high ionic strength. The molecular structure of $\{POM_6L_9\}^{36-}$ can be described as two basal POM_3L_3 triangles connected by three $(Mo_2O_2S_2)$ pillars. The nine $(Mo_2O_2S_2)$ groups have their axial Mo=O bonds directed inside the cage forming the cluster. Experimental observations supported by crystal structure analysis and solution studies confirm that sodium or potassium cations develop close

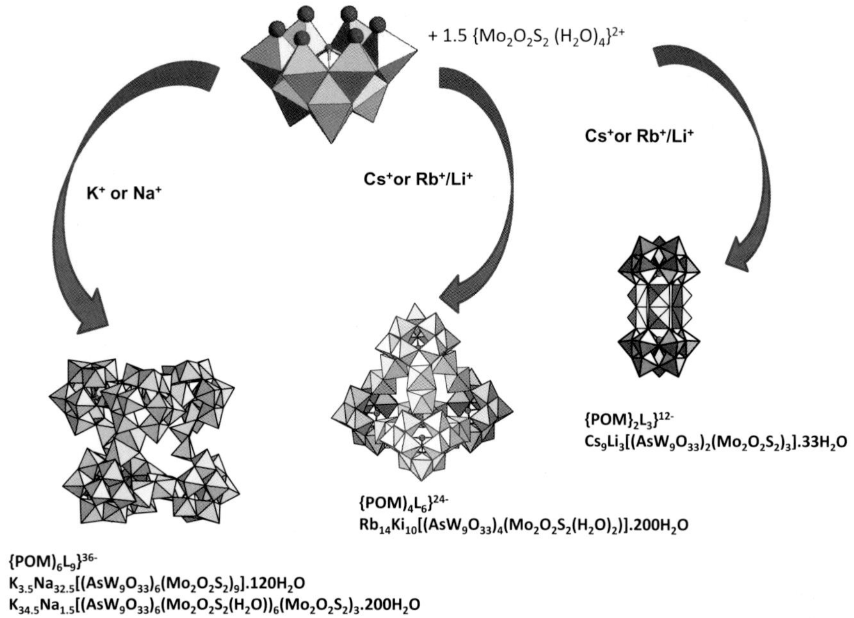

Fig. 32. Diagrammatic representation of the conditions of formation of $\{POM_2L_3\}^{12-}$, $\{POM_4L_6\}^{24-}$ and $\{POM_6L_9\}^{36-}$, with their molecular geometries and chemical compositions, (M^{VI} in grey, M^{V} in yellow). **See Color Plate 31**.

interactions with the $\{POM_6L_9\}$ scaffolding. In spite of its apparent density, $\{POM_6L_9\}^{36-}$ presents surprising properties of flexibility in the solid state illustrated by the evidence of two conformations, either eclipsed (D_{3h} symmetry), either staggered (D_3 symmetry), as represented in Fig. 33.

This effect results of the relative rotations of the two triangular POM_3L_3 basal planes. In the eclipsed geometry, the large cluster $[Na_6(H_2O)_{14}]^{6+}$ trapped inside the molecular cage plays the role of structurating agent for filling the central cavity of the structure.

The substitution of the sodium aquo-cluster by the less bulky $[K_8[H_2O)_8]^+$ cation forces the system to minimize the empty volume created by the twisting of the POM_3L_3 basal planes.

The modularity of the molecular structure of $\{POM_6L_9\}^{36-}$ is directly related to the idea of "*angle of approach*" introduced by E. Cadot[85] and is defined as the angle between two vectors, one representing the orientation of the $\{Mo_2O_2S_2\}$ linkers and the other that of the POM subunits.

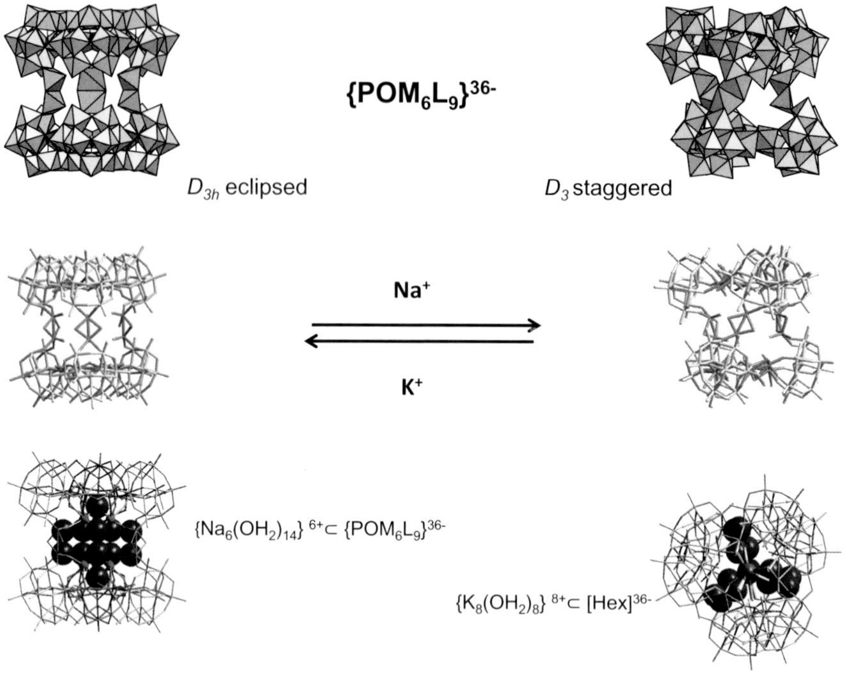

Fig. 33. Representations of the two conformations eclipsed and staggered of $\{POM_6L_9\}^{36-}$ in stick and polyhedra models. View of the aquo clusters $[Na_6(H_2O)_{14}]^{6+}$ and $[K_8(OH_2)_8]^{86+}$ trapped in the respective structures, (M^{VI} in grey, M^{V} in yellow, Na^+ and K^+ cations as blue spheres). **See Color Plate 32.**

$\{Mo_2O_2S_2\}$ is represented by a vector aligned along the Mo-Mo bond while the POM is characterized by a vector perpendicular to the plane delimiting the hexagonal vacancy as represented in Fig. 34. In $\{POM_2L_3\}^{12-}$ the three α angles have a value close to zero while they are significantly open for $\{POM_4L_6\}^{12-}$, $\alpha = 35.2 \pm 2°$, very close to the idealized value of 35.3° expected for a regular tetrahedron. For the regular D_{3h} $\{POM_6L_9\}^{36-}$ the angles within the two basal $\{POM_3L_3\}$ planes are significantly larger ($\alpha = 40 \pm 1°$) and the angle of connectivity between the three $\{Mo_2O_2S_2\}$ pillars and the two basal $\{POM_3L_3\}$ subunits is increased to $62 \pm 1°$.

The values of the α angle depend on the alkali cations present inside the molecular structure as aquoclusters. Thus, $\{POM_2L_3\}$ which exhibits a molecular structure empty of any cation has the most compact molecular arrangement.

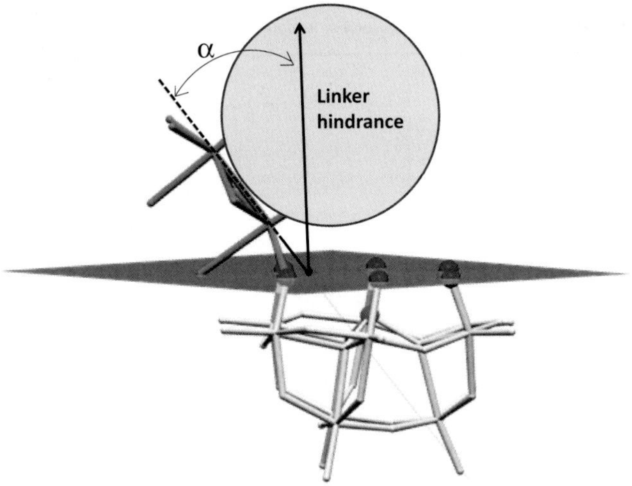

Fig. 34. Definition of the angle of approach between a $\{Mo_2O_2S_2\}$ linker and the vector perpendicular to the hexagonal vacancy of the $[A\text{-}\alpha\text{-}AsW_9O_{33}]^{9-}$ POM.

In $\{POM_4L_6\}^{12-}$ the α angle is increased because of the steric hindrance introduced by the presence of inner Rb^+ cations. $\{POM_6L_9\}^{36-}$. $\{POM_6L_9\}$ is characterized by the largest angle of approach probably since it contains the hindering cluster $[Na_6(H_2O)_{14}]^{6-}$.

The mixed lithium salt $KLi_{35}\{POM_6L_9\}$ spontaneously converts in aqueous solution into $Li_{12}\{POM_2L_3\}$ accompanied with the release of the water clustered K^+ cation.

The conversion is kinetically slow enough to be quantitatively followed by ^{183}W and ^{39}K NMR experiments as represented in Fig. 35. The experimental data have been supported by DFT calculations.

Geometries and energies of the three assemblies $\{POM_6L_9\}^{36-}$, $\{POM_4L_6\}^{24-}$ and $\{POM_2L_3\}^{12-}$ have been computed by Poblet and Lopez[86] to tentatively understand the role of the cations in the stabilization of the host-guest supramolecular assemblies.

3.1.4. *Tetravacant-trivacant POMs conversion induced by $\{Mo_2O_2S_2\}$ coordination*

The tetravacant anion[87] $[\beta\text{-}HAs_2W_8O_{31}]^{7-}$ quantitatively reacts with the $[Mo_2O_2S_2(H_2O)_6]^{2+}$ oxocation to form the triangular $\{POM_3L_3\}^{15-}$ arrangement

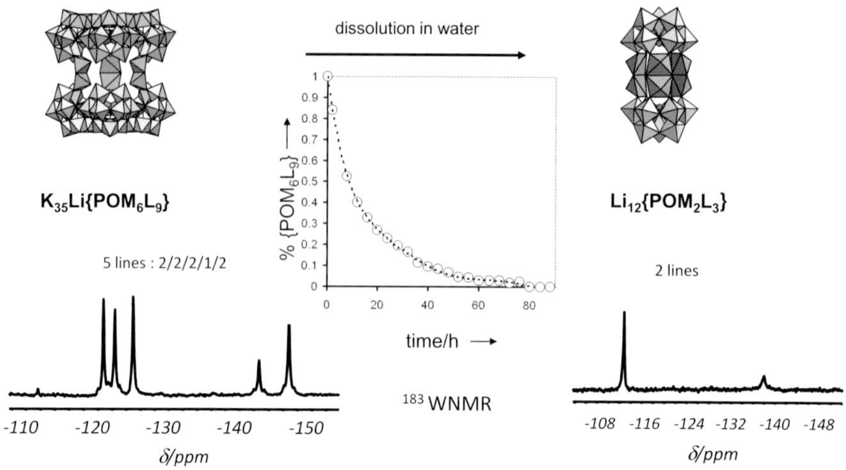

Fig. 35. Conversion of $\{POM_6L_9\}^{36-}$ into $\{POM_2L_3\}^{12-}$ under cation exchange followed by ^{183}W NMR spectroscopy.

represented in Figs. 36(a) and 36(b).[88] The POM part of $[(HAs_2W_9O_{34})_3(Mo_2O_2S_2(OH_2)_2)_3]^{15-}$ surprisingly corresponds to the trivacant anion $\{AsOH(\beta\text{-}AsW_9O_{33})\}^{7-}$. That means that during the reaction, the parent β-tetravacant anion has been transformed into a trivacant POM which has retained the β conformation. The molecular structure is described as three $\beta\text{-}\{HAs_2W_9O_{34}\}$ subunits mutually linked by three $\{Mo_2O_2S_2\}$ groups, the triangular arrangement delimiting a large "hexagonal" cavity, lined at its periphery by three {As-OH} groups and closed by a small oxygen atoms pocket.[88]

The central hexagonal cavity contains potassium, rubidium or cesium cations depending on the type of salt used for crystallization. The outer {As-OH} pendent group adopts two positions called *up* and *down*. With potassium the three {As-OH} groups are in *up* configurations characterizing the {up, up, up} isomer while with rubidium the two {up, up, up} and {down, up, up} isomers are disordered over the same crystallographic site. With dimethyl ammonium, only the {down, up, up} isomer is detected in the structure. In solution, ^{183}W NMR experiments carried out on the sodium salt result in a complex spectrum consistent with the presence of the four {up, up, up}, {down, up, up}, {down, down, up} and {down, down, down} isomers. The pendent {As(III)-OH} groups act as protecting agent against the coordination of additional $\{Mo_2O_2S_2\}$ linkers. This role has been evidenced by the oxidation of As(III) into As(V) which becomes

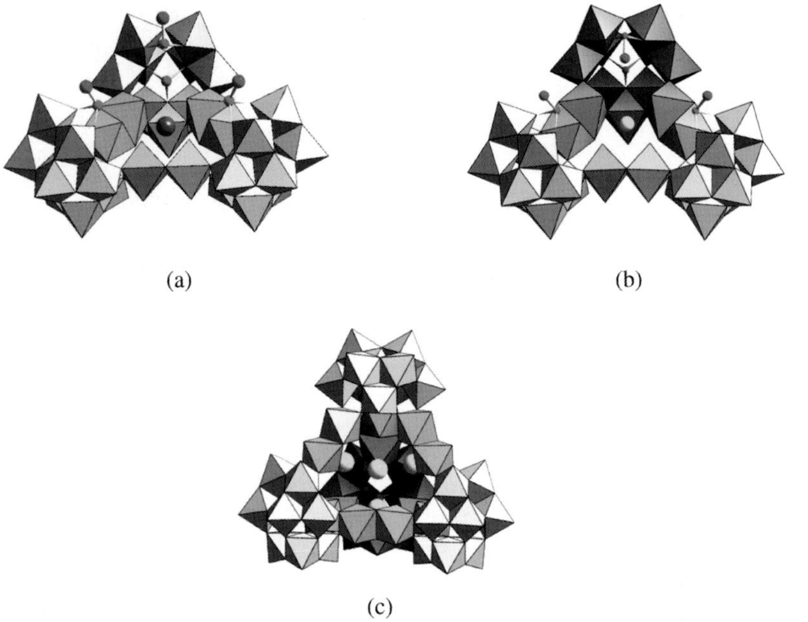

Fig. 36. Views in polyhedra models of (a) (up, up, up) and (b) (up, up, down) isomers of $[(HAs_2W_9O_{34})_3(Mo_2O_2S_2(OH_2)_2)_3]^{15-}$; (c) $[(\beta\text{-}AsW_9O_{33})_4(Mo_2O_2S_2(OH_2)_2)_6]^{24-}$ obtained after oxidation by iodine and reaction with $\{Mo_2O_2S_2\}$, (M^{VI} in grey, M^{V} in yellow, cations as colored spheres). **See Color Plate 33**.

labile enough to permit the condensation process to continue. After addition of three equivalents of iodine and reaction with $\{Mo_2O_2S_2\}$, $[(\beta\text{-}AsW_9O_{33})_4(Mo_2O_2S_2(OH_2)_2)_6]^{24-}$ has been isolated as mixed Cs-Na salt.

It is described as four $\beta\text{-}\{AsW_9O_{33}\}$ trivacant POMs mutually linked by six $\{Mo_2O_2S_2\}$ oxocations, see Fig. 35c. This result confirms that the three outer {As-OH} groups have been selectively removed by iodine oxidation, this reaction consisting in a real deprotecting process allowing the size and shape of the molecular arrangement extended from triangular to tetrahedral.

4. $[Mo_2O_2S_2(OH_2)_2\}^{2+}$ a Building Block Involved in Mo- and W-giant Capsules

Many isopolyoxometalates differing by their size, shape and structure have been obtained by acidobasic condensation and characterized in solution and in the solid state.[89] Their condensation process can be modified

(a) (b) (c)

$[Mo_8S_8O_8(OH)_8Mo(W)(O)OH_2)]^{2-}$ $[Mo_{40}S_8O_{120}]^{24-}$ $[\{W_6O_{21}(H_2O)_6\}_{12}\{Mo_2O_2S_2(CH_3COO)\}_{20}\{Mo_2O_2S_2(H_2O)_2\}_{10}]^{32-}$

Fig. 37. Representations in polyhedra of (a) $[M_9S_8O_{10}(OH)_{10}(H_2O)]^{2-}$, (b) $[Mo_{40}S_8O_{120}]^{24-}$ and (c) common backbone of the oxo $[Mo_{132}O_{372}(SO_4)_{30}(OH_2)_{72}]^{72-}$ and thio $[Mo_{132}S_{60}O_{312}(SO_4)_{15}(OH_2)_{102}]^{42-}$ Kleperates (M^{VI} in grey, M^{V} in yellow). **See Color Plate 33.**

or oriented in a different way by addition of exogenous species such as chelating ligands, reducing agents, transition metal ions, specific counter-ions, participating to the formation or the stabilization of transient species susceptible to open toward giant or nanosized systems.

Historically, the first example combining the $\{Mo_2O_2S_2\}$ core and oxo-metalates within a polycondensed backbone is $[M_9S_8O_{10}(OH)_{10}(H_2O)]^{2-}$ represented in Fig. 37(a).[90]

In this compound the overall geometry of the cyclic arrangement $\{M_8S_8O_8(OH)_8\}$ (M= Mo or W) is based on four $\{Mo_2O_2S_2\}$ clusters double-bridged by hydroxo groups and encapsulating a central $O=M{-}OH_2$ octahedron. This compound can be also considered as a member of the family of templated wheels described in Sec. 2.3, the main difference coming from the replacement of the negatively charged central organic guest by the positively charged $\{W{=}O\}^{4+}$ inorganic group.

4.1. *Sulfurated Kleperates*

Generally, giant systems are very sensitive to the *modus operandi* of the syntheses: order of addition of reagents, nature of the buffer, type of counter ions and obviously the $[M^{VI}O_4]^{2-}/[M_2O_2S_2(H_2O)_6\}^{2+}$ stoichiometry. This

(a) (b)

$\{Mo_6O_{21}\}^{6-}$ $\{Mo_6O_{21}\}_3\{M_2O_2X_2\}_3$

Fig. 38. (a) Polyhedra representation of the $\{M_6O_{21}\}^{6-}$ pentagon$^-$, M = Mo, W, formed of a central MO_6 pentagon surrounded by five MO_6 octahedra; (b) view of the structural motif formed of 3 pentagons (in grey) and 3 $\{Mo_2O_2X_2\}$ X = O or S linkers ensuring the structural connectivity in oxo and thio Mo_{132} Kleperates (M^{VI} in grey, M^{V} in yellow). **See Color Plate 34**.

versatility is as the origin of the characterization of highly condensed anions as for example $[Mo_{40}S_8O_{120}]^{24-}$[91] resulting from the condensation of four octamolybdate building blocks with four $\{Mo_2O_2S_2\}$ linkers as represented in Fig. 37(b). $[Mo_{40}S_8O_{120}]^{24-}$ is poorly stable in solution and rapidly transforms into complex mixtures as recently reported by Cronin *et al.*[92]

One of the most important oxo directing building block is $\{M_6O_{21}\}^{6-}$ (M = Mo or W) at the origin of the spectacular molecular molybdenum oxides discovered by A. Müller and known as Keplerate capsules.[93] $\{M_6O_{21}\}^{6-}$ represented in Fig. 38(a) is a soluble transient species, resulting from the acidification of MO_4^{2-} M=Mo, W and permitting the 3D-edification of the Kleperate derivatives.

Even if no mechanism is presently available explaining the formation of this precursor in solution, it seems obvious that the presence of weakly coordinative groups such as sulfates or acetates is necessary for its formation and stabilization, and thus the exchange properties.

4.1.1. *Sulfates as stabilizing ligands*

With sulfates, the oxo $[\{Mo_6O_{21}(H_2O)_6\}_{12}\{Mo_2O_4(SO_4)\}_{30}]^{72-}$[94] has been first reported by A. Müller while the sulfurated analogue $[\{Mo_6O_{21}(H_2O)_6$

$\}_{12}\{(Mo_2O_2S_2)_{30}(SO_4)\}_{15}]^{42-}$ has been obtained only recently[95] by Cadot *et al.* Both compounds exhibit the same global backbone composed of 12 pentagonal motives connected to 30 $\{Mo_2O_2X_2\}$ X = O or S linkers as represented in Fig. 37(c), but differ mainly by their metal-metal bond distances within the $\{Mo_2O_2X_2\}$ fragments which varies from 2.57 Å for X=O to 2.83 Å for X=S. The replacement of oxygen by sulfur induces a significant reduction of the radius of the inner pore from about 1.80 Å to 1.35 Å which probably modifies the host-guest interactions. The presence of sulfur in the $\{Mo_2O_2S_2\}$ fragments is expected to bring softness, pore flexibility and lability compared with the related oxo species as previously observed in templated-$\{Mo_2O_2S_2\}$ rings[75] or POM-$\{Mo_2O_2S_2\}$ system.[83,84] The control of the lability within the pores is a crucial a parameter since it determines the pore openings and closings.

4.1.2. *Acetates as stabilizing ligands*

A. Müller recently shown that the $\{Mo_2O_2S_2\}$ fragment combines with the $\{W_6O_{21}\}$ pentagonal building block formed *in situ* to form the spherical Kleperate analogue $[\{W_6O_{21}(H_2O)_6\}_{12}\{Mo_2O_2S_2(CH_3COO)\}_{20}\{Mo_2O_2S_2(H_2O)_2\}_{10}]^{32-}$, in which the replacement of sulfates for acetates does not induce significant structural differences except for the affinity for external cations and exchange properties.[96]

In concentrated acetate buffer and pH=3.6, E. Cadot, succeeded to isolate, besides the well known Kleperate type $[Mo_{132}S_{60}O_{312}(SO_4)_{15}(OH_2)_{102}]^{42-}$, a striking compound with a cuboidal arrangement he qualified of *inverse* Kleperate.[97] The anion $[H_{32}Mo_{84}S_{48}O_{200}(AcO)_{24}]^{46-}$ is represented in Fig. 39. It can be described as cube whose eight corners are occupied by trigonal $\{Mo_6S_6O_6(OH)\}^{5+}$ units and the 12 edges by $\{Mo_3O_{11}\}^{4-}$ oxomolybdates.

The arrangement of the trigonal $\{Mo^V_6\}$ units results of the mutual connections of three $\{Mo_2O_2S_2\}$ fragments *via* a μ_3-OH central group *trans* to the three terminal oxygen atoms of the {Mo = O} groups. The presence of the proton is deduced from the elongated Mo–O distance (Mo–O = 2.32 Å) yielding a bond valence sum (BVS) of 1.02 for this oxygen atom.

Three sulfur atoms occupy an equatorial position of the neighboring Mo^V octahedra which generate long Mo-S contacts (Mo-S = 2.58 Å) while the

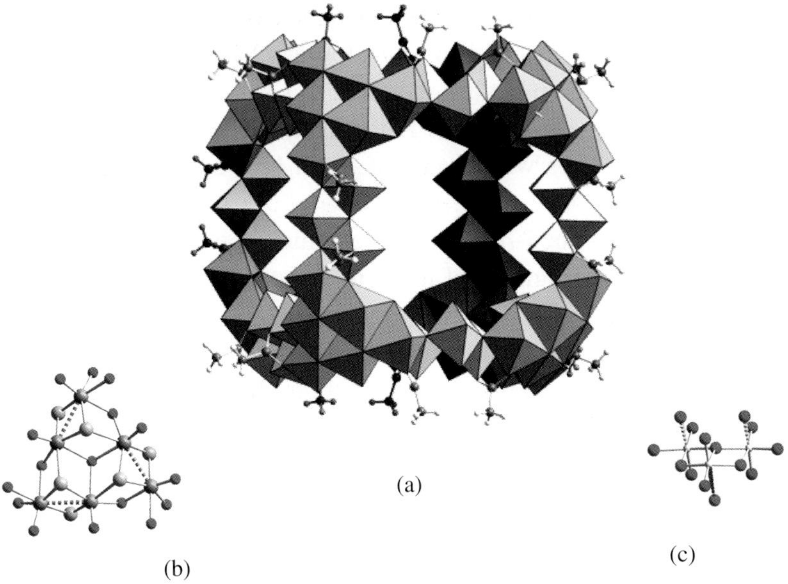

Fig. 39. (a) Polyhedra representation of $[H_{32}Mo_{84}S_{48}O_{200}(AcO)_{24}]^{46-}$; $\{Mo_6S_6O_6(H)\}$ units located at the corners of the cube (in yellow); $\{Mo_3O_{11}\}$ linkers located at the edges of the cube (in grey); (b) ball and stick model of $\{Mo_6S_6O_6(H)\}$ units; (c) ball and stick model of $\{Mo_3O_{11}\}$ linkers (Mo in green, S in yellow, and O in red). **See Color Plate 34**.

Mo–S bond distances in each $\{Mo_2O_2S_2\}$ subunit have the usual 2.29–2.36 Å value.[98] The resulting $\{Mo^V{}_6\}$ core has a local C_3 symmetry generated by the helicoidal arrangement of the three $\{Mo_2O_2S_2\}$ fragments around the central μ_3-OH (crossed by the C_3 axis). The eight $\{Mo^V{}_6\}$ cationic clusters are mutually connected through twelve $\{Mo_3O_{11}\}^{4-}$ anionic fragments acting here as nearly linear linkers to generate the cuboidal arrangement. In the $\{Mo_3O_{11}\}$ fragments three edge-shared octahedra are perpendicularly arranged around a μ_3-O oxygen atom. A water molecule is bonded to the Mo^{VI} atom through a long Mo-OH_2 distance (Mo-OH_2=2.55 Å). The $\{Mo^V{}_6\}$ building blocks and the $\{Mo^{VI}{}_3\}$ linkers are connected through edge sharing junctions involving two μ_3-oxo ligands. Strikingly, 24 acetate ligands, outside-orientated complete the coordination sphere of the neighboring Mo^{VI} and Mo^V atoms, thus bridging the $\{Mo^V{}_6\}$ building block and the $\{Mo^{VI}{}_3\}$ linker. The helicoidal $\{Mo^V{}_6\}$ building block is chiral (C_3 symmetry), but both enantiomers are symmetrically related by mirror planes and are alternately located on the eight corners of the achiral cubic box (T_d symmetry). As usually observed for very large clusters, the remaining constituents, sodium cations, acetates and water molecules are disordered within the crystal lattice.[99] Remarkably, the cuboidal arrangement exhibits six open faces with a gate of about 9 Å opening to an internal volume of 2100 Å.[3]

That makes the cube is expected to chemically behave as a *molecular* MOF: molecules or fragment of molecules, catalysts, and other chemical objects could be confined in the cage and *in situ* reacted with gas or liquids. This can be reasonably considered regarding the very good stability of the molecular cube in aqueous solution. The 24 acetates at the surface of the cube are labile enough to be easily exchanged for specific organic groups given the giant cube the possibility of post functionalization for a specific application.

5. Functionalization of Vacant POMS by $[Mo_3S_4(H_2O)_9]^{4-}$

5.1. *Saturation of a monovacant Dawson POM*

As seen in Sec. 2.2, $[Mo_3S_4(H_2O)_9]^{4-}$ exhibits a geometry quite similar to that of the $\{W_3O_{13}\}^{8-}$ building unit at the origin of Keggin and Dawson structures. But up to now, Keggin or standard $\{X_2W_{18}O_{62}\}^{n-}$ Dawson POMs functionalized by a Mo_3S_4 group replacing a W_3O_{13} group remain unknown. This is not surprising such a situation would suppose that a terminal oxygen atom of the central XO_4 tetrahedron of the POM has been substituted by a sulfur atom of the entering Mo_3S_4 group which is not chemically satisfactory. In fact, $\{XW_{18}O_{60}\}$ Dawson-like architectures[100] with X = Se(IV), Te(IV), As(III), Sb(III), Bi(III), Sn(II)) are better candidates because the central heteroelement X present in the first $\{XW_9\}$ half of the POM possesses an hindering lone pair preventing the presence of a second X-heteroelement in the center of the second half of the POM. In this situation, the X-free half of the POM is composed of only 3 $\{W_3O_{13}\}$ units and the replacement of one of them by a M_3S_4 fragment becomes easier.

Five POMs functionalized by $\{Mo_3S_4\}^{4+}$ or $\{Mo_3S_2O_2\}^{4+}$ thio-fragments have been reported so far.[101] In these structures, a $\{W_3O_4\}^{10+}$

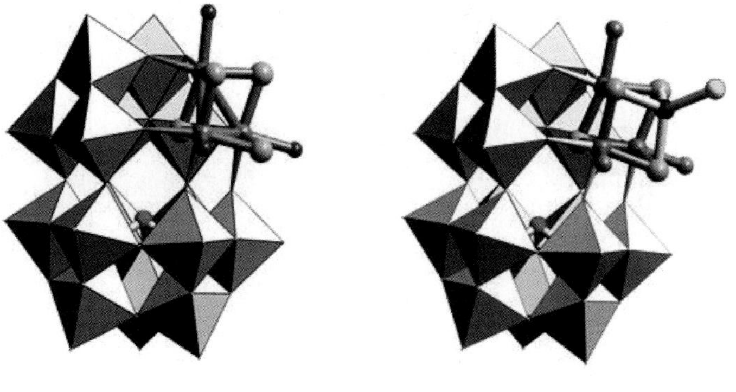

(a) (b)

Fig. 40. Representations of (a) $[AsW_{15}Mo_3S_4(H_2O)_3O_{53}]^{9-}$ showing the substitution of a $W_3O_{13}\}$ group by Mo_3S_4 and (b) $[AsW_{15}Mo_3S_4CuCl(H_2O)_3O_{53}]^{9-}$, As^{III} in pink, W^{VI}, in grey, Mo^{IV} in small green spheres, S in yellow, Cu in dark purple, oxygen of terminal water molecules in blue. **See Color Plate 35**.

group has been replaced by a topologically similar chalcogenide $\{Mo_3S_4\}^{4+}$ or $\{Mo_3S_2O_2\}^{4+}$ fragment leading to $[XW_{15}Mo_3S_4(H_2O)_3O_{53}]^{9-}$ (X=As, Sb), $[TeW_{15}Mo_3S_4(H_2O)_3O_{53}]^{8-}$, $[TeW_{15}Mo_3S_4(H_2O)_2ClO_{53}]^{9-}$ and $[AsW_{15}Mo_3O_2S_2(H_2O)_3O_{53}]^{9}$.[101]

$[AsW_{15}Mo_3S_4(H_2O)_3O_{53}]^{9-}$ represented in Fig. 40 can be viewed as issued from the parent[102] $[H_2AsW_{18}O_{60}]^{7-}$ by replacing one of the six $\{W_3O_{13}\}$ units of the As-free hemisphere by a $\{Mo_3S_4O_6(H_2O)_3\}$ unit. A Mo atom of $\{Mo_3S_4\}$ is located in one of the polar belts while the other two are in the equatorial belt. The Mo atoms in the POM have retained their +IV oxidation state and the 3 Mo–Mo bonds (2.764 (3)–2.763 (3) Å) present in the parent Mo_3 subunit. The main interest of these compounds is that the soft centers introduced in the framework of the POMs (sulfur atoms) can be reacted with metals with low oxidation states. The resulting clusters are expected to exhibit redox potentialities mainly due to the added low valent metal. Thus, $[As\ W_{15}Mo_3S_4(H_2O)_3O_{53}]^{9-}$ easily reacts with Cu^+ to form $[AsW_{15}Mo_3S_4CuCl(H_2O)_3O_{53}]^{9-}$ containing three metals in three different oxidation states, W^{VI}, Mo^{IV} and Cu^I respectively.[103]

If Mo_3S_4 cannot directly substitute a W_3O_{13} group of the Keggin or Dawson structures, the related monovacant types $[SiW_{11}O_{39}]^{8-}$ and $[P_2W_{17}O_{61}]^{10-}$ have

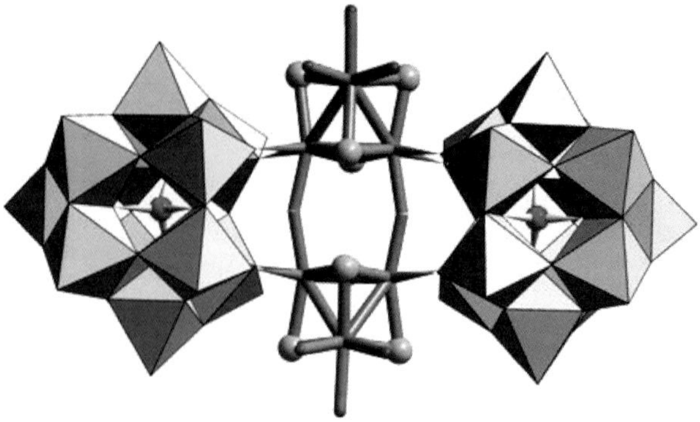

Fig. 41. Stick representation of $[(SiW_{11}O_{39})Mo_3S_4(H_2O)_3(\mu\text{-}OH))_2]^{10-}$ showing the two Mo_3S_4 groups blocked between two POMs subunits. W^{VI} in grey polyhedra, two Mo_3S_4 groups in ball and stick model. **See Color Plate 35.**

been successfully used as ligands. Thus, the reaction of $[Mo_3S_4(H_2O)_9]^{4+}$ with the Keggin and Dawson monovacant anions led to $[(SiW_{11}O_{39})Mo_3S_4(H_2O)_3(\mu\text{-}OH))_2]^{10-}$ and $[(P_2W_{17}O_{61})Mo_3S_4(H_2O)_3(\mu\text{-}OH))_2]^{14-}$, respectively, which have been fully structurally characterized.[46] Two central $\{Mo_3S_4(H_2O)_3\}^{4+}$ cores are sandwiched by two POMs subunits *via* OH bridges resulting in the ellipsoidal molecular structure represented in Fig. 41 for the Keggin derivative.

Such architectures are quite comparable to that obtained from the reaction of monovacant Keggin or Dawson POMs with $[Mo_2O_2S_2(H_2O)_6]^{2+}$. The two clusters are interesting for being stable in solution in a wide domain of acidity extending from pH = 1 to about pH = 7. Below pH = 1 the two clusters are hydrolyzed restoring the $[Mo_3S_4(H_2O)_9]^{4+}$ precursor stable in strong acid medium.

5.2. *Mo_3S_4 as supramolecular linker*

In Secs. 3.1.3 and 3.1.4, it has been shown that the product of the reaction of $\{Mo_2O_2S_2\}^{2+}$ with the tri-vacant $[B\text{-}AsW_9O_{33}]^{9-}$ POMs, owing to the large size of the POM vacancy and the presence of the lone pair of As^{III}, is not the corresponding saturated POM, but supramolecular architectures in which $\{Mo_2O_2S_2\}^{2+}$ acts as a linker. Regarding the relative sizes of $[Mo_3S_4(H_2O)_9]^{4+}$ and $[B\text{-}AsW_9O_{33}]^{9-}$ a similar behavior is expected of their mutual reaction keeping in mind the specific chemical properties brought by $[Mo_3S_4(H_2O)_9]^{4+}$. The result is the amazing supramolecular $[(H_4AsW_9O_{33})_2(Mo_3S_4(H_2O)_5)]_2^{12-}$ anion[104] represented in Fig. 41(b) which can be viewed as resulting of the dimerization of the two $[(H_4AsW_9O_{33})_2(Mo_3S_4(H_2O)_5)]^{6-}$ units depicted in Fig. 42(a).

The stability of the dimer is ensured by hydrogen bondings between the terminal hydroxo groups of the $[H_4AsW_9O_{33}]^{5-}$ POM and Mo-coordinated water molecules, and by S---S contacts between two Mo_3S_4 cores. In solution the monomer and the dimer are in equilibrium, the thermodynamic constant being $K_f = 10^{-3}$ at 25°C. The set of weak bondings participating to the dimer stabilization is still present in solution as confirmed UV-vis experiments.

The reaction of $[(H_4AsW_9O_{33})_2(Mo_3S_4(H_2O)_5)]_2\}^{12-}$ with a solution of arsenite permits the grafting of two $\{AsOH\}^{2+}$ groups on two $\{AsW_9\}$ subunits related by the inversion center of the molecule.[104] The molecular geometry of $\{[(H_2As_2W_9O_{34})(H_4AsW_9O_{33})(Mo_3S_4(H_2O)_5)]_2\}^{14-}$ obtained after the As-OH addition is represented in Fig. 43(a). As observed for the parent anion, $\{[(H_2AsW_9O_{34})(H_4AsW_9O_{33})(Mo_3S_4(H_2O)_5)]_2\}^{14-}$ dissociates

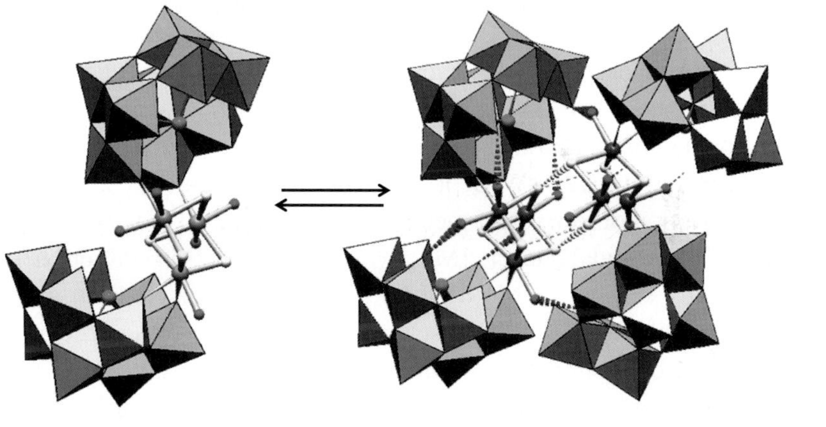

(a) (b)

Fig. 42. Representations of (a) the $[(H_4AsW_9O_{33})_2(Mo_3S_4(H_2O)_5)]^{6-}$ "monomer" and (b) the $[(H_4AsW_9O_{33})_2(Mo_3S_4(H_2O)_5)]_2^{12-}$ "dimer". W are grey octahedra, Mo are green spheres, S in yellow and O in red. Hydrogen bonds as blue dotted lines and S-S contacts in yellow ones. **See Color Plate 36**.

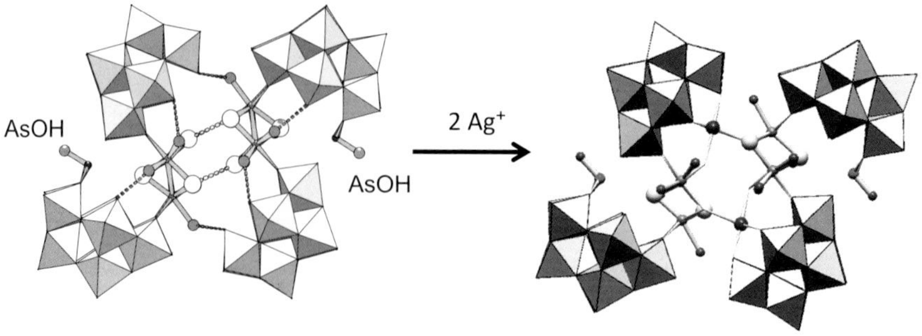

(a) (b)

Fig. 43. (a) Representation of $\{[(H_2As_2W_9O_{34})(H_4AsW_9O_{33})(Mo_3S_4(H_2O)_5)]_2\}^{14-}$ showing the two AsOH groups grafted on AsW9 units; (b) view of $\{[Ag_2(H_2As_2W_9O_{34})(H_2AsW_9O_{33})(Mo_3S_4(H_2O)_5)]_2\}^{16-}$, Ag^+ ions are dark purple spheres linked to sulfur atoms of Mo_3S_4 and oxygen atoms of {AsW9} POMs units. **See Color Plate 36.**

in solution and is in equilibrium with the monomeric $[H_2AsW_9O_{33})(H_4AsW_9O_{33})((Mo_3S_4(H_2O)_5)]^{7-}$ anion, $Kd \sim 930$.

The dimeric $\{[(H_2AsW_9O_{34})(H_4AsW_9O_{33})(Mo_3S_4(H_2O)_5)]_2\}^{14-}$ anion reacts with low valent metal cations such as Cu^+ and Ag^+.[104] With silver, two Ag^+ cations are directly coordinated to the Mo_3S_4 group to form $\{[Ag_2(H_2As_2W_9O_{34})(H_2AsW_9O_{33})(Mo_3S_4(H_2O)_5)]_2\}^{16-}$ represented in Fig. 43(b).

The addition of the two Ag^+ cations proceeds in two steps characterized by potentiometric and UV-vis experiments, $K_1 = 4.1 \cdot 10^6$ and $K_2 = 2.3 \cdot 10^5$. With copper the addition is limited to one Cu^+ cation which leads to $\{[Cu(H_2As_2W_9O_{34})(H_4AsW_9O_{33})(Mo_3S_4(H_2O)_5)]_2\}^{13-}$ anion.

The insertion of low valent metals has been extended to nickel and palladium. The cuboidal clusters $[Mo_3S_4(MCl)(H_2O)_9]^{3+}$ ($M = Pd^0$ or Ni^0) are topologically similar to $[Mo_3S_4(H_2O)_9]^{4+}$ and for that have been directly reacted with the trivacant POM subunit $[B\text{-}AsW_9O_{33}]^{9-}$. As expected, the reaction proceeds similarly to $\{Mo_3S_4\}$ leading to $[(H_2AsW_9O_{33})_4\{Mo_3S_4M(H_2O)_5\}_2]^{20-}$ which presents a dimeric arrangement geometrically comparable to $\{[(H_2AsW_9O_{34})(H_4AsW_9O_{33})(Mo_3S_4(H_2O)_5)]_2\}^{14-}$ represented in Fig. 42(a).

In fact the pseudo-cube Mo_3S_4 has been replaced by a complete Mo_3S_4M cube, the interest coming from the insertion within the POMs of "zerovalent" metals, Ni^0 or Pd^0 expected to be redox actives. The

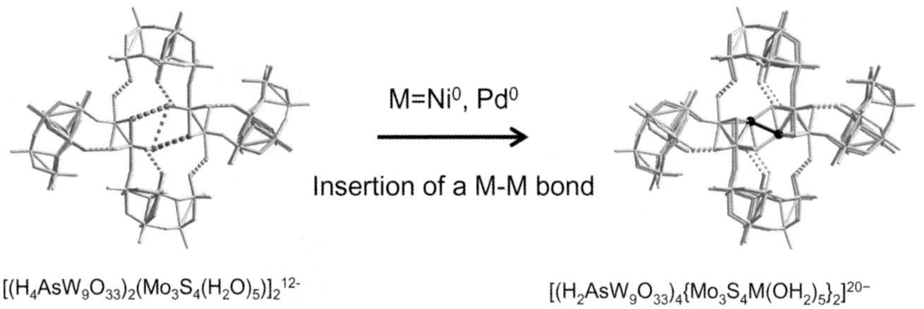

Fig. 44. Insertion of a M–M metal bond, M=Ni, Pd, in the central core of the $[(H_2AsW_9O_{33})_4\{Mo_3S_4M(H_2O)_5\}_2]^{20-}$ anion.

molecular structures consists of four $\{AsW_9O_{33}\}$ bidentate ligands bound to the Mo_3S_4M cube as represented in Fig. 44. Both compounds are isostructural their molecular structures consisting in two $\{(H_2AsW_9O_{33})_2\{Mo_3S_4M(H_2O)_5\}$ moieties linked by a M–M metal bond created in the central $\{M_2S_2\}$ core and by internal hydrogen bonds.

The compound is stable in solution in the studied concentration range of 2.10^{-4} to 2.10^{-3} mol. L^{-1}. ^{183}W NMR experiments carried out in solution with the Pd-derivative confirm the structure observed in the solid state is retained in aqueous solution. Addition of iodine or pyridine provokes the dissociation of the anion by creation of a Pd-I or Pd-Py bonds.[105]

Conclusion

The introduction of sulfur in molybdenum-oxygen based combinations drastically modifies their structures and their chemical behaviors. Because of the softness of sulfur compared to oxygen, the formation of Mo–S bonds introduces flexibility in the overall molecular structure that permits the edification of architectures which are unknown with solely oxygen. In addition to structural hard-soft consequences, redox properties of sulfur favor the localization of electrons resulting in stable metal-metal bonds while with oxygen delocalization is common. The acidobasic condensation of $[Mo_2O_2S_2]^{2+}$ in the presence of a templating group is at the origin of exciting wheels based on the host-guest concept. In a different way, $[Mo_2O_2S_2]^{2+}$ adapts its coordination modes to the shapes and sizes of the vacancies in archetypal POMs to generate non archetypal POTMs.

References

1. Pope M. T., *Heteropoly and Isopoly Oxometalates,* Springer-Verlag, New York, 1983.
2. Souchay P., *Polyanions et polycations,* Gauthier-Villars, Paris, (1963).
3. Pope M. T. and Müller A., *Angew Chem Int Ed Engl* **30** (1991) 34.
4. Pope M. T. and Müller A. (Eds.) *Polyoxometalates: From Platonic Solids to Anti-Retroviral Activity,* Kluwer, Dordrecht, The Netherland, 1994.
5. Gouzerh P. and Proust A., *Chem Rev* **98** (1998) 77–111.
6. Khan I. M. and Zubieta J., Oxovanadium and Oxomolybdenum Clusters and Solids Incorporating Oxygen-Donor Ligands; *Progress in Inorganic Chemistry*, Karlin. K. D. Ed. John Wiley and Sons, Inc. Vol. 43, (1995)
7. Berzelius J., *Poggendorff's Ann Phys* **6** (1826) 369, 380.
8. Miolati A. J., *Prakt Chem* **77** (1908) 439.
9. Rosenheim, A. and Bilecki R. *Berichte* **4** (1913) 543.
10. (a) Gouzerh P., *et al.*, *Angew Chem Int Ed Engl* **28** (1989) 1363; (b) Proust A. Gouzerh P., Robert F., *Inorg Chem* **32** (1993) 5291; (c) Stark J. L. Rheingold A. L. and Maatta E., *J Chem Soc Chem Commun* (1995) 1165; (d) Clegg W., *et al.*, *J Chem Soc Chem Commun* (1995) 455; (e) Strong J. B., *et al.*, *J Am Chem Soc* **116** (1994) 3601.
11. Grasselli R. K. and Burrington J. D., *Adv Catal* **30** (1981) 133.
12. (a) Baker L. C. W., *Advances in the Chemistry of Coordination Compounds*; Kirschner, S. (Ed.); Macmillan, New York, p. 604 (1961); (b) Day V. and Klemperer W. G., *Science* **228** (1985) 533.
13. Sundermeyer J. and Runge D., *Angew Chem Int .Ed Engl*, **22** (1994) 1255.
14. Wingley D. E., *Prog Inorg Chem* **42** (1994) 239.
15. Kang H. and Zubieta J., *J Chem Soc Chem Commun* (1988) 1192.
16. (a) Du Y., Rheingold A. L. and Maatta E. A., *J Am Chem Soc* **114** (1992) 345; (b) Stark J. L., Young V. G. and Maatta E. A., *Angew Chem Int Ed Engl* **34** (1995) 2547; (c) Proust A., *et al.*, *Inorg Chim Acta* **224** (1994) 81; (d) Errington R. J., *et al.*, *in ref 4*; p 105.
17. Strong J. B., *et al.*, *J Am Chem Soc.* **116** (1994) 3601.
18. (a) Jeannin Y., Hervé G. and Proust A., *Inorg Chim Acta* 198–200 (1992) 319; (b) Gouzerh P., *et al.*, *In Ref, 4*; p. 115 *Mol Engineering* **3** (1993) 79.
19. (a) Chilou V., *et al.*, *Inorg Chim Acta* **33** (1987) 205; (b) Hunter A. D. and Legzdins P., *Inorg Chem* **23** (1987) 4198; (c) Proust A., *et al.*, *Inorg Chem* **34** (1995) 4106.

20. Proust A., *et al.*, *Inorg Chem* **32** (1993) 5299.
21. Chae H. K., *et al.*, *Coord Chem Rev* **128** (1993) 209.
22. Rehder D., Priebsch W. and Von Oeynhausen M., *Angew Chem Int Ed Engl* **28** (1989) 1221.
23. (a) Kwak W., Pope M. T. and Scully T. F., *J Am Chem Soc* **97** (1975) 5735; (b) Stalick J. K. and Quicksall C. O., *Inorg Chem* **15** (1976) 1577; (c) Lowe M. P., Lockhart J. C., Clegg W. and Fraser K. A., *Angew Chem Int Ed Engl* **33** (1994) 451.
24. Müller A., *et al.*, *Chem Rev* **98** (1998) 1.
25. Müller A., *et al. Angew Chem Int Ed Engl* **33** (1994) 849.
26. (a) Müller A., *et al.*, *Angew Chem Int Ed Engl* **34** (1995) 2122; (b) Müller A. and Beuholt C., *Nature* **383** (1996) 296; (c) Plenio H., *Angew Chem Int Ed Engl* **36** (1997) 348; (d) Müller A., *et al.*, *Angew Chem Int Ed Engl* **37** (1998) 1220; (e) Müller A., *et al.*, *Angew Chem Int Ed Engl* **38** (1999) 3238; (f) Müller A., *et al.*, *Angew Chem Int Ed Engl* **38** (1999) 3241.
27. Müller A., *et al.*, *Angew Chem Int Ed Engl* **29** (1990) 926.
28. Müller A., Reuter H. and Dillinger S., *Angew Chem Int Ed Engl* **34** (1995) 2328.
29. Berzelius J. J., *Poggensdorffs Ann Phys Chem* **7** (1826) 262; **8** (1826) 269.
30. Müller A., *et al.*, *Chem Int Ed Engl* **20** (1981) 934.
31. Weisser O. and Landa S. (Eds.) *Sulphide Catalysts Their Properties and Application* Pergamon Press New-York (1973)
32 (a) Smith B. E., *et al.*, *Biochem J* **250** (1988) 299; (b) Garner C. D. *Molybdenum: An Outline of Its Chemistry and Uses* Braithwaite E. R. and Haber J (Eds.) Elsevier (1994) p 403.
33. Dilworth J. R. *Molybdenum: An Outline of Its Chemistry and Uses* Braithwaite E. R. and Haber J., (Eds.) Elsevier (1994) p 185.
34. (a) Hou Y., *et al.*, *Nature Mat* **10** (2011) 434–438; (b) Laursen A. B., *et al.*, *Energy & Env Sci* **5** (2012) 5577–5591.
35. (a) Diemann E. and Müller A., *Coord Chem Rev* **10** (1973) 79; (b) Müller A., *et al.*, *Anorg Allg Chem* **461** (1980) 91; (c) Müller A., *et al.*, *Anorg Allg Chem* **472** (1981) 69; (d) Müller A., *et al.*, *Inorg Chim Acta* **37** (1979) L493; (e) Sécheresse F., *et al.*, *Inorg Chem* **21** (1982) 1311.
36. (a) Müller A., *et al.*, *Inorg Chim Acta* **37** (1979) L493.; (b) Sécheresse F., Lavigne G. and Jeannin Y., *J Coord Chem* **45**. (1980).
37. Sécheresse F., *et al.*, *Inorg Chem* **21** (1982) 1311.
38. (a) Coucouvanis D., Toupadakis A. and Hadjikyriacou A., *Inorg Chem* **27** (1988) 3272; (b) Coucouvanis D., *Adv Inorg Chem* **45** (1997) 1.

39. (a) TopsØe H.; Clausen B. S.; Massoth F. E., In *Catalysis-Science and Technology* Anderson J. R. and Boudart M. (Eds.) Springer Berlin Vol. 11 (1996) pp 1–130; (b) Curtis M. D. and Druker S. H., *J Am Chem Soc* **119** (1997) 1027; (c) Feliz M., *et al.*, *Polyhedron* **24** (2005) 1212.
40. (a) Burgess B., *Chem Rev* **90** (1990) 1377; (b) Kim J. and Rees D. C., *Science* **257** (1992) 1677; (c) Kim J. and Rees D. C., *Nature* **360** (1992) 553.
41. (a) Keita B., *et al.*, *J Phys Chem* **112** (2008) 1109; (b) Appel A. M.; DuBois D.L. and Rakowski-DuBois M., *J Am Chem Soc* **127** (2005) 12717; (c) Lopez L. L., *et al.*, *Organometallics* **11** (1992) 2424; (d) Lopez L. L. Godziela G. and Rakowski-DuBois M., *Organometallics* **10** (1991) 2660; (e) Birnbaum J., *et al.*, *Organometallics* **9** (1990) 394; (f) Laurie J. C. V., *et al.*, *J Am Chem Soc* **108** (1986) 6234.
42. (a) Rauchfuss T. B., *Science* **316** (2007) 553; (b) Ogo S., *et al.*, *Science* **316** (2007) 585; (c) Justice A. K., *et al.*, *Inorg Chem* **46** (2007) 1655; (d) van der Vlugt J. I., *et al.*, *J Am Chem Soc* **127** (2005) 7012; (e) Borg S. J., *et al.*, *Inorg Chem* **46** (2007) 384; (f) Pardo A., *et al.*, *J Biol Inorg Chem* **11** (2006) 286; (g) Liu T. B. and Darensbourg M. Y., *J Am Chem Soc* **129** (2007) 7008; (h) Tye J. W, Darensbourg M. Y. and Hall M. B., *J Mol Struct THEOCHEM* **771** (2006) 123; (i) Gao W., *et al.*, *Inorg Chem* **46** (2007) 1981; (j) Cheah M. H., Borg S. J. and Best S.P., *Inorg Chem* **46** (2007) 1741; (k) Jiang S., *et al.*, *Dalton Trans* **896** (2007); (l) Oudart Y., *et al.*, *Eur J Inorg Chem* **2613** (2007); (m) Henry R. M., *et al.*, *J Am Chem Soc* **128** (2006) 3002; (n) Perra A., *et al.*, *Chem Commun* **1103** (2006); (o) Capon J-F., *et al.*, *J Electroanal Chem* **595** (2006) 47; (p) Artero V. and Fontecave M., *Coord Chem Rev* **249** (2005) 1518; (q) Ott S., *et al.*, *Angew Chem Int Ed* **43** (2004) 1006; (r) Dong W., *et al.*, *Chem Commun* **305** (2006).
43. Shibahara T., *et al.*, *Inorg Chem* 31, (1992) 640.
44. Jaramillo T. F., *et al.*, *J Phys Chem C* **112** (2008) 17492.
45. Llusar R. and Uriel S., *Eur J Inorg Chem* **7** (2003) 1271.
46. (a) Müller A., *et al.*, *Chem Commun* **1189** (1999); (b) Izarova N.V., *et al.*, *Russ Chem Bull Int Ed* **53** (2004) 1503.
47. Sokolov M., *et al.*, *Angew Chem Int Ed* **47** (2008) 1465.
48. Klemperer W. G. and Schwartz C., *Inorg Chem* **24** (1985) 446.
49. (a) Cadot E., Béreau V. and Sécheresse F., *Inorg Chim Acta* **239** (1995) 39; (b) Béreau V., Thèse de doctorat de l'Université de Versailles St Quentin en Yvelines (1997).

50. Cadot E., *et al.*, *Angew Chem Int Ed* **37** (5) (1998) 611.
51. (a) Cadot E., *et al.*, *Chem Commun* (2000) 261–262 ; (b) Cadot E., *et al.*, *J Phys Chem Solids* **62** (8) (2001) 1533.
52. Salignac B., Thesis University of Versailles St Quentin chap. IV p. 67 (2001).
53. Lemonnier J. F., *et al.*, *Inorg Chem* **46** (23) (2007) 9516.
54. Lemonnier J. F., *et al.*, *Eur J Inorg Chem* **34** (2009) 5233.
55. Cadot E., *et al. J Am Chem Soc* **126** (29) (2004) 127.
56. Cadot E., *et al.*, *Inorg Chim Acta* **350** (2003) 414.
57. Salignac B., Thesis University of Versailles St Quentin chap. II p. 21 (2001).
58. Salignac B., *et al.*, *J Am Chem Soc* **122** (42) (2000) 10381.
59. Dolbecq A., Cadot E. and Sécheresse F., *C R Acad Sci* **3** (3) (2000) 193.
60. (a) Cadot E., Marrot J. and Sécheresse F., *Angew Chem Int Ed* **40** (2001) 774; (b) Floquet S., Marrot J. and Cadot E., *C R Chim* **8** (2005) 1067–1075; (c) Lemonnier J. F., *et al.*, *Dalton Trans* (2007) 3043–3054.
61. Lemonnier J. F., *et al.*, *Chem Eur J* **13** (2007) 3548–3557.
62. Martinez-Viviente E., *et al.*, *Chem Eur J* **10** (2004) 2912–2918.
63. Terazzi E., *et al.*, *J Am Chem Soc* **127** (2005) 888–903.
64. Greenwald M., *et al.*, *New J Chem* (1999) 337–344.
65. Fernandez I., Martinez-Viviente E. and Pregosin P. S., *Inorg Chem* **44** (2005) 5509–5513.
66. Böhme U. and Scheler U., *Macromol Chem Phys* **208** (2007) 2254–2257.
67. Auge S., *et al.*, *J Phys Chem B* **113** (2009) 1914–1918.
68. Pluth M. D., *et al.*, *Inorg Chem* **47** (2008) 1411–1413.
69. Giuseppone N., *et al.*, *Angew Chem Int Ed* **47** (2008) 2235–2239.
70. LemonnierJ. F., *et al.*, *Dalton Trans* (2008) 4565–4574.
71. Muller A., *et al.*, *Angew Chem Int Ed* **37** (1998) 3360–3363.
72. Cadot E., *et al.*, *Inorg Chem* **35** (1996) 3099.
73. Cadot E., Béreau V. and Sécheresse F., *Inorg Chim Acta* **252** (1996) 101–106.
74. Cohen S. A. and Stiefel E. I., *Inorg Chem* **24** (1985) 4657.
75. Tézé A., *et al.*, *Inorg Chem* **35** (1996) 1001.
76. Rohmer M. M., *et al.*, Pope M. T. and Müller A., (Eds.) *Polyoxometalates from Topology to Industrial Applications* (2001) p. 117–133; Kluwer Academic Publishers the Netherlands.
77. Marrot J., *et al.*, *Inorg Chem* **42** (2003) 3609–3615.
78. Pilette M. A., *et al.*, *Eur J Inorg Chem* (2011) 3523.

79. Randall W. T., Weakley T. J. R. and Finke R. G., *Inorg Chem* **32** (1993) 1068.
80. Termes S. C. and Pope M. T., *Trans Met Chem* **3** (1978) 103.
81. Izarova N. V., Banerjee A. and Kortz U., *Inorg Chem* **50** (2011) 10379–10386.
82. Anderson T. M. and Hill C., *Inorg Chem* **41** (2002) 4252.
83. Cadot E., *et al.*, *Angew Chem Int Ed* **42** (2003) 2173–2176.
84. Lemmonier J. F., *et al.*, *Is J Chem* **51** (2011) 290.
85. Cadot E., *et al.*, *Chem Soc Rev* **41** (2012) 7335.
86. Marrot J., *et al.*, *J Am Chem Soc* **134** (2012) 1724.
87. Leyrie M., Tézé A. and Hervé G., *Inorg Chem* **24** (1985) 1275.
88. Pilette M. A., *et al.*, *Inorg Chim Acta* **363** (2010) 4253–4261.
89. (a) Hagrman P. J., Hagrman D. and Zubieta J., *Angew Chem Int Ed* **98** (1999) 2638; (b) Xi R. M., *et al.*, *Inorg Chem* **33** (1994) 833.
90. (a) Dolbecq A., Cadot E. and Sécheresse F., *Chem Commun* (1998) 2293–2294; (b) Cadot E., Marrot J. and Sécheresse F., *J Clust Sci* **13** (2002) 303–312.
91. Korenev V. S., *et al.*, *Inorg Chem* **49** (2010) 9740–9742.
92. Zang H. Y., Long D. L. and Cronin L., *Eur J Inorg Chem* (2011) 5105–5111.
93. Kortz U., *et al.*, *Coord Chem Rev* **253** (2009) 2315 and references therein.
94. (a) Müller A., *et al.*, *Angew Chem* **116** (2004) 4566–4570; *Angew Chem Int Ed* **43** (2004) 4466–4470; (b) Merca A., *et al.*, *Chem Eur J* **13** (2007) 7650–7658.
95. Korenev V. S., *et al.*, to be published.
96. Schäffer C., *et al.*, *Angew Chem Int Ed* **48** (2009) 149–153.
97. Bannani F., *et al.*, *J Am Chem Soc* (2012) accepted for publication.
98. (a) Schäffer C., *et al.*, *Angew Chem* **123** (2011) 12534–12537; (b) *Angew Chem Int Ed* **50** (2011) 12326–12329.
99. Müller A., *et al.*, *Angew Chem Int Ed* **41** (2002) 3604–3608.
100. (a) Jeannin Y. and Martin-Frère J., *Inorg Chem* **18** (1979) 3010. (b) Long D. L., *et al.*, *J Am Chem Soc* **130** (2008) 1830. (c) Rodewald D., *et al.*, *C. R. Acad Sci Paris Ser IIc* **63** (1999) (d) Krebs B., *et al.*, *C. R. Acad Sci Paris Ser IIc* **3** (2000) 205.
101. (a) Sokolov M. N., *et al.*, *Angew Chem Int Ed* **47** (2008) 1465–1468; (b) Sokolov M. N., *et al.*, *Eur J Inorg Chem* (2010) 5446–5454.
102. JeanninY. and Martin-Frère J., *Inorg Chem* **18** (1979) 3010–3014.
103. Cadot E., *et al.*, *Angew Chem Int Ed* **42** (2003) 2173.
104. (a) Duval S., *et al.*, *Chem Eur J* **14** (2008) 3457–3466; (b) Kalinina I. V., *et al.*, *J Struct Chem* **51** (2010) 303–307.
105. Duval S., *et al.*, *Dalton Trans* **41** (2012) 3174–3184.

INDEX